STATISTICS for the SOCIAL SCIENCES

To my wife Debbie and my daughters Edie and Farah.
"All my pretty chickens and their dam."

STATISTICS for the SOCIAL SCIENCES

R. Mark Sirkin

SAGE Publications
International Educational and Professional Publisher
Thousand Oaks London New Delhi

For information address:

SAGE Publications, Inc.
2455 Teller Road
Thousand Oaks, California 91320
E-mail: order@sagepub.com

SAGE Publications Ltd.
6 Bonhill Street
London EC2A 4PU
United Kingdom

SAGE Publications India Pvt. Ltd.
M-32 Market
Greater Kailash I
New Delhi 110 048 India

Printed in the United States of America

Library of Congress Cataloging-in-Publication Data

Sirkin, R. Mark.
Statistics for the social sciences / R. Mark Sirkin.
p. cm.
Includes index.
ISBN 0-8039-5144-2 (cl). — ISBN 0-8039-5145-0 (pb)
1. Social Sciences—Statistical methods. 2. Statistics.
I. Title.
HA29.S5763 1995
519.5—dc20 94-18897

98 99 00 01 10 9 8 7 6 5

Sage Production Editor: Diane S. Foster

Brief Contents

Detailed Contents

Preface

To the students and the instructors using this text, welcome! This book is designed to teach introductory statistics primarily to undergraduates majoring in the social sciences. I have tried to use a wide variety of examples that are both relevant to the social and behavioral sciences and of interest to today's undergraduate students. This book may be used as a text in a statistics course geared to any of the social sciences, or it may be used as part of a course or sequence of courses in research methodology.

Why another statistics text? After years of teaching students, many of whom claim to be victims of math anxiety, I wanted to provide a teaching device that could be used by the nonmathematically inclined, but at the same time would cover all relevant topics thoroughly enough to meet the needs of all students. To do this, (a) I am assuming that the only recent math courses that readers have had did not go much beyond introductory algebra and (b) many of the more onerous calculations encountered can be done on a computer. So, while all relevant calculations are presented here, emphasis—particularly in the later chapters—is also placed on the analysis of computer printouts.

Another thing that I have done is to begin with as little computational work as possible and move slowly into the math. This approach should enable students to gradually overcome their fear of numbers, build confidence in their ability to handle quantitative work, and (who knows?) even come to enjoy what they are doing. Note that many of the earlier topics, such as those on the scientific method, levels of measurement, and interpretation of tables, are given far less attention in many other statistics texts than is given here. By including them, it is my hope that students will see statistics as linked to the more comprehensive field of research methodology, rather than just as an entity unto itself. My emphasis is on the analysis and interpretation of data, rather than on how those data are collected. However, I do want the reader to have a feel for the way interpretation of data is related to the methods whereby the data were obtained. This approach also guards the student from immediate inundation in calculation.

Examples and exercises are designed to mirror the subject matter explored in all the social science disciplines. Easily spotted throughout the text are examples that can be identified with sociology, political science, communications, psychology, social work, management, education, and other disciplines. In selecting examples, I have chosen topics that should be of general interest to undergraduates in each field or to all undergraduates in the social sciences. I have also sought to include examples that reflect applied research as well as basic research. Such examples and exercises should help students retain interest in the course material. This has been my experience with my own students after having "test marketed" drafts of these chapters.

Each chapter begins with an introduction and a list of key concepts that are introduced for the first time in that particular chapter. In the body of each chapter, key concepts are presented in boldface type with their accompanying definitions in italics. Boxes are used to provide supplemental information reinforcing certain topics presented in the chapter. The exercises at the end of the chapter are presented in the same order in which the material is presented within the chapters, so that they may be undertaken prior to the completion of all topics presented in the chapter. There are ample exercises so that instructors may assign at their discretion a subset of the problems and still cover all the appropriate statistical procedures presented in the chapter. Thus more homework problems are included than one may need to assign.

A word about the ordering of the chapters: Chapters 1 through 6 are designed to introduce students to concepts of empirical research and the basic working vocabulary of statistics. These chapters cover the scientific method, levels of measurement and formats for manipulating and presenting data, operational definitions and index construction, central tendency and dispersion measures, and contingency tables. Although all statistics introductions cover central tendency and dispersion, few give as much emphasis to the other topics I have done here. This additional coverage will be of particular value if you are using this text in a combined methods/statistics course or if students have not already taken a separate methods course.

Chapters 7 through 10, together with Chapter 12, cover inferential statistics. Chapter 7 is an overview of the entire area and should be read first, followed by Chapter 11. From that point, it is possible to cover the remaining inferential statistics chapters in any order. Although generally one presents the two-sample *t* test prior to covering analysis of variance, it is possible, for instance, to cover chi-square without having first covered ANOVA.

Chapters 11 (association measures) and 13 (linear regression) cover additional topics in descriptive statistics and could be presented prior to the chapters on statistical inference. Chapter 14, however, can only be fully utilized if students have already had several of the inferential statistics topics in addition to linear regression. If you have a two-course sequence, it is possible to group the descriptive statistics chapters in the first course (Chapters 1-6, 11, and 13) and the inferential chapters in the second course (Chapters 7-10 and 12), culminating with Chapter 14, which interweaves the two threads. In short, I have designed the chapters with the knowledge that there are many possible sequences of topics and that we all march to different drummers.

For the instructor, I will save more of my comments for the instructor's manual that accompanies the book. In that manual, I will give suggestions gleaned from years of teaching this material. In turn, I hope that you will share your observations—both positive and negative—with me. You may contact me in care of Sage Publications. Best wishes for a positive teaching and learning experience.

Note to Students

Unlike many other courses, the material presented here is often cumulative in nature. This means that to understand today's assignment, you need to understand the material previously presented. If you do not understand today's topic, you may not be able to understand tomorrow's. Accordingly, do not try to "cram" this material. Learn it at a regular pace. If something confuses you, stop and reread it. If you still do not understand the topic, ask your instructor. Don't feel self-conscious about raising your hand in class and asking. You need to understand the material! Moreover, if you are confused, the odds are that others are likewise having trouble. Never assume that you are the only one "snowed." One other suggestion: attend class! You will learn better with the reinforcement provided by your instructor in class and you will have your instructor at hand to help explain any material that is causing you difficulty. Statistics calls for attendance.

It is my hope that this book will help contribute to a worthwhile educational experience for you. Best wishes with it!

Acknowledgments

There are many people who contributed directly or indirectly to this book. Indirectly, all those professors and colleagues who helped me develop my interest in statistics and methodology deserve my thanks. Likewise, my family deserves my gratitude for their patience and support during a period when the demands of the book interfered with my family activities and responsibilities. Finally, I owe a debt to all of my colleagues in the Wright State University Department of Political Science and the College of Liberal Arts for their encouragement and support.

Of those who contributed directly, nobody deserves greater praise than Joanne Ballmann, who typed, retyped, and often re-retyped the manuscript. I don't think that Joanne realized how difficult a job it would be to prepare a text so full of symbols, Greek letters, and algebraic equations. However, she tackled the project with much patience and good humor. Thanks! My gratitude also goes to Bruce Stiver and Gloria Sparks, in the graphic arts office of Wright State's Media Services, who prepared the many figures and graphs found throughout this book.

Thanks also to the College of Liberal Arts at Wright State and its dean, Perry Moore. The college provided me with funds for travel, graphics, and manuscript preparation. In the process of gaining this seed money, I received support and assistance also from Jim Jacob and Charlie Funderburk, my department chairs, Bill Rickert, associate dean of our college, and of course, my colleagues on the college's Faculty Development Committee. And to my many friends at Wright State who provided me with ideas and examples from their various social science disciplines, my gratitude goes out to you.

Many faculty members at universities throughout the country read and commented on drafts of chapters. Specifically, I would like to thank professors Rick Brown, California State University at Fresno; Alfred DeMaris, Bowling Green State University, Ohio; David Dooley, University of California at Irvine; Donald Gross, University of Kentucky, Lexington; Carl J. Huberty, University of Georgia, Athens; Garth Lipps, Statistics Canada, Ottawa; and John P. McIver, University of Colorado, Boulder. Many other helpful reviewers provided useful assistance during the course of this manuscript's development.

Thanks also to C. Deborah Laughton and Diane Foster, my editors; their assistants, Nancy Hale and Tricia Howell Bennett; and Andrea Swanson and Christina Hill, my typesetters; as well as all of the many other fine people at Sage who worked with me on the project.

Also, one learns from one's students. I wish to thank my students at Wright State University who took my classes in quantitative methods during the past 2 years and used earlier drafts of the chapters as text material. Their comments and feedback contributed greatly to improvements I was able to make in these chapters. In particular, Marge Gibson, Ann Koch, and Connie Weber, three of my students, were kind enough to supply detailed commentary—and proofreading—for several draft chapters. Thanks also to Chang Li for assisting in the index preparation and to Farah Sirkin for helping type the index.

I am grateful to the Longman Group UK Ltd., on behalf of the Literary Executor of the late Sir Ronald A. Fisher, F.R.S., and Dr. Frank Yates, F.R.S., for permission to reproduce tables II1, III, IV, V, and VII from *Statistical Tables for Biological, Agricultural and Medical Research* 6/e (1974).

Thanks to the SAS Institute, Inc., whose software is used to generate printouts incorporated into this text and the accompanying Instructor's Manual. Also, in the Instructor's Manual, several printouts are reproduced using SPSS Release 3.0. SPSS is a registered trademark of SPSS, Inc.

I wish to thank CQ Press for permission to reprint material from H. Stanley and R. Niemi (1992), *Vital Statistics on American Politics* (3rd ed.) and to Americans for Democratic Action for allowing me to use their ratings of members of Congress.

Yale University Press permitted my use of data and excerpts from C. L. Taylor and D. Jodice (1983), *World Handbook of Political and Social Indicators* and Freedom House, Inc. allowed the use of data from R. Gastil (1980), *Freedom in the World* as initially reprinted in Taylor and Jodice, above. I appreciate their cooperation.

Data were also utilized from *Global Studies: The Middle East*, 2nd ed., by William Spencer. Copyright © 1988 by The Dushkin Publishing Group, Inc., Guilford, CT 06437. I thank them for providing the permission.

Others too numerous to mention have contributed to this textbook. To all of them, thanks! Any errors to be found are, of course, not theirs but mine.

KEY CONCEPTS

empirical
normative
scientists
hypothesis/hypotheses
social sciences
scientific method
table/cross-tabulation/ contingency table
marginal totals
grand total
cell (of a table)
association
concept
variable
induction
deduction
experiment
scientific law
data (pl.)/datum or piece of data (sing.)
necessary condition
sufficient condition
theory
main diagonal
positively related/ a positive relationship
off diagonal
inversely related/ an inverse relationship
causation
temporal sequence
dependent variable
independent variable
criterion variable
predictor variable
unit of analysis
statistics

1 How We Reason

INTRODUCTION

This is a textbook on statistics and data analysis for the social sciences. Its techniques apply whenever data that involve counting or measuring have been collected. A standard logical process—the scientific method—underlies the collection and interpretation of data and applies to all the sciences.

The scientific method is the procedure whereby we propose possible relationships among characteristics of phenomena under study and then test to see whether those relationships actually exist. Though the techniques for doing this vary from one discipline to another, the logical sequence remains the same. Therefore the scientific method provides a good jumping off point for the chapters to come. This, in other words, is how we reason.

SETTING THE STAGE

All social sciences, indeed all sciences in general, have many goals. Essentially, we seek to understand some relevant phenomenon in order to make predictions or provide explanations about that phenomenon, and possibly to gain some control over it. To do this, we generally find ourselves concerned with two fundamental kinds of questions—What is? and What ought to be? *Questions pertaining to knowing "what is" we call* **empirical**, *and questions about "what ought to be" we call* **normative**. These two types of questions are found side by side in all academic disciplines, with the normative predominating in the humanities, the empirical predominating in the laboratory sciences, and both being components of the social sciences.

Questions about what ought to be form a core essential to understanding our society. We ask such questions as these: What is a good society? What is justice? Is a democratic government an ideal form of political system? Is it moral for opinion leaders to use deceit in dealing with the public? Should we employ the death penalty in the case of premeditated murder? Is racial discrimination morally acceptable? The answers to these questions depend on our values, our priorities and preferences, and our feelings of what is right or just. When we ask, "What are the components of an ideal society?" we are really asking ourselves to specify what we as individuals desire or think contributes to the common good. Normative questions form the background of social and political philosophy. They extend as far back as human preference can be traced, and some of the greatest minds have sought their answers: Plato, John Locke, John Stuart Mill, Thomas Jefferson, Adam Smith, and others—the giants of normative theory.

The empirical questions pertain to establishment of facts rather than values. We may accept the normative ideal that freedom of the press is a necessity for a free society. The related empirical question would be: Is there freedom of the press in a given nation? and/or How much freedom of the press actually exists there? Suppose we define freedom of the press as the existence of at least one newspaper or TV station not owned or controlled by the government and not subject to censorship by the government. Whatever we may like or dislike about that definition of freedom of the press, once we tentatively agree to use it as our working definition, we are in a position to examine data about each country and in doing so (perhaps with the aid of experts) to determine whether or not freedom of the press really exists there. The existence of freedom of the press as we defined it is subject to empirical verification. The question of whether or not freedom of the press is good and should exist in a democracy is a normative question, a matter of personal preference.

Both normative and empirical considerations enter into any definition of freedom of the press. For instance, we have many examples where countries with wide peacetime press freedoms must impose military censorship in time of war for security purposes. Our values will determine whether we would still categorize that country as having a free press. What about restrictions resulting from libel laws? What about nudity? What kinds of restrictions may we accept and still consider the press to be free? Empirical considerations pertain to our need to be able to clearly measure what we study. How are we going to obtain the information needed to categorize a country's press freedom? If a free press, in our definition, allows "soft core" but not "hard core" pornography to be published, how are we to determine where one ends and the other begins?

The normative and the empirical coexist and complement one another in a simple way: Empirical facts tell us how close we are to the normative ideal or how far we still must go to achieve the ideal. Suppose we have normative agreement that poverty should be eliminated. If we know that 10 years ago 15% of all families were legally defined as living below the poverty level and this year only 12% can be so defined, we might conclude, at least initially, that poverty is declining and we are moving toward our normative goals.

The art of empirical analysis—how we establish facts, how we determine what is, what actually does exist—is the subject matter of this text. While many techniques are available for studying "what is," an underlying logical process exists to enable us to make use of these analytical techniques.

EXERCISE

For each of the following, indicate which is normative and which is empirical:

1. Abortion should be outlawed.
2. There were 3 abortions per 100 live births in the United States last year.

3. The right to vote is essential to a free society.
4. Only 40% of all registered voters cast votes in the last presidential election.
5. Children should not be physically abused.
6. Last year, the rate of reported child abuse cases rose 3%.
7. A person who is old enough to fight for his or her country and vote in elections should be able to purchase liquor legally.
8. In this state the minimum age to purchase hard liquor is 21, although the voting age and minimum age for military service is 18.

Obviously, the odd-numbered questions are normative and their even-numbered counterparts are empirical.

A final observation: That something is normative and therefore exists as a goal rather than an established fact does not make it either more or less valid to scholars than an empirical fact. Both have a place in the way we learn about society. Consider the following:

> We hold these truths to be self-evident: That all men are created equal; that they are endowed by their creator with certain unalienable rights; that among these are life, liberty, and the pursuit of happiness . . .

These are Thomas Jefferson's words from the U.S. Declaration of Independence. "Self-evident" or not, what do we mean by men (persons) being created equal? Equal in size? Intelligence? Strength? Equal before the law? Economically? What about slaves? (Jefferson owned them.) The term "unalienable" means that which cannot be taken away. But your right to life may be taken away by the electric chair and your right to liberty may be taken away by a prison term (or a required college course).

Does this mean that what is normative is nonsense? Only if we ignore its value in providing guidelines and goals for what ought to be or if we ignore its value in inspiring and motivating us.

By the same token, the following empirical statement may be true factually, but hardly inspirational.

> A recent survey showed that four out of five consumers prefer Cleeno detergent to the next leading brand.

Unless I were the owner of the company making Cleeno, I would probably be more inspired by the Declaration of Independence than by the detergent statement.

SCIENCE

When we focus our attention on empirical knowledge, we concern ourselves with knowledge based on observation and experimentation. To the extent that we are engaged in discovering and categorizing such empirical knowledge, we are being scientists. **Scientists** *are those engaged in collecting and interpreting empirical information.* They do so in order to formulate and test hypotheses. **Hypotheses**, as we use them here, *are statements positing possible relationships or associations among the phenomena being studied.* These relationships, which will be elaborated on throughout this chapter, suggest that when some attribute or quantity of one phenomenon exists, a specific attribute or quantity of another phenomenon is also likely to occur. Since communications, cultural anthropology, political science, psychology, social work, and sociology, among other disciplines, concern themselves with aspects of societies, they are often termed the **social sciences**. *They empirically study social phenomena.* Their goals are to formulate and test hypotheses or suppositions about relationships and possible causes and effects among various aspects of a society, a culture, or a political system.

Social sciences share with all sciences two common aspects. The first is a commitment to the **scientific method**, *a series of logical steps that, if followed, help minimize any distortion of facts stemming from the researcher's personal values and beliefs.* The second is the use of *quantitative techniques*, measuring and counting, for the gathering and analysis of the factual information that is collected.

The scientific method is really a series of intellectual steps. It is not so much the actual techniques whereby the research is performed as it is the thought process whereby hypotheses are formed, tested, and verified (or not verified). If followed, the scientific method provides a basis for acquiring knowledge that will be eventually accepted by the scientific community. This accepted "truth" would be independent of the values and preferences of the researcher or any other observer. A properly conducted study of attitudes on abortion might find that in a particular group, your class for instance, 55% of the students are pro-choice and 45% are pro-life. This result would be accepted as fact regardless of the researcher's personal preference on the abortion issue.

As we will see, the scientific method involves the formulation of hypotheses, the testing of hypotheses via observation or experimentation, and the ultimate verification (or disconfirmation) of the hypotheses in a manner that will enable other scholars to draw the *same* conclusion as did the investigator. Thus facts are separated from values.

A major challenge for both the researcher and the consumer of research is to distinguish between what is established fact and what is a moral or ethical value judgment about some aspect of human behavior. This is often not an easy task because both the selection of facts and the interpretation of the selected facts filter through the researcher's own normative screen of values and preferences.

Among scholars there has long been a philosophical debate over whether a science can truly be value free. (This is a question posited against the claims of those empiricists who desire a value-free science.) Although it is probably true that we can never totally eliminate the effects of value and preference, adherence to the scientific method certainly helps minimize these effects and helps keep those effects from clouding our conclusions. For instance, a medical researcher convinced that cigarette smoke harms nonsmokers may publish a report citing the evidence from studies showing such harmful effects but ignoring those studies showing no damaging effects. Despite the intervention of personal values in this example, further empirical studies of passive smoking will either verify existing dangerous effects or show that these effects do not exist. Eventually, the issue will be put to rest, despite the occasional encumbrance of personal preference. "Facts are facts."

To enable us to take a closer look at the scientific method, I have chosen a fundamentally simple example. The example starts with an observation and proceeds to the establishment of theory.

It should be understood that from theory, further observations are generated, leading in time to more elaborate theory. Thus the process is a cyclical one: observation to theory to observation. Few scientific studies begin without at least some theoretical foundation. In this example, however, we assume no prior theory exists.

Now join me in a literal walk through the scientific method, but bring an umbrella—it might rain.

EXAMPLE

Suppose each morning I take an hour's walk. On Sunday, the sun shown during my walk and I admired the blue, cloudless sky. The same was also true on Monday. On Tuesday and again on Wednesday, it was raining and the skies were gray and overcast. On Thursday it only sprinkled and the sky, while generally blue, was broken here and there

Table 1.1

Presence of Rainfall	*Sky Conditions*			*Total*
	Cloudy	*Partly Cloudy*	*Clear*	
Rain				15
No Rain				15
Total	10	10	10	30

with a dark cloud. Assuming I lack education and prior awareness, I note something obvious to you: Rainfall appears to be associated with the presence of clouds in the sky, whereas sunshine generally means fewer clouds. In any event, whenever it did rain there were inevitably clouds in the sky. I conclude that rainfall is associated with cloudy conditions. At least that was the case for this 5-day period.

I think to myself that if for these 5 days rain is associated with cloudiness, then that same pattern should persist over a longer period. I decide to keep a log, indicating for each day whether I considered it cloudy, partly cloudy, or clear. I also note for each of these days whether or not it rained. At the end of 30 days, I take stock of my results. I note that of the 30 days, 10 were cloudy, 10 were partly cloudy, and 10 were clear. Also, I note that on 15 days it rained and on the other 15 it did not. *I put together a chart, Table 1.1, to summarize this. We term this chart* either a **contingency table**, a **table**, or a **cross-tabulation**. *The totals in the margins of my table* are called **marginal totals**. *The three marginal column totals* (10 cloudy, 10 partly cloudy, 10 clear) *add up to a* **grand total** of 30 days. The two marginal row totals (15 rain, 15 no rain) also add up to the same grand total of 30 days.

Now I review my records and tally my results for the 30-day period, noting for each day what the sky conditions were and whether or not it rained. (See Table 1.2.) I count up my tallies and put the appropriate number in each **cell** (*e.g., "cloudy, rain" or "clear, no rain")* in Table 1.3.

The results for the 30-day period are consistent with the observations for the initial 5 days: on cloudy days it always rained, on partly cloudy days it sometimes rained, and on clear days it never rained. The presence of rainfall is **associated** with the presence of clouds, and without clouds it appears that no rain will fall. Similar observations over other 30-day periods of time yield similar results and reinforce my initial conclusions. After a while, I take the conclusion that clouds are associated with rain for granted.

Table 1.2

Presence of Rainfall	*Sky Conditions* Cloudy	*Partly Cloudy*	*Clear*	*Total*
Rain	||||| |||||	|||||		15
No Rain		|||||	||||| |||||	15
Total	10	10	10	30

Table 1.3

Presence of Rainfall	*Sky Conditions* Cloudy	*Partly Cloudy*	*Clear*	*Total*
Rain	10	5	0	15
No Rain	0	5	10	15
Total	10	10	10	30

THE SCIENTIFIC METHOD

Aside from my simplistic observational techniques and my simplistic lack of even a basic understanding of weather, in this little example I have followed the scientific method. This is the thought process that is the underpinning of empirical research. In the first 5 days, I noticed a possible relationship between two **concepts** *or ideas:* sky conditions and the presence of rainfall. Both of these concepts are known as **variables** because *they vary, or change, from one observation to another*. Sky conditions vary from cloudy to partly cloudy to clear. Rainfall (as I observed it) varies in the sense that on some days it rains and on other days it does not. (I could also have classified my days more specifically if I had wanted to, for instance: heavy rain, moderate rain, light rain, drizzle, no rain. This time I chose to keep it simple.)

The concepts or ideas that we call variables are the phenomena of particular interest in our social science disciplines. They are called variables because they vary in amount or attribute for each individual (or group, or society, or state, or culture—whatever we happen to be observing). Some of the variables (and their categories or amounts) that we may be trying to better understand might include social class (upper, middle, lower), occupational status (white collar, blue collar), political party (Democrat, Republican), status (ascribed, achieved), government (democratic, authoritarian, totalitarian), or population density (high, medium, low). They parallel the presence of rainfall during my little walks. Other variables may explain the differences in the categories of the variables I wish to better understand: income (high, medium, low—or expressed in actual dollars), education (elementary school, high school, college—or expressed in total years of schooling), religion (Catholic, Protestant, Eastern Orthodox, Jewish, etc.), type of community (urban, suburban, rural nonfarm, farm). And of course there may be times when a variable from the first list, such as social class, might be explaining variables in the second list, such as income or education.

The relationship that I noticed on my walk was that certain categories of one variable were associated with specific categories of the other: cloudy sky conditions with the presence of rain, clear sky conditions with no rain. At the end of my initial 5 days of observation, I could have stated what I saw as follows:

> There is a relationship between sky conditions and the presence of rain, such that cloudy sky conditions are associated with the presence of rainfall and clear sky conditions are associated with no rainfall.

The above statement we term a **hypothesis**. *The hypothesis names the two variables that appear to be related and indicates the nature of that relationship* (clouds with rain, no clouds with clear weather). Note that another way of thinking about the relationship in the hypothesis is an "if . . . then" format: if clouds, then there will be rain; if clear, then there will be no rain. Another but wrong hypothesis could have been that clear skies are associated with rain; cloudy skies, with no rain. That would have been an alternative hypothesis, but not one consistent with my initial observations.

Here are some examples of hypotheses similar to those one would expect to find in the social sciences. These examples make use of some of the concepts presented above:

- ✦ There is a relationship between one's income and one's social status, such that the higher one's social status, the higher will be one's income and the lower one's status, the lower will be one's income.
- ✦ There is a relationship between the status of one's occupation and one's level of education such that indi-

viduals with higher status occupations are more likely to have college educations and individuals with lower status occupations are more likely to have only grade school educations.

✦ There is a relationship in the United States between religious preference and partisan identity, such that Democrats draw a greater proportion of support from Catholics and Jews than do Republicans, and Republicans draw a greater proportion of their support from Protestants than do the Democrats.

In each of these cases, despite variations in the wording, the two variables are named and the relationship between the categories of each variable is specified.

EXERCISE

Try forming some hypotheses that you think may be true using this same basic format.

TESTING HYPOTHESES

Based on a small number of observations during my morning walks (5 days), I noticed a relationship that I then assumed would hold over one or more 30-day periods. In using a small number of observations to assume that the relationship should hold for most or all observations, I was undertaking a process we call **induction**, *going from the specific to the general*. I **induced** my hypothesis from five specific observations and then assumed that the hypothesis would apply in all cases.

Once the hypothesis was induced, I set out to substantiate the hypothesis by acquiring information over a specific 30-day period. I reasoned that if the hypothesis was true in general, it should be true for the specific 30-day period that I had chosen. Here, I reversed my reasoning and went not from the specific to the general but the other way, *from the general to the specific; part of a logical process known as* **deduction**. I **deduced** that if the hypothesis were true all the time, it should be true for the 30-day period I had selected to study. In other areas of research, I might have been able to design an **experiment** *to test the hypothesis under laboratory-like conditions*. Here, my "experiment" was to select the 30-day period that I did and keep records on cloudiness and rainfall.

As the information in Table 1.3 indicated, my hypothesis was verified. If many subsequent studies had produced similar results, so that the relationship between presence of clouds and rainfall was widely accepted by the scholarly community and rarely if ever questioned, then my hypothesis would become a **scientific law**.

Scientific laws *are simply hypotheses with a high probability of being correct.* There is never absolute certainty about them, despite our use of the word "law." Two facts that emphasize this point will be covered in some detail later on. First, since most of our data come from sample surveys or randomized experiments, we are generalizing from a smaller group, the sample, to a larger one, called a population. However, we never can be absolutely sure that what was true for our sample is true for the population as a whole. We only estimate that probability. Second, by the rules of formal logic, we never in sampling actually prove anything. Instead, we demonstrate that all other possible alternatives are unlikely to be true, thus leaving us with only one remaining possibility—the thing that we are proving.

Despite these problems, scientific laws differ from hypotheses in general, since they are accepted as having a high probability of being correct by the scholarly community actively pursuing research in the field to which that scientific law pertains. The law of gravity would be an example. A scientific law is a law because experts in the appropriate area or discipline have reached that conclusion. It is neither the general public nor scholars in nonrelated fields that determine what scientific laws have validity. Physicists, not sociologists or theologians, determined the validity of gravity.

But the real world rarely cooperates with the researcher as neatly as indicated in Table 1.3. For example, suppose that after 30 days I tallied my results and what appears in Table 1.4 emerged.

Here, the hypothesis is not true; it is *disconfirmed* rather than confirmed. There appears to be *no* relationship between sky conditions and the presence of rain. Fifty percent of the cloudy days produced rain (5 out of 10 days), but so did 50% of the partly cloudy days and the same percentage of clear days. Regardless of sky conditions, it rained half of the time. Moreover, knowing a day's sky conditions gives us *no* useful information for predicting rainfall. Compare this to the information in Table 1.3. There, if we know that a given day is cloudy, we can predict rain and be 100% correct (all 10 cloudy days produced rain). If the day is clear we can perfectly predict no rain (on no clear day did it rain). The partly cloudy days have rain 50% of the time only—this is the only category of sky conditions that does not give us perfect predictability.

Table 1.4

Presence of Rainfall	*Sky Conditions* Cloudy	*Partly Cloudy*	*Clear*	*Total*
Rain	5	5	5	15
No Rain	5	5	5	15
Total	10	10	10	30

Table 1.5

Presence of Rainfall	*Sky Conditions* Cloudy	*Partly Cloudy*	*Clear*	*Total*
Rain	8	8	8	24
No Rain	2	2	2	6
Total	10	10	10	30

If on a partly cloudy day we predict rain, we know that we can expect to be correct half of the time. But half of the partly cloudy days produced no rain. We have an equal likelihood of being wrong in predicting rain. By contrast, in Table 1.4 cloud conditions are not useful at all for predicting the weather.

At this juncture, note that Table 1.4 indicates a 50–50 chance of rain, regardless of sky conditions. The 50–50 ratio is the result of the fact that there are an equal number of rainy and no-rain days, 15 days each, in this example. One does not need all equal entries to conclude that no relationship exists. Rather, the cell entries need only be proportionate to the marginal totals. Suppose that out of 30 days studied, it rained 24 days. That would be 80% of all days studied. If within each category of sky conditions it rains 80% of the time, then we also would have no relationship. Assuming 10 days for each of the three weather conditions, that would amount to 8 rainy and 2 no-rain days in each category of sky conditions. This is illustrated in Table 1.5.

The information found in these tables we call **data**. Data is the plural form. One single piece of information should be called a **piece of data** or a **datum**. Often we forget to differentiate singular from plural, but grammatically, we should say "these data," and so on. In Table 1.3, the data confirm the hypothesis; in Tables 1.4 and 1.5, they do not.

Table 1.6

Presence of Rainfall	Sky Conditions: Cloudy	Partly Cloudy	Clear	Total
Rain	0	5	10	15
No Rain	10	5	0	15
Total	10	10	10	30

Table 1.7

Presence of Rainfall	Sky Conditions: Cloudy	Partly Cloudy	Clear	Total
Rain	8	5	0	13
No Rain	2	5	10	17
Total	10	10	10	30

From time to time a relationship may be found that is *not* in the predicted direction, as shown in Table 1.6.

Here there is a relationship and high predictability, but the relationship does not follow what was logically anticipated by the hypothesis. Clear days produce rain; cloudy days do not. As in the case of Table 1.4, the original hypothesis was not verified, but unlike Table 1.4, there is a relationship between the variables in Table 1.6, only the relationship is the opposite of the one predicted.

One should note that the relationship found in Table 1.3 is very clear-cut. Rarely do results appear so clear-cut. More likely it is a case where a trend is noticeable, even though there are clear examples of days inconsistent with the hypothesis. Note the illustration in Table 1.7. Only 8 of the 10 cloudy days resulted in rain; 2 days were inconsistent with the anticipated results. Nevertheless, the partly cloudy and clear categories remain unaffected. (The marginal totals for the rows have also changed in this example.) The hypothesis has still been verified even though the results of the study do not produce perfect predictability for cloudy days. We may conclude that if the clouds appear before the rain (clouds come first in time), then cloudy sky conditions are a **necessary** but not **sufficient** condition for rain. No rain falls without the presence of clouds, but the presence of clouds does not *always* result in rain.

We should be aware of this distinction between necessary and sufficient. A **necessary condition** *is a condition that must be present in order for some outcome (in this case, rain) to occur*. Its presence, however, does not guarantee that the outcome will occur. By comparison, *if a* **sufficient condition** *exists, the predicted outcome will definitely take place*. For example, one could argue that poverty is a cause of communist revolutions. Indeed the presence of poverty motivated Marx, Lenin, and Mao in their writings and strategies, and there was great poverty in prerevolutionary Russia and China. Yet, many impoverished nations have not undergone Marxist revolutions. Why a revolution in Cuba but not in Haiti? Perhaps poverty is necessary but not sufficient for such a revolution. Then, in addition to poverty, one or more other factors may be needed for a revolution, such as a perception of inequality, unmet rising expectations of an end to poverty, or an organized revolutionary movement. If the presence of poverty alone always led to leftist revolution, then it would be both necessary and sufficient. It is also possible that any of several conditions when accompanying poverty can cause revolution; for example, either poverty plus a perception of inequality or poverty plus a charismatic revolutionary leader is sufficient to bring about revolution. When we study hypotheses containing more than two variables, we take the necessary versus sufficient aspect of relationships into particular consideration.

FROM HYPOTHESES TO THEORIES

Much in the same manner as my rainfall study, I could design studies to test my social science hypotheses. Perhaps I am interested in the possible relationship between religion and political party preference. I could prepare an attitude questionnaire and administer it to a randomly selected group of people. One question would ask each person his or her party preference and another question would tap religious preference. From the results we could put together a table similar to those in the rainfall example.

	Religious Preference				
Party Preference	*Protestant*	*Catholic*	*Jewish*	*Eastern Orthodox*	*Total*
Republican					
Democrat					
Total					

We would then look for discrepancies in the proportion of each religious group expressing preference first for the Republicans and then for the Democrats.

Let us assume that the findings show what we anticipated—Catholics and Jews have a greater tendency to express preference for the Democratic party than do Protestants, and Protestants have a greater tendency to identify with the Republican party than do the two other religious groups. This would motivate me to expand my study. Perhaps I might use as an explanatory variable family origin (for example, the country from which the respondents' forebears came when immigrating to the United States). I might look at the communities where the immigrants settled and the freedom and social mobility available to those immigrants. I might look at the impact of the Great Depression or the civil rights movement on the voting patterns of the families of my respondents. *If I can establish a number of interrelated hypotheses, each of which partly explains or accounts for differing partisan identities, I would be developing a* **theory** *of partisan identity.* The theory must also account for as many exceptions to the rule as possible. The better the theory, the fewer exceptions there will be.

Into my theory would pour dozens of observations and hypotheses, some already known and others undiscovered. For example:

1. African Americans were pro-Republican following the Emancipation Proclamation. As a result, however, of post-Reconstruction Jim Crow legislation, African Americans in the South abandoned the Republican party.
2. Southern Whites, opposed to Lincoln's policies, identified overwhelmingly with the Democrats until the "Solid South" began to crumble in the 1960s due to the civil rights movement and the perceived liberalism of the Kennedy and Johnson administrations.
3. Irish and Italian immigrants (mainly Catholic) as well as Eastern European Jews immigrated to the United States and settled in large cities, most of which were controlled by the Democrats. Their offspring remained Democrat. By contrast, Catholic and Jewish immigrants who settled in Republican-controlled communities (a smaller number) became Republicans.

Eventually, I would have enough explanatory capability to explain why more Catholics were Democrats than Republicans and why those Catholics who were exceptions identified with the G.O.P. I would know why Protestants had a higher probability of being Republicans than people from the other religion categories. I would know why African and Jewish Americans were generally Democrats. Finally, I might be able to predict and explain future trends, such as an impending change in party affiliation by one of these groups. In short, I would have a theory of partisan identity.

A similar process could take place in the rainfall example. With the hypothesis tested, I would be inclined to elaborate on it and expand my study. I might decide to replace the variable sky conditions by one such as type of clouds present (e.g., cumulus, cumulonimbus, cirrus, etc.) or humidity level. Perhaps I would also examine temperature or other atmospheric conditions. If I can establish a number of interrelated hypotheses, each of which partly explains or accounts for the presence of rainfall, I would have a theory of rainfall. Going to Table 1.7, I would need to develop hypotheses that would, for instance, explain why those 2 cloudy days produced no rain. What else besides just clouds needs to be present if rain is to fall? Other hypotheses in my theory would explain such exceptions and more about the nature of rainfall.

TYPES OF RELATIONSHIP

We worded our original hypothesis this way:

> There is a relationship between sky conditions and the presence of rain, such that cloudy sky conditions are associated with the presence of rainfall and clear sky conditions are associated with no rainfall.

We named the two variables and went on to specify the nature of the relationship. When both variables are measured in quantities—as amounts rather than as differing attributes—it is possible to simplify the specification of the relationship in our hypothesis. Both variables must be measuring more or less of an amount. Examples would be net income, either in exact dollars or categorized as high, medium, or low; age, in years or categorized as old, middle aged, young; or liberalism (high, medium, low). Variables that measure attributes rather than amounts, such as gender, religion, region, or race, require us to word the hypotheses as we have done so far. To illustrate such a simplification of our hypothesis, let us recast the categories of our variables so that both clearly indicate amounts or quantities.

Table 1.8

Amount of Rainfall	*Amount of Cloudiness* Very Cloudy	Partly Cloudy	Not Cloudy	Total
Heavy	7	1	0	8
Moderate	2	4	0	6
Light	1	4	0	5
None	0	1	10	11
Total	10	10	10	30

Now in Table 1.8, Amount of Cloudiness ranges from most (very cloudy) to least (not cloudy) and Amount of Rainfall ranges from most rainfall (heavy) to least rainfall (none). The categories of both variables describe differing amounts and they are in logical sequence, ordered from largest to smallest amounts. Note that the vast majority of the 30 days cluster in the table along a diagonal line from upper left to lower right, indicating that heavier rains are associated with greater amounts of cloudiness and lighter rains are associated with lesser amounts of cloudiness. Finally, as shown in Figure 1.1, "no rain" is associated with "no cloudiness."

We call this diagonal line from upper left to lower right the **main diagonal**. In a table with an equal number of rows and columns, the main diagonal would be a straight line; here it only approximates one.

Amount of Rainfall	**Amount of Cloudiness** Very Cloudy	Partly Cloudy	Not Cloudy
Heavy	7	1	0
Moderate	2	4	0
Light	1	4	0
None	0	1	10

Figure 1.1

	Amount of Cloudiness		
Amount of Rainfall	Very Cloudy	Partly Cloudy	Not Cloudy
Heavy	0	1	10
Moderate	1	4	0
Light	2	4	0
None	7	1	0

Figure 1.2

When most cases cluster on or near the main diagonal, indicating that greater amounts of one variable are associated with greater amounts of the other, and conversely, less of one with less of the other, we can describe the nature of the relationships with the expression **positively related**. *A* **positive relationship** *is one where greater is associated with greater; less with less.*

Our hypothesis may now be reworded in simplified form.

> The amount of cloudiness and the amount of rainfall are positively related.

Please note that in each instance the terms "positively" pertains to the nature of the actual relationship (high with high, low with low). It is *not* used as a description of how certain we are that a relationship exists. We are not saying that we are "positive" that the two variables are associated. We are not saying that cloudiness and rainfall are "absolutely, positively" related. Rather, we are hypothesizing that the nature of that association is that the more cloudiness there is, the more rain there will be.

The opposite of a positive relationship occurs when the variables are related in a way that the more of one variable there is the less of the other there will be. Suppose more clouds mean less rain and fewer clouds mean more rain, as shown in Figure 1.2.

In Figure 1.2 the clustering is on a diagonal line going from the upper right-hand side of the table to the lower left-hand side. We refer to this as the **off diagonal** and *when most cases cluster about the off diagonal, we*

Amount of Rainfall	Amount of Cloudiness: Not Cloudy	Partly Cloudy	Very Cloudy
Heavy	0	1	7
Moderate	0	4	2
Light	0	4	1
None	10	1	0

Figure 1.3

say that the variables are **inversely related**. The term *negatively related* is sometimes used, but the term *inversely related* is preferred.

If we were initially inclined to believe clouds were associated with a lack of rain and clear days were associated with rainfall, our hypothesis could have been worded:

> The amount of cloudiness and the amount of rainfall are inversely related.

Warning: The nature of a relationship, positive versus inverse, is taken from the logic of the hypothesis and corresponds to the stated diagonals in a table only if the table is set up so that the main diagonal represents "more with more" and "less with less."

Suppose Figure 1.1 had been recast as in Figure 1.3.

Although the categories of rainfall amount remain as before, the other variable, amount of cloudiness, was entered in the sequence opposite the way it was done in Figure 1.1. Now the clustering is on the off diagonal, but the relationship is *positive*, not inverse. Heavy rainfall is still associated with very cloudy sky conditions; low rainfall, with clear days. While the general convention is to set up tables so that clustering in the main diagonal indicates a positive relationship, there are exceptions to every convention. One must be on guard for such exceptions.

Finally, to reiterate a point made earlier, the positive versus inverse relationship terminology cannot be used unless both variables have categories representing *amounts* of the variables! Suppose we have a

Table 1.9

	Hair Color		
Gender	*Brown or Black*	*Blonde*	*Total*
Male	35	15	50
Female	15	35	50
Total	50	50	100

hypothesis that states that an individual's gender is related to his or her hair color such that women are more likely to be blondes than men and men are more likely to have dark hair than women. A study of 100 people yields the data shown in Table 1.9.

Since 70 of the 100 people appear to be consistent with the hypothesis (men with dark hair; women with blonde hair), the hypothesis is verified. Yet, we could not say that hair color and gender are "positively associated." There is no quantification in either variable in the sense of the categories implying more or less of the variable. Male and female are two types of gender; neither category possesses more or less gender. The same applies to hair color. Blonde is perceived by most people as a different color than dark hair, but a blonde has neither more nor less *an amount* of hair color than a dark-haired person. (Ignore the fact that physicists do view colors in amounts—the frequency of light from one end of the spectrum to the other.) Thus the terms "positive" or "inverse" would be unclear in characterizing this relationship and should not be used.

Let us examine how some social science hypotheses might be worded in this new manner:

- Income and social alienation are inversely related.
- Occupational status and educational level are positively related.
- Support for the existing political system and expectations of impending improvement in one's living standards are positively related.
- Time spent viewing television and time spent reading print media are inversely related.
- Income and support for organized labor are inversely related.

We could not word the hypothesis relating religion to partisan identity in this manner, however. Such a hypothesis relates attributes—

Republican, Democrat—to other attributes—Protestant, Catholic, and so on. These attributes are not quantitative in nature. They do not represent amounts of the variables.

Association and Causation

When we do research, we are ultimately seeking to infer the **cause** of the phenomenon we study. What variables when changing in value cause the variable we are studying to change? What factors account for the amount of rainfall? If levels of cloudiness are associated with levels of rainfall, and we assume for a moment that the actions of other variables such as temperature and humidity play no role in the direct relationship between cloudiness and rainfall, we are tempted to suggest a *causal relationship* between the two. We might assume that changes in levels of cloudiness bring about changes in the amount of rainfall. In effect, we are saying that it is the clouds that cause the rain. Given what is known about meteorology and climate, it is a logical assumption that clouds cause rain. If we knew nothing else about weather, however, we might just as likely conclude that it is the rainfall that causes the cloudiness level. Which of the two directions of causation we choose will often depend on two things: the logic of the situation and the temporal sequence of the variables, or which variable came first in time.

As a point of departure, it should be borne in mind that in general it is illogical to talk about a pattern of causation between two variables unless it can be first demonstrated that there is association between the variables. Without association it is meaningless to consider causation. As demonstrated in Table 1.4, the same proportion of rainy days exists under cloudy conditions as under partly cloudy and as under clear conditions. Knowing sky conditions does *not* improve our ability to predict or explain rainfall. Conversely, knowing whether this is a rainy day or not does not help us predict or explain cloud level. If variation in one variable does not relate to variation in the other variable there is no evidence to suggest that either variable causes, or brings about a change in, the other variable. Thus, with the exception of a few instances presented in subsequent chapters, *association is a necessary condition for causation.*

Once association has been established it may be possible to infer a causal direction between the two variables. A simple tool for doing this, particularly when we want to explain and not just predict the change in a variable, is **temporal sequence**. *If one variable changes earlier in* *time than does the other variable, we might assume that the former may cause the latter.* In the case of the rainfall problem, we may observe that the buildup in the cloud level takes place prior to the rainfall. In fact, once the rains stop, the cloud level often dissipates. Since the cloudiness precedes the rainfall, we may assume that clouds cause the rain. Since the rain follows the cloudiness in time, it would make no sense to suggest that the rainfall causes the cloudiness. The cause precedes the effect temporally.

Unfortunately, we are not always able to ascertain the time ordering of the variables. The gender versus hair color problem presented in Table 1.9 is an example of this dilemma. We receive through inheritance of genetic factors both our gender and hair color predispositions before we are born. From this perspective, it would be just as illogical to assume that gender causes hair color as it would be to assume that hair color causes gender. Even if there is association found between the two variables, there is no evidence that one either preceded the other in time or could logically have caused the other.

Given this fact, it would seem reasonable not to concern ourselves with the causation question at all unless we had evidence for making a causal inference. Unfortunately, many of the statistical techniques used in data analysis require that we designate, *in advance* of calculating the statistic, which variable is doing the causing and which variable is being caused. Because of this problem, there will be times when we may have to arbitrarily select one variable to be doing the causing or explaining and by default assume that the other variable's changes are being "caused" by changes in the former. In the case of the gender versus hair color problem, if we are primarily interested in studying hair color, we are likely to assume that for our purposes, gender "causes" hair color. If our goal were to account for gender differences, we may have to treat differences in one's hair color as leading to differences in one's gender.

Likewise, in American politics there has been a relationship between religion and partisan identity. Though it has been dissipating in recent years, Protestants have had a slightly higher affinity for the Republicans, whereas Catholics and Jews were more likely to vote Democrat. What is the variable doing the causing? Probably religious identity comes first in time, although not by much, so it would be logical to consider religion the cause and partisanship the effect. This would certainly make sense to a political scientist who would be trying to account for party identification. However, if one's field is the sociology of religion, and religious identity is the subject of inquiry, it would be perfectly logical to treat partisan identity as causing religious identity.

When we must differentiate the variable presumed to do the causing from the variable being caused—whether the selection is based on logic or based on an arbitrary decision—*we usually call the variable being described, caused, or explained the* **dependent variable**, *and the variable doing the causing or explaining, the* **independent variable**. The dependent variable is the "causee"; the independent variable is the "causer." Changes in the dependent variable *depend* on changes in the independent variable, but not necessarily the other way around, just as increases in rainfall depend on increases in cloud level but rainfall does not cause the clouds to form.

If we hypothesize that social inequality leads to revolution, then the occurrence of a revolution depends on prior social inequality. Revolution is the dependent variable (being caused), and social inequality is the independent variable (doing the causing). If we believe that air pollution (occurring first in time) causes certain forms of cancer, then level of air pollution is the independent variable and the cancer rate is the dependent variable. Implicitly, we assume in the latter instance that cancer levels do not cause air pollution.

Some people take issue with the use of the terms dependent and independent variable in cases where no logical causal ordering can be inferred. *They prefer the term* **criterion variable** *as a substitute for dependent variable and* **predictor variable** *as a replacement for independent variable.* With this alternate terminology, there is no connotation of causality nor implication of change in one variable being dependent on change in the other. Nevertheless, while one should be aware of this alternative to the terms dependent and independent variable, this text will continue to utilize the more traditional dependent/independent variable terminology.

THE UNIT OF ANALYSIS

There remains one other item of discussion in this review of the scientific method—the unit of analysis. The **unit of analysis** *is what we actually measure or study to test our hypothesis. It is not the variable being studied but rather the entity being studied—the person, place, or thing from which a measurement is obtained.* In the rainfall problem, *days* were the units of analysis. For each of the 30 days we took two "measurements," the presence (or amount) of rainfall and the presence (or amount) of clouds. In the hair color/gender problem individual *people* were the units of analysis. For each person we determined two things, that individual's gender and that individual's hair color. In the problem asking whether social inequality led to revolution, we would have to design a study in which we collected information from a number of countries. Thus *country* or *nation-state* would be the unit of analysis. For each country in our study, we would then seek to ascertain its people's level of inequality and also whether or not revolution had taken place in that country during some specified time interval. That would likely be the way a comparative political scientist would handle the problem. By contrast, a social psychologist might hypothesize that for an individual, his or her self-perception of being the victim of social inequality would influence his or her tendency to be supportive of revolution. Here, individuals, not countries, would be studied, so individuals would be the units of analysis. An urbanist might assume that where murder rates are high, so too would robbery rates be high. Data might be collected from a number of cities, finding each one's murder and robbery rates for a given year. Cities would be the units of analysis.

Since so many social and behavioral scientists study aspects of human attitudes and behavior, we often encounter an individual of one kind or another being studied as the unit of analysis. Thus many examples in this book also utilize the individual as the unit of analysis. It must not be assumed that this is always the case, however. Depending on availability of data, the units of analysis could be individuals; business firms or social organizations; counties, states, or provinces; nations or international alliances. Care must be taken to avoid assuming that what holds true for one unit of analysis holds true for others. Conclusions about the behavior of business firms do not necessarily carry over to individuals or counties or other units of analysis.

EXAMPLE

Suppose we were doing a study of some state's criminal justice system. During the process of collecting data, we notice that in the case of public defenders, those people employed in urban areas seem to have higher case loads than those employed in rural areas. Assuming that we wish to understand *case load* (amount of cases per public defender) and that a community's *population size* (urban, implying large population and rural, implying small) may account for the variations in size of public defender case load, then case load is our dependent variable and population size is our independent variable.

The hypothesis we induce from our observations is:

> There is a relationship between the size of a public defender's case load and the size of the community where that public defender is employed, such that public defenders in urban communities have higher case loads than their colleagues in rural communities.

Noting that both case load and population size are quantifiable variables, we may simplify our hypothesis as follows:

> The size of a public defender's case load and the size of the community employing that individual are positively related.

To test our hypothesis, suppose we have access to data for each county in that state or province showing the county's average public defender case load and also that county's population. County is our unit of analysis.

To keep our example *very* simple, assume that we establish a cutoff point in terms of case load and another cutoff point in terms of population size, such that each county is classified as either high or low in terms of public defender case load, and urban or rural in terms of population size.

If the hypothesis we induced is assumed to be true universally, it is assumed to be true—we deduce—for this particular province or state. We categorize each county in terms of case load (high versus low) and population (urban versus rural).

Note that our hypothesis is empirical; it can be tested from the data at hand. Whether or not the hypothesis is true is kept apart—as much as we can—from our own normative judgments. The facts will hold, regardless of our own normative opinions and beliefs about what *should* be the case. These normative beliefs could be any of a number of possible attitudes:

> Public defenders *ought* to have equal case loads, regardless of population density, since it is *unfair* for urban case workers to have greater work loads than their country cousins.
>
> Public defenders in cities *ought* to have higher case loads than their rural counterparts, since cities have higher concentrations of poor and disadvantaged, and thus more crime. Therefore urban personnel *should* be working harder.

> Public defenders in rural areas *ought* to have case loads as high as their more assertive colleagues in the cities.
>
> Public defenders in cities *ought* to have higher case loads than rural public defenders, since the former were foolish enough to select jobs in the cities.

Regardless of what we think ought to be, the empirical study will tell us what actually is the case.

Assume that we are studying 40 counties, of which half are classified as urban and half are rural. A tabulation of our results might look like this:

	Size of County	
Case Load	*Urban*	*Rural*
High	17	5
Low	3	15
Total	20	20

We note that out of 40 counties, 32 are consistent with the hypothesis (high load/urban or low load/rural). Only eight counties are classified in a manner inconsistent with the hypothesis. The evidence suggests that our hypothesis is confirmed.

But what about the eight inconsistent counties? Why did three urban counties have low case loads and five rural counties have high case loads? To move toward a *theory* of case loads, we need to test the impact of other independent variables on case load. Which ones might they be? Perhaps the three urban counties with low case loads are prosperous counties with a high tax base. We could study the impact of median income or tax revenue on case load. Perhaps the political party controlling the county's government has an impact. We could make the governing party our independent variable. Perhaps the nature of the crimes committed in the county can explain some of the inconsistencies. Maybe the five rural counties with high case loads have a large rate of misdemeanors—easily disposed of by the judicial system. Perhaps the other 15 rural counties have a larger occurrence of felonies. The cases may be fewer, but their adjudication might be more complex and time consuming. What other independent variables might you want to test to build our theory?

Over time, if we repeat these studies in other places and keep getting similar results, confidence will grow as to the validity of our hypothesis.

We have, to the extent that science allows, verified our hypothesis empirically and moved along the road of theory building.

CONCLUSION

In this chapter we have discussed the scientific method. Scientific reasoning is by no means the only way to understand the world. We could view the world through more traditional ways, such as through theology, a political ideology, facts or myths generated by our cultural environment, or simply what those in authority tell us. All of these, however, require faith in the sources telling us about the world and faith in those who interpret those sources for us. Scientific reasoning also requires faith, but it is a faith in ourselves and our colleagues. This is a faith based not on outside authority but on our own ability to collect and interpret data and our ability to scrutinize the research of others and to be able to reach the same conclusions they reached.

The scientific method provides us with logical steps for formulating and testing hypotheses. This thought process parallels the process used in all scientific research and remains stable. What does not remain so stable are the techniques of observation and experimentation used to verify hypotheses, research techniques, and the data analysis techniques used in reaching conclusions.

Research techniques vary with the field of study. In many of the social sciences we use some observation and experimentation techniques, but we also depend quite a bit on survey research through interviews and questionnaires. Other social sciences such as psychology may use the same techniques but put more emphasis on experimentation.

Although this chapter focuses on the scientific method, the chapters that follow concentrate on the techniques of quantitative analysis; not so much on the research design but on the techniques of measuring and counting for the purposes of analyzing data and showing how the numbers come to tell us what the facts are. All of these topics are part of the field of **statistics,** *the study of how we describe and make inferences from data*. Our aim will be to learn how to make the numbers make sense.

The techniques employed in analyzing data depend in part on the type of research design utilized and in part on other factors that we will encounter in Chapter 2, "Levels of Measurement and Forms of Data." Before going on, try working the following exercises.

EXERCISES

Exercise 1.1

1. Develop a hypothesis appropriate to your major field of interest. Word the hypothesis using one of the formats presented in this chapter.
2. Identify the dependent and independent variables. What was the reasoning behind your decision as to which was which?
3. Identify logical categories for each variable. Are you measuring amounts or attributes?
4. What is your unit of analysis that you will need to study, and how will you collect the data?
5. Put together a table similar to the one in the caseload/population example (see page 26). Simulate the numbers in the table to resemble the results you would expect to get if your hypothesis is verified.

Exercise 1.2

For the three hypotheses on pages 9-10, repeat steps 2, 3, and 4 as in Exercise 1.1.

Exercise 1.3

Formulate hypotheses useful for explaining or for measuring progress toward attaining the following normative goals. For each hypothesis, state the dependent and independent variables, their categories (or how you will measure amounts of each variable), and the units of analysis.

1. Women and men should receive equal salaries for similar occupations.
2. Housing restrictions on nonwhites should be eliminated.
3. Students studying foreign languages should study the languages of the major minority groups in their country (e.g., French in Canada or Spanish in the United States).
4. Campaign contributions by interest groups or political action committees should be limited by law.
5. Chemical and biological weapons should be eliminated.
6. The death penalty should be applied in a timely manner.
7. All who are mentally ill should receive treatment.

8. Use of illegal drugs should be eliminated.
9. All citizens should receive a college education.
10. All gun control laws should be repealed.

Exercise 1.4

Each of the following hypotheses has a flaw in either its format or its logic. Identify the flaw and correct the hypothesis.

1. There is a relationship between women and math anxiety such that women have math anxiety.
2. Are age and need for social services positively related?
3. There is a relationship between birth weight and smoking such that mothers who smoke have lower birth weight.
4. There is a relationship between British political parties and support for national health insurance such that Labour, Conservative, and Liberal Democrats support national health insurance.
5. Religion and support for church tax exemption are positively related.
6. Liberals run for public office.
7. Communication graduates earn less than public administration graduates and thus drive cheaper cars.
8. "Right-brained" people are more likely to vote for conservative candidates.

Exercise 1.5

Identify the appropriate unit of analysis for each of the following. Be as specific as possible.

1. Levels of censorship are greater among countries at war than those at peace.
2. Urban areas have higher juvenile deliquency rates than do rural ones.
3. Per capita income is higher in English counties than in the rest of Britain.
4. Two thirds of the kindergarten students in Mrs. Smith's class at Apple Valley Elementary School were absent last February 7, due to the flu.
5. Managers are more likely to contribute to charities than are technicians.
6. First-degree murder rates tend to be higher in southwestern U.S. states than in southeastern ones.

7. Of all NHL teams that year, Detroit averaged the most goals per game.
8. Elvis Presley sold more albums than Jerry Lee Lewis, Bo Diddley, Chuck Berry, or any other musicians of that era.
9. Mexico City and Cairo are the largest and second largest cities, respectively.
10. The United States had more submarine-launched ballistic missiles than did the U.S.S.R.

KEY CONCEPTS

measurement
levels of measurement/scales
nominal level of measurement
attributes versus quantities
individual nominal data
grouped nominal data
frequency distributions
dichotomies
n-category variables
ordinal level of measurement
individual ordinal data
grouped ordinal data
Likert scale
frequencies
scores
interval level of measurement
ratio level of measurement
absolute zero
demographic variables
individual interval data/raw scores
ungrouped frequency distribution
grouped interval data
class interval
closed-ended versus open-ended class intervals

2 Levels of Measurement and Forms of Data

INTRODUCTION

To do research using the scientific method, we must first collect data. The data we collect are actually measurements or counts of characteristics of the entities we are studying. There are differing kinds of measurements that we may make in our study. These measurements also fall into categories of mathematical sophistication known as levels of measurement.

Levels of measurement range from assignments of attributes, such as identifying a person's ethnic background, to assignments of numerical scores. Scores may be naturally occurring, such as a respondent's age, or may involve some scale developed by the researcher for the measurement of some characteristic, such as the respondent's attitude on some social issue. The level of measurement used by the researcher is important in that only certain techniques of data analysis should be used on data measured at specified levels. Thus the level of measurement of our data determines what we may or may not do in our analysis of the data.

MEASUREMENT

When we hear the term **measurement,** we usually think of *a very specific process such as measuring length* by means of a ruler or yardstick *or measuring weight* by means of a scale. We then assign a number corresponding to that indicated by the measuring device. Thus an object is determined to be 10 inches long and 6 inches wide, or someone weighs 61 kilograms. *In the social sciences,* **measurement** *indicates the above, but also other simpler actions such as assignment of a person to a particular category of a variable,* such as to Catholic rather than Protestant or Jewish. When we are able to assign a number with a specific meaning, we are also able to make precise comparisons. The person weighing 61 kilograms weighs 2 kilograms more than the person weighing 59 kilograms. In other instances we may just be assigning subjects to categories. If one person is an Australian and another person is a New Zealander and we categorize them according to nationality, we are also "measuring" them, but we usually do not use numerical scores. Even if we coded Australians with the number 1 and New Zealanders with the number 2, as we often do for computer data entry, the numbers could not be compared the way weights are compared. Here numbers are stand-ins for names but do not indicate amounts the way weights do.

Depending on the meaning of the variable chosen and our available methods of researching that variable, we select a particular pattern of measurement. *The kind of measurement selected will fall into one of four general categories known as* **levels of measurement** or **scales:** *nominal, ordinal, interval,* or *ratio.*

NOMINAL LEVEL OF MEASUREMENT

The term **nominal** *pertains to the act of naming. Here we are assigning our subjects, our units of analysis, to a particular category of a variable. These categories represent differing* **attributes,** *not quantities.* People differ from one another not only in quantifiable ways such as weight and height, but also in nonquantifiable ways based on attributes or characteristics possessed.

An example would be the variable gender, which contains two categories, male and female. Neither category reflects more or less "gender." There are simply two different categories. Jim happens to be male and Jill happens to be female. Also, there is no particular reason to list the categories in a particular order. The variable "religious preference" may have three categories for a particular study: Catholic, Protestant, and Jewish. There is *no* reason to place Catholic first or Protestant second. We could have as easily said Protestant, Jewish, and Catholic. Each is a different type of religious preference. The categories reflect differing types of religious preference, not differing amounts. Also, there is absolutely no reason to order these categories in terms of what we anticipate will be the ultimate number of people in each group. We are not yet concerned with the question of the size of each category. Thus each of our variables could be presented in any order of categorization.

Gender	*Religion*
Female	Jewish
Male	Protestant
	Catholic

We do require that there be a category to accommodate each subject or person studied, and that each subject go into only one category. If among the group being studied there is a Muslim and an atheist, we must either add those specific categories or add one or more residual categories to accommodate them. All three of the following sets of categories would meet our criteria.

A. *Religion*	B. *Religion*	C. *Religion*
Jewish	Jewish	Jewish
Protestant	Protestant	Protestant
Catholic	Catholic	Catholic
Muslim	Other	Muslim
Atheist	None	Other or None

If one of our subjects is a Hindu, set A above would be inappropriate unless we added the category "Hindu," but set B would be appropriate (we would put the Hindu in the category labeled "Other") and so would set C (the Hindu would be listed in the category labeled "Other or None"). The choice of what religions to list would depend on where we are doing the study. If we were doing the study in New Delhi, our categories might appear as follows:

Religion	
Hindu	Christian
Muslim	Other
Jain	None
Parsi	

In other words, include the specific denominations that we would logically expect to find.

Here are some other nominal level variables that we might encounter:

Region of Birth (U.S.)	*Race*	*Major*
New England	African American	Anthropology
Middle Atlantic	Asiatic	Communications
South	Caucasian	History
Midwest	Native American	Political Science
Great Plains	Polynesian	Sociology
Far West	Other	
None of the Above		

An Australian Aborigine now residing in the United States might be placed in "None of the Above" under Region of Birth and in "Other" under Race.

We often learn nominal categories in a particular order just for the ease of learning:

Direction	*Gender*	*Religion*
North	Men	Catholic
South	Women	Protestant
East		Jewish
West		

The category ordering helps us learn them, but means nothing in terms of the variable itself. We could list "West" before "North" or "Women" before "Men" and lose no information.

FORMS OF NOMINAL LEVEL DATA

There are two general ways that nominal data may be presented. In **individual** *form there is a listing of each individual subject in the study (often these are persons) and his or her category assignment for one or more variables.* For instance:

Name of Subject	*Gender*	*Religion*	*Region of Birth (Canada)*
Allen Jones	Male	Protestant	Prairie Provinces
Susan Smith	Female	Catholic	Atlantic Provinces
Robert Blondel	Male	Catholic	Quebec
Lisa Goldberg	Female	Jewish	Ontario

Although listing these data individually may be a first step for data entry in a computer or code sheet, rarely are we interested in the information on an individual-by-individual basis. Generally the data are presented in another form known as **grouped nominal data.** *Each category of the variable is listed and the subjects are not named but are counted (grouped) in the category into which each subject falls.* Tabulation of these numbers is in the form of a **frequency distribution.** *We list the variable, its categories, and a frequency column.* If for a group of 10 people, 4 are men and 6 are women, we would display the information as follows:

Gender	$f =$
Male	4
Female	6
$n =$	10

The letter f stands for frequency and the letter n stands for the total number of cases. The sum of the f column yields our n $(4 + 6 = 10)$. When you do a study and someone asks, "What is your n?" that is the same as asking you, "How many cases are included in your study?"

Two-category variables, such as gender, are often called **dichotomies.** Gender is a two-category nominal scale, or more simply, a nominal dichotomy. When there are more than two categories, we may specify that category number:

Religion (A Four-Category Nominal Scale)

Protestant
Jewish
Catholic
Other/None

Religion (A Seven-Category Nominal Scale)

Protestant
Catholic
Jewish
Muslim
Hindu
Other
None

Often, if our variable is *not* a dichotomy, we simply say it is measured with an ***n*-category nominal scale.** *Here, n-category means any number more than two.* We use this simplification since certain characteristics of dichotomies differentiate two-category variables from variables with more than two categories. Thus the crucial thing we need to know usually is, "Is it or is it not a dichotomy?" If it is not a dichotomy, it really does not matter to us how many categories it has. We shall return to the subject of dichotomies a bit later.

ORDINAL LEVEL OF MEASUREMENT

The term **ordinal** *refers to order or ordering.* This suggests a quantifiable ranking from most to least or some other logical sequence or ordering of a variable's categories. When the quantification is by rank order, it suggests a sequence but not yet an exact amount of a variable. To say that in terms of population size China ranks first; India, second; and the former U.S.S.R., third gives us an ordering. But we do not know how much larger China is than India. We still do not know the exact population sizes.

FORMS OF ORDINAL LEVEL DATA

This becomes a bit more complex than was the case with nominal data. By **individual ordinal** *format, we mean that we rank each individual subject from highest to lowest along the variable.* The variable is such that not just different attributes but also amounts are implied. Suppose we rank our subjects from tallest to shortest:

Name of Subject	*Height*
Allen Jones	Tallest (i.e., first tallest)
Lisa Goldberg	Second Tallest
Robert Blondel	Third Tallest
Susan Smith	Shortest (i.e., fourth tallest)

The rankings do *not* show us specific heights. We know that Jones is taller than Goldberg, but we do not know how much taller. Allen could be 1 centimeter taller than Lisa, or 6 centimeters, or a half meter. Likewise, we know that Goldberg is taller than Blondel, but not by how much.

When we actually collect our data, we generally collect it in individual format as above. In fact, we would collect it at the highest level of measurement available to us. So, ideally, we would like to have the exact height in centimeters or inches for Jones, Goldberg, Blondel, or Smith. If that is not possible, as would be the case when we rank only according to our visual observation, we could still order each one, as above, according to height, and assign the rankings of tallest, second tallest, and so on.

Often, however, when it comes to presenting data in a paper or report and we have a large number of subjects to be accounted for, we are far more likely to encounter a format known as **grouped ordinal** data. *Here, it is not the individuals who are ordered highest to lowest. Rather, individuals are placed into ranked categories ordered highest to lowest* (or lowest to highest).

Height	$f =$
Very Tall	3
Tall	7
Medium	10
Short	6
Very Short	4
$n =$	30

As with grouped nominal data, the $f =$ column (frequency =) indicates the number of people in each category of height. The sum of the frequency column is 30, indicated by n, which stands for "number," meaning the total number of people. Another example:

Economic Status	$f =$
Wealthy	10
Upper-middle income	20
Lower-middle income	30
Modest income	20
Poor	10
$n =$	90

The categories represent a sequence from highest economic status (wealthy) to second highest (upper-middle) to third highest (lower-middle) and so on down to fifth highest, or lowest (poor).

To be grouped ordinal data, the categories must represent a ranking (such as high, medium, low) or some other sequence that appears logical (such as liberal, moderate, conservative), and the categories must be presented in that sequence. We can completely reverse the sequence and the variable is still grouped ordinal.

Economic Status	$f =$
Poor	10
Modest income	20
Lower-middle income	30
Upper-middle income	20
Wealthy	10
$n =$	90

If, however, the categories are presented out of sequence, which a good researcher would never do, the variable is no longer ordinal but must be treated as if it were *grouped nominal* data.

Economic Status	$f =$
Poor	10
Upper-middle income	20
Lower-middle income	30
Wealthy	10
Modest income	20
$n =$	90

To be grouped ordinal, we would have to reorder the categories back into their logical sequence.

LIKERT SCALES

In the social sciences, particularly in survey research involving administration of a questionnaire, we often encounter an item whereby respondents are given a statement and then asked their level of agreement.

> Statement: U.N. troops should be permanently stationed in the Persian Gulf.
> Do you: Strongly Agree?
> Agree?
> Disagree?
> Strongly Disagree?

This is known as a **Likert scale** (named for its inventor) *and may be considered to be ordinal, going from most agreement to least agreement.* Actually, we are tapping both the nature of the respondent's opinion (for or against the permanent stationing of U.N. troops in the Persian Gulf) and how strongly that respondent feels on the issue. The reasoning behind treating this as a single ordinal variable is assumption of a logical sequence. The "strongly agrees" are more in agreement than those who agree not as strongly. The "agrees" are more in agreement than the "disagrees," and "disagrees" do not disagree as much as the "strongly disagrees." Therefore, "disagrees" are more in agreement than the strongly disagreeing respondents. Thus there is an ordering from most to least agreement.

What if a respondent is unsure of his or her opinion? The Likert scale is still ordinal as long as we add an "unsure" response and place it in the middle.

This is ordinal: Strongly Agree
Agree
Unsure
Disagree
Strongly Disagree

The "unsures" are less in agreement than the "agrees" but more in agreement with the proposition than those who disagree.

BOX 2.1

Level of Measurement and Dichotomies

Consider the following two dichotomies:

Gender	*Income*
Male	High
Female	Low

Of the two, gender is nominal and income is ordinal. We could reverse the sequence of their categories as follows:

Gender	*Income*
Female	Low
Male	High

Gender, of course, is still nominal. Income remains ordinal because the categories are still in their logical sequence. Note, though, that with a dichotomy, there is *no* way to take the categories out of sequence. All we can do is reverse their ordering, but even in doing that the sequence is not broken.

Consequently, a nominal dichotomy may be treated as if it were an ordinal dichotomy, even though it really is not ordinal in terms of the logic of its categories. For this reason, techniques of analysis that apply only to ordinal level data may also be applied to nominal dichotomies. While we differentiate nominal from ordinal dichotomies, many methodologists do not make that differentiation and subsequently treat all dichotomies as a single category of data.

If, however, we put the unsure response at the bottom (out of logical sequence), we only have a *nominal* scale.

Strongly Agree
Agree
Disagree
Strongly Disagree
Unsure

Those who are unsure are not more in disagreement with the proposition than those strongly disagreeing with it. So the logical sequence is broken and the ordinal nature of this variable disappears.

In studying U.S. politics, we often list the variable party identification in a way that resembles a Likert scale.

Party Identification
Strong Democrat
Democrat
Independent
Republican
Strong Republican

Though it may be debated, we treat such a variable as being an *ordinal* level of measurement by assuming that Democrat and Republican are logical opposites and there are no other meaningful "third parties." Since so few U.S. citizens belong to such "third parties," we lose only a tiny number of cases by doing so. Thus in terms of identification with the Democratic Party, we assume that the "Strong Democrat" will most often support that party's candidates, a "Democrat" usually will, and an "Independent" may or may not support the Democrat but is not predisposed either in favor or against Democrats. The "Republican" is less likely to support a Democrat, but is not as strongly predisposed against Democrats as is the "Strong Republican."

SCORES VERSUS FREQUENCIES

Up to this point, most numbers that have appeared in this text are **frequencies.** *They are headcounts or tallies indicating the number of cases in a particular category or the total number of cases measured.* The tables in Chapter 1 are composed of a series of frequencies. Thus in Table 1.7, there were 8 cloudy days on which it rained, there were 17 days having no rainfall, and so on. In Table 1.9, there were 35 blonde females out of a total of 50 blondes. In this chapter we found frequencies such as the number of males versus females or the number of individuals studied whose economic status could be classified as wealthy. *When we came to ordinal level of measurement, however, we encountered, in addition to frequencies, numbers being used to represent rankings or* **scores.** Jones ranked 1st in height, Goldberg 2nd, and Blondel ranked 3rd. These rankings were not frequencies but rather represented relative amounts of the variable being measured: tallest, second tallest, third tallest, and so on. Once we move into the interval level of measurement, we will encounter whole number scores, which are more than mere rankings.

INTERVAL AND RATIO LEVELS OF MEASUREMENT

The next level of measurement is known as an **interval level of measurement** *or an interval scale. Here the subject receives a numerical score rather than a ranking.* Imagine that our subjects undertook an hour of strenuous physical exercise, after which their body temperatures were recorded:

Name of Subject	*Temperature (individual, ordinal)*	*Temperature (individual, interval)*
Allen Jones	1st (highest)	98.8°F
Lisa Goldberg	2nd	98.7°F
Robert Blondel	3rd	98.5°F
Susan Smith	4th (lowest)	98.3°F

The interval data may be added and subtracted. They enable us to define the distance between subjects' scores. With rank orderings we knew that Jones had a higher temperature than Goldberg but not what the actual difference was. With interval level data, we know that since Jones measures 98.8°F and Goldberg 98.7°F, Jones is one-tenth of a degree warmer than Goldberg, the person with the next highest temperature. Similarly, Jones is 0.3°F warmer than Blondel and Blondel is 0.2°F warmer than Smith. Interval levels of measurement correspond to what people generally mean when they use the term *measurement.*

Suppose in the problem where we examined the height of our subjects, we listed each individual's actual height as well as the rank order. The height as measured in feet and inches (or in centimeters) would appear to be measured at the interval level, but it is really an even *higher level of measurement known as* **ratio level.**

Name of Subject	*Height (individual, ordinal)*	*Height (individual, ratio)*
Allen Jones	1st	6′0"
Lisa Goldberg	2nd	5′11"
Robert Blondel	3rd	5′7"
Susan Smith	4th	5′3"

Interval and ratio levels are nearly identical. The difference between the two is the nature of the meaning of zero. In interval data, zero is an arbitrary point, whereas in ratio data, zero is an **absolute zero,** *meaning a complete lack of the variable being measured.* An example would be temperature. In the commonly used Fahrenheit and Celsius scales, zero is arbitrarily selected and it is possible to have below-zero temperatures. In the Kelvin scale, however, zero °K (–273°C) is called *absolute zero.* In theory, it cannot get any colder than absolute zero, so there are no below-zero readings on the Kelvin scale. The distinction between interval and ratio levels of measurement leads to other mathematical implications, but for our purposes these implications are not important.

In the social sciences, many variables above the ordinal level of measurement are actually ratio level. We encounter few examples where zero is arbitrarily assigned. In fact, many statistics books do not even differentiate interval from ratio levels. We should be aware of the distinction between the two levels of measurement, but we should also be aware that throughout this text, when references are made to interval level data, we really mean both interval and ratio levels (unless told otherwise) and most examples of techniques requiring interval level of measurement will actually be applied to ratio level data.

Our levels of measurement may be thought of as increasing in sophistication, as we move from nominal to ordinal to interval to ratio. At the interval and ratio levels we assign an exact measurement with a numerical score indicating the quantity measured. Height (in inches or centimeters), weight (in pounds or kilograms), income (in dollars or pounds sterling), age (in years), academic performance (test scores) — all are examples of such data used in the social sciences.

In the social sciences *we also use interval and ratio level background information on human subjects studied:* age, years of schooling, income, and so on. These are known as **demographic variables.** For other units of analysis, a wide variety of interval scales also exist. Examples include crime rates for metropolitan areas, counties, or larger political units;

other kinds of census data such as mortality rates; percentage of people in some region who are employed in agriculture; per capita gross national product; and so on. In addition, many of us are constantly examining public opinion and voting statistics pertaining to elections at all levels of government.

As well as using existing census or other collected data, many of our endeavors involve the actual creation of interval or ratio level scores. This is done, as you will see in the next chapter, by creating indices that purport to provide interval level scores for social or political attitudes. These scores might measure religious tolerance, attitudes toward capital punishment, abortion, military intervention in the Middle East, censorship, drug use—whatever issues are salient at that time and place. Such indices are often constructed from scales of ordinal level of measurement.

We also create interval scales based on examining roll call voting by elected representatives. What is a particular senator's roll call voting score on the issue of free trade in North America? Past studies have examined judicial decisions in a similar manner. What is a particular judge's record of decisions on cases pertaining to pornography? Voting behavior studies have also been done at the international level. How often is Israel's vote in the U.N. General Assembly the same as that of the United States? How often is it the same as Egypt's?

Since the interval level of measurement is a more sophisticated level, we can do more things with interval data than with nominal or ordinal

BOX 2.2

Changing Levels of Measurement

Because certain mathematical assumptions are violated when interval level data are "created" from lower levels, there is some lingering criticism of this practice. Nevertheless, all the social and behavioral sciences make use of this.

One should also be careful of going the other way and coding an interval variable such as income as if it were only ordinal (for example, high, medium, or low income). This is a good way to present data but beyond presentation, statistical techniques should be performed on the original interval level variable and not its ordinal clone.

data. In fact, when we arrive at techniques that manipulate large numbers of variables at once, virtually all such techniques in use assume interval level data. (You will meet some of these in Chapter 14.)

FORMS OF INTERVAL LEVEL DATA

Interval level data may appear as **individual** *listings, often called* **raw scores,** as was also the case with ordinal data. For purposes of computation this is an appropriate format, since we are best served when calculations are based on all available scores. For a group of our subjects whose ages are given, if we wanted to average the scores (later we'll call this average the arithmetic mean) we would need to add all scores together and divide by the total number of cases:

Name	*Age*
Aaron	6
Bryan	15
Edie	23
Farah	16

Adding the scores yields a total of 60, and there are 4 subjects ($n = 4$). Dividing 60 by 4 we get 15.0, so the average age for this group is 15 years.

Suppose we had a larger n, say 30 people. While we (or our computer) could add up all 30 scores and divide the sum by 30, a listing of 30 scores by themselves would be difficult to interpret until we had calculated averages or other measures. By contrast, when we examine the 4 cases above it is not hard to see that we have 4 young people ranging in age from 6 to 23. It would be harder to ascertain this kind of trend from a larger listing of scores.

For this reason, we often make use of an **ungrouped frequency distribution** format for presenting interval level data to the reader. Here, *we list the scores in sequence (usually highest to lowest), making sure to include every score that actually appears in our results.* (Scores that could occur but do not actually appear in the final results could be listed with a frequency of zero or deleted from the listing, as we see fit.)

Suppose we have a group of 30 people whose scores along some variable are: 29, 28, 26, 25, 23, 22, 21, 21, 21, 20, 20, 20, 20, 20, 19, 19, 19, 18, 18, 17, 17, 16, 16, 15, 14, 14, 12, 12, 11, and 10. (These scores have been listed in descending order for our convenience.) We can see that the maximum score is 29 and the minimum score is 10. We begin listing the scores for our frequency distribution at 29 and list all scores through 10. (We need not put in scores from 9 to 0, since their frequencies will all be zero.) Also, although the scores 27, 24, and 13 do not occur in the study and their frequencies are zero, we list them for purposes of clarity.

Score	*f* =
29	1
28	1
27	0
26	1
25	1
24	0
23	1
22	1
21	3
20	5
19	3
18	2
17	2
16	2
15	1
14	2
13	0
12	2
11	1
10	1
n =	30

Later on, we will consider some techniques for calculating averages and other statistics from frequency distributions. Primarily, though, frequency distribution format is utilized in presenting results when statistics have already been calculated for the data.

A third form for interval level data is known as **grouped interval data.** This is an additional simplification of a frequency distribution and is most useful when in addition to a large number of cases, there is also a large number of possible scores, making even an ungrouped frequency distribution seem unwieldy. *The following is an example of how the data in the above frequency distribution may be grouped:*

Score	*f* =
25-29.9	4
20-24.9	10
15-19.9	10
10-14.9	6
n =	30

Each of the score ranges, 25-29.9, 20-24.9, 15-19.9, 10-14.9, is known as a **class interval** *since it indicates the space between two end points.* (Try not to confuse that usage with "interval" level of measurement.) Each class interval has both a lower and an upper limit (e.g., 25 = lower limit; 29.9 = upper limit).

The way we group the data can be tricky. Technically, for our grouped data to remain at the interval level of measurement, two criteria must be met. First, the class intervals must all be equal in size. To see this, subtract the lower limit from the upper limit of the topmost class interval: 29.9 − 25 = 4.9. That class interval covers 4.9 units in magnitude. Making the same subtraction for the other class intervals shows us that all of them are 4.9 units in size. Since that is the case, the first of the two criteria for grouped interval level data has been met.

By contrast, suppose we had grouped our data this way:

Score	*f* =
25-29.9	4
15-24.9	20
10-14.9	6
n =	30

The upper and lower class intervals are each 4.9 units in magnitude, but the class interval in the middle has a range of 9.9 units (24.9 − 15 = 9.9). The class intervals are *not* equal in magnitude, so the first criterion for grouped interval level data has not been met. What we have now is treated as if it were *grouped ordinal* level of measurement, despite the fact that the information originated from individual scores and frequency distribution data that were indeed interval level of measurement. When we grouped the data into unequal-sized class intervals we technically dropped down a level of measurement to ordinal.

The second criterion for grouped interval level data is that all class intervals must be **closed-ended.** *This means that each class interval must have both an upper and a lower limit.* Consider the following:

Score	*f* =
25 and above	4
20-24.9	10
15-19.9	10
10-14.9	6
5- 9.9	0
0- 4.9	0
n =	30

Notice the topmost class interval: It has a lower limit of 25 but no upper limit. It is an **open-ended** rather than a closed-ended class interval. Since we do not know its upper limit, we cannot assume that it is the same size as the other class intervals, all of whom are 4.9 units in magnitude. Thus as presented this variable is only grouped ordinal level data.

This does not mean that the data should *not* be presented this way, particularly if any relevant statistics can be calculated from the data in their original individual interval level format. Instead of the people in the top class interval having scores of 29, 28, 26, and 25, as in the original problem, so that they are easily grouped into a class interval of 25-29.9, suppose those 4 scores were 29, 36, 55, and 72. To preserve equal-sized class intervals, we would have to continue our groupings from 25-29.9 all the way up to 70-74.9 just to accommodate 4 subjects.

Score	$f =$
70-74.9	1
65-69.9	0
60-64.9	0
55-59.9	1
50-54.9	0
45-49.9	0
40-44.9	0
35-39.9	1
30-34.9	0
25-29.9	1
20-24.9	10
15-19.9	10
10-14.9	6
5- 9.9	0
0- 4.9	0
$n =$	30

The results are clumsy looking and hard to interpret. In this instance, it is worth our while to present the data as grouped ordinal, with a top class interval of 25 and above.

Technically, the lower limit of each class interval must also be provided for the same reason that we need to know the upper limit. The following is also only grouped ordinal level data.

Score	$f =$
25-29.9	4
20-24.9	10
15-19.9	10
14.9 and below	6
$n =$	30

We don't know the size of the lowest class interval.

If, however, in an open-ended lower-level class interval we can logically assume that the lower limit is zero, and if making this assumption gives the lowest class interval a size equal to the other class intervals, we may treat the data as *grouped interval* level.

Score	*f* =
25-29.9	4
20-24.9	10
15-19.9	10
10-14.9	4
5- 9.9	1
Below 5	1
n =	30

The "below 5" class interval may be assumed to be the same as 0-4.9.

From time to time, we may find class intervals listed as follows:

Score	*f* =
25-30	4
20-25	10
15-20	10
10-15	4
5-10	1
0- 5	1
n =	30

Where would we place a respondent whose score falls at one of the interval's limits? Suppose someone's score is 25—do we count that score in the 20 to 25 interval or in the 25 to 30 interval? If the score is *exactly* 25, we include that person in the *higher* (25-30) class interval. If the score is 25 by rounding off (e.g., 24.9—not quite 25) but we rounded it up to 25, we place that person in the lower (20 to 25) class interval. This is a needlessly confusing way of presenting class intervals and should be avoided!

TABLES CONTAINING NOMINAL LEVEL OF MEASUREMENT VARIABLES

When one or both of the variables is nominal level of measurement, care must be taken in interpretation of that table, since we may not base our

Table 2.1

	Catholic	*Protestant*	*Jewish*
Agree	78.8%	34.5	25.0
Unsure	11.2	40.5	25.0
Disagree	10.0	25.0	50.0
	100.0%	100.0%	100.0%

Table 2.2

	Catholic	*Jewish*	*Protestant*
Agree	78.8%	25.0	34.5
Unsure	11.2	25.0	40.5
Disagree	10.0	50.0	25.0
	100.0%	100.0%	100.0%

interpretation on whether or not there is clustering on a diagonal in the table. This is because the "ordering" of a nominal variable's categories is arbitrary. Suppose we ask respondents whether they agree, are unsure, or disagree with the statement that a law should be passed severely limiting a woman's right to an abortion. The attitude on abortion legislation—our dependent variable—remains ordinal as long as we keep the "unsure" response in between "agree" and "disagree." Suppose, however, our independent variable is religious preference measured by a three-category nominal scale: Catholic, Protestant, Jewish. Each of the tables (2.1 and 2.2) is an equally valid presentation of the data.

Table 2.1 shows a relationship because there appears to be clustering on the main diagonal, but it is really the *change in percentages* as one moves across the row (along the top row, 78.8% to 34.5% to 25.0%) that indicates the relationship. If we reorder the categories for religion, there is no longer a discernible clustering on a diagonal, but percentage changes are still apparent (in the top row, 78.8% to 25.0% to 34.5%). See Table 2.2.

Note that the reordering of religious categories rearranges percentages for each category of the dependent variable, but in either case, the relationship is equally strong. Since religion is nominal, and therefore the ordering of its categories is arbitrary, we look for any change in the

Table 2.3

	Catholic	*Protestant*	*Jewish*
Agree	30.0%	30.0	30.0
Unsure	40.0	40.0	40.0
Disagree	30.0	30.0	30.0
	100.0%	100.0%	100.0%

percentages to indicate a relationship. We do not examine only the diagonals.

Finally, note that in Table 2.3, if there are *no* percentage changes along the rows, there is *no relationship* between the variables. The same percentage of each group agrees (30%), is unsure (40%), and disagrees (30%). Knowing a person's religious preference, in this case, gives us *no* additional aid in predicting or explaining a person's attitude toward abortion legislation.

CONCLUSION

Statistical and analytical techniques are often geared to specific levels of measurement and forms of data. Thus to select the appropriate analytical procedure we must usually identify each variable's level of measurement—nominal, ordinal, interval, or ratio—as well as the form in which that variable appears—individual (raw score), grouped, or (for interval level variables) ungrouped frequency distribution. Some of the exercises that follow will help you to refine your ability to make these identifications.

EXERCISES

Exercise 2.1

All of the following variables are *grouped.* Indicate for each its level of measurement (see Examples).

Examples:

(1)

Residence	*f* =
House	7
Tent	3
Apartment	5
Total	15

(2)

Residence	*f* =
Palace	1
Mansion	3
Hut	11
Total	15

Both variables are called "Residence," although Example (1) is really type of residence whereas Example (2) suggests size, status, or cost of residence. Thus Example (1) is *nominal*; differing types of residence are listed but no sequencing of the categories is apparent. By contrast, Example (2) is *ordinal,* going from largest (and presumably most expensive) to smallest (and least expensive).

1.

Cost of Residence	*f* =
Above $1,000,000	3
$250,000 – $999,999	5
$100,000 – $249,999	7
$ 75,000 – $ 99,999	20
Below $75,000	18
Total	53

2.

Cost of Residence	*f* =
$75,000 – $99,999	15
$50,000 – $74,999	20
$25,000 – $49,999	10
0 – $24,999	5
Total	50

3. *The United States should withdraw all of its citizens from a known war zone.*

Response	*f* =
Strongly Agree	25
Agree	20
Unsure	5
Strongly Disagree	5
Disagree	15
Total	70

EXERCISES

4\.

Idealism	f =
Very Idealistic	3
Moderately Idealistic	5
Somewhat Idealistic	7
Not Idealistic	4
Total	19

5\.

Weight (in pounds)	f =
150 – 200	200
100 – 150	180
50 – 100	25
0 – 50	5
Total	410

6\.

Weight (in Kilograms)	f =
Over 75	200
50 – 75	180
40 – 50	15
0 – 40	15
Total	410

7\.

Media Censorship	f =
Applied to all topics	25
Applied to most topics	25
Applied only to military topics	85
No Censorship	15
Total	150

8\.

Race	f =
Automobile	6
Foot	3
Speedboat	2
Ski	1
Horse	5
$n =$	17

9\.

Education	f =
Vocational	7
Technical	8
College Preparatory	10
$n =$	25

10.

Education	*f* =
Graduate School	20
Undergraduate School	50
High School	45
Junior High or Middle School	15
Elementary School	5
Total	135

11.

Ideology	*f* =
Conservative	35
Moderate	45
Liberal	30
n =	110

12.

Eye Color	*f* =
Blue	5
Green	5
Brown	15
Hazel	5
n =	30

13.

Religion	*f* =
Protestant	200
Roman Catholic	150
Eastern Orthodox	50
Other Christian	25
Jewish	20
Muslim	30
Hindu	20
Other non-Christian	5
Total	500

14.

Religiosity	*f* =
Very High	15
High	25
Moderate	50
Low	40
Very Low	15
n =	145

15. *Religiosity (Number of times per year respondent attends religious services)*

	f =
50 – 59	0
40 – 49	5
30 – 39	20
20 – 29	50
10 – 19	20
Below 10	5
Total	100

EXERCISES

Exercise 2.2

Look at the following and determine its level of measurement.
Low Temperature (°F) on January 1 of last year:

Dayton, OH	10
New York, NY	15
Vancouver, BC	40
Sydney, NSW	70
Fairbanks, AK	–10

Despite the fact that this looks somewhat like the examples in Exercise 2.1, it is really *individual interval* level data. The variable is Low Temperature (the coldest registered temperature on January 1 of last year). The unit of analysis is city, and the cities listed are not some variable's categories. The number to the right of each city is *not* a frequency, but rather a *score*—that city's low temperature for the day. (See why it is so important to differentiate frequencies from scores?)

For each of the following, indicate the *level of measurement* and also the probable *unit of analysis.*

1. *Income*

D. Smith	$ 24,000
R. Jones	$ 60,000
M. Jackson	$500,000
P. Roberts	$ 15,500

2. *Senior Class Ranking*

J. Thomas	1st
P. Roberts	2nd
A. Albertson	3rd
C. Chen	4th

3. *Political Party (Canada)*

A. Jones	Liberal
S. Smith	Progressive Conservative
R. Blondel	Bloc Québécois
L. Goldberg	Liberal

4. *The Most Populous Countries (1990)*

Name	*Population*
China	1
India	2
U.S.S.R.	3
U.S.A.	4
Indonesia	5
Brazil	6
Japan	7

5. *Per Capita GNP - 1988 (U.S. dollars)*

Switzerland	$28,660
Japan	$23,290
Sweden	$20,883
W. Germany	$19,900
U.S.A.	$19,840
Canada	$18,090
France	$16,490

6. *Net Exports (Exports Minus Imports)-1989*

	(Billions of Dollars, U.S.)
W. Germany	71.6
Japan	64.4
Netherlands	3.6
Canada	2.1
Belgium	1.5
Italy	–13.2
France	–17.8
U.K.	–45.0
U.S.A.	–109.6

7. *Political Ideology*

D. Smith	Moderate
R. Jones	Conservative
M. Jackson	Conservative
P. Roberts	Liberal

KEY CONCEPTS

- demographic data
- operational definitions/working definitions
- conceptual definitions
- items (on an index or a scale)
- index (scale) construction
- validity
- face validity
- content validity
- criterion validity
- construct validity
- reliability
- split-half reliability
- test-retest reliability

3 Defining Variables

INTRODUCTION

In this chapter we are dealing with the way in which we develop systems of measurement for the variables we are studying. We begin by determining how we will make a measurement and what specific criteria we will use for assigning our subjects or respondents to specific categories of each variable. Attention is given to the creation of numerical scales or indices of opinions or attitudes and how we determine the validity and reliability of such scales. Selected examples of variable measurement are also presented.

GATHERING THE DATA

Let us assume that a researcher has identified one or more hypotheses to be tested in a study. The selection of the hypothesis also generally locks a researcher into studying a particular unit of analysis. If the generalizations are about individual behavior or attitudes, we normally choose human subjects as our units of analysis. If the hypothesis refers to characteristics of cities, cities or metropolitan areas become the units of analysis, and so on. In each hypothesis there are usually two variables to be studied, a dependent variable whose variation the researcher is trying to explain or predict and an independent variable that hypothetically accounts for the change in the dependent variable. To study these variables, we must be able to measure the characteristic or amount possessed by each of our units of analysis. Before we can make measurements, however, we must determine exactly what we want to measure and how we are going to take these measurements.

Suppose we have chosen individuals as our units of analysis and we intend to administer a questionnaire to each of our subjects. Also suppose, as is often the case, that we intend to begin our survey by asking each respondent for some basic background information. We would not want the subject's name if we wanted an anonymous survey, but *we might like to know such things as the subject's age, sex, marital status, religion, ethnic background, and so on.* These social characteristics may be related to many of the variables in our study. *Such background information is called* **demographic data.** If one of our hypotheses is that liberalism and age are inversely related, one of our variables—age—will be among the demographic data in the early part of our questionnaire. For our purposes, what do we mean by age and how do we propose to measure it? Do we want the respondent's age at the time he or she fills out the questionnaire? Suppose the survey is to be conducted over several weeks and to several groups of people. We might select to have the respondent indicate his or her age as of some specific date; for example, "How old were you as of February 1st of this year?" Let us assume that we are satisfied with the respondent's age at the time the survey document is filled out. Are we satisfied to know the respondent's age only in years? This is usually the case, but there are instances when we might opt for more specific information. In studying children of elementary school age, for instance, we might want the age in years plus months if we have reason to believe that, for example, a 7-year-old child may respond to certain items quite differently from a 7 1/2-year-old child.

Suppose for our study that age in years only is sufficient. How shall we get our age data? With an adult respondent we may simply use the following format:

Age: ________ years old

The respondent just fills in the blank with the appropriate year.

Most of the time this is adequate for social research, but imagine a situation in which we have reason to suspect that the respondent may misrepresent his or her age. We might want to obtain the age from documentation provided by the respondent, such as a birth certificate. What if someone said he was 18 years old but his birth certificate indicates that he is only 17 1/2 years old? The age we record depends on what we have decided in advance. If we had decided to accept whatever age the respondent gave, then this person will be listed as 18 years old. If we wanted the age as indicated on the birth certificate, we would record 17 1/2.

Likewise, in studying voting behavior, we often find instances of people claiming they had voted in a particular election when they had not. (After all, we learn that voting is a civic duty.) In this case, the respondent's answer to the question of having voted in that election may be a far less accurate operational definition than one requiring the researcher to confirm the answer by examining public voting records.

OPERATIONAL DEFINITIONS

In making such decisions we are formulating a **working definition** or an **operational definition.** For demographic data from an adult sample, we are usually satisfied to operationalize these concepts by accepting whatever response the subject provides. We would list the subject as 18 years old because that was what the subject said, and we assume that he or she is telling the truth.

The idea behind the operational definition is that once formulated and applied, there would be no disagreement as to the respondent's score or category assignment. In a particular room, some occupants might find the temperature too hot, whereas others are comfortable. Because there is disagreement among the occupants, we cannot characterize the room temperature as being either too hot or comfortable. Suppose, though, that we agree in advance to measure room temperature with a thermometer and operationally define "too hot" to be any temperature equal to or greater than 78°F. If the thermometer reads 77°F, we consider the room to be comfortable even though several occupants feel it to be too hot; if the thermometer reads 78°F, we consider the room to be too hot even though several occupants consider the room to be comfortable. Thus the operational definition, by virtue of its arbitrary specificity, eliminates for our purposes any disagreement as to whether or not the room is too hot. The disagreement comes in advance of our measurement when we decide arbitrarily that 78°F is our cutoff point.

When we move from demographic concepts to other social or political variables, the problems of operationalization may become more difficult. All of these must be addressed before we can continue our study.

In the case of research involving human subjects, we are likely to face conflicts between attributes (what we say we are), attitudes (the way we actually feel), and behaviors (what we actually do). Suppose ideology (liberal to conservative) is our variable. We could ask the respondent for a self-assignment to an ideological attribute as follows:

Do you consider yourself to be: (please indicate)

______ a Liberal?
______ a Moderate?
______ a Conservative?

Suppose the respondent checks liberal. We then ask a series of questions designed to tap attitudes that would reflect ideology, such as attitudes on abortion, aid to antidictatorial insurgencies in Latin America, censorship of "adult" magazines, and so on. Suppose the same respondent who said he or she was a liberal then gives consistently conservative responses to these attitude questions. Assuming that we have used attitude questions that reflect current major differences between liberals and conservatives so that our questions are valid, we obviously have a conflict between the respondent's self-assigned attribute (liberal) and that person's political attitudes (conservative). In designing our study, we need to know what will be most germane and useful to us—the attribute or the attitude.

A similar conflict between an attribute and a behavior could occur. Take a respondent who when asked his or her political party identification says Democrat. We then discover that in the last five elections the same respondent consistently voted for the Republican candidate. What will be most useful for us in our study, to assign that subject by attribute

(Democrat) or by behavior (Republican)? Subjects are rarely as consistent as we would like them to be, particularly when they do not perceive the topic that interests the researcher as having much direct importance in their own daily lives. Because we must live with such inconsistencies, we as researchers must make decisions about what we are trying to find out and what we will do with the information. If, for instance, our goal is to predict a respondent's vote in the next election, that person's past voting behavior is likely to be a better predictor of future voting than is the self-assigned attribute of party identification.

A related problem in forming our operational definition is that before we operationalize, we must have consensus on at least the major parts of our **conceptual definition.** *The conceptual definition is the more general definition of that concept such as one would find in a textbook or dictionary.* As an extreme example, note the term *democracy.* As we currently use it in the West, a democracy is a political system that governs based on a popular consent determined by free elections. By our standards, prior to reunification West Germany was more democratic than East Germany. Yet, the formal name for East Germany was the German Democratic Republic, and at least to a Marxist ideologue, East Germany was democratic in that the representatives of the workers and peasants, through the Communist Party, controlled the government. Clearly, we have two very different views and definitions of the word *democracy.* Any operational definitions that stem from the Western concept of a democracy will be far different from the operational definitions based on Communist interpretations.

The above case is extreme. More commonly there are agreements as to the general, conceptual definition, but disagreements as to what aspects of that conceptual definition compose the essence of the concept essential to the operational definition. An example is the attempt to operationally define a concept such as *freedom.* Is a particular country free (its citizens possess freedom) and if so, how free? Suppose we begin by looking up the dictionary definitions of freedom and selecting the portion of those definitions most germane to political freedom.

> *Freedom:* Possession of civil rights; immunity from arbitrary exercise of authority.[1]

There are two general parts to the definition: (1) civil rights and (2) exercise of authority. Should our operational definition be based on one of these? Which one? Or should we use both?

Suppose we decide to include possession of civil rights. What is a civil right and which rights should we include in the operational definition? Civil rights are rights granted to an individual based on citizenship or national residency. We might begin with the "four freedoms" in the First Amendment to the U.S. Constitution:

- freedom of religion
- freedom of speech
- freedom of the press
- freedom of assembly

To this list we could add other civil rights gleaned from the U.S. Constitution's Bill of Rights:

- the right to bear arms
- freedom from unreasonable searches and seizures
- the right to a jury trial
- freedom from double jeopardy
- freedom from cruel and unusual punishment

If we examine other documents such as the U.N. Charter or other bills of rights, we could add additional items such as the rights of certain linguistic groups to have their language used as an official national language or the rights of citizens to a specified economic standard of living.

What we include in our operational definition will reflect our individual values and levels of knowledge. Once we agree on what to include, further clarification must be undertaken to tighten our definitions. Suppose we had decided to use the four freedoms of religion, speech, press, and assembly. We still have to clarify what these mean. In the U.S. Bill of Rights, for instance, freedom of religion really referred to the government's not making laws *establishing* a particular religion. In modern times, many nations have "established" religions, even though they are, by our definition, democracies (examine the status of the Church of England in the United Kingdom). The real issue for us to examine is not whether there are official religions in a country but whether adherents to the other religions are restricted in their freedom of worship or in other civil rights.

A second consideration is that all freedoms are limited even in the most democratic of countries. For example, your religion may believe in ritual human sacrifice, but that does not mean that the state allows you to practice that ritual. Likewise, freedom of speech is limited. Recall Justice Oliver Wendell Holmes's dictum that freedom of speech does not give one the right to shout "Fire!" in a crowded theater. We limit freedom of the press through libel laws and antipornography legislation. We limit freedom of assembly by requiring permits to hold public meetings. Therefore our operational definition cannot be so tight as to disallow these kinds of limitations.

A final but crucial problem in forming operational definitions is whether or not there exist available data that will enable us to code each country in terms of the specific civil liberties chosen for inclusion in our operational definition. Is there any source of data available to us that would enable us to determine, say, the existence and level of freedom of assembly in each country? Economic and social statistics are available from several sources, but do they contain the information we need? In the case of our civil rights scores, we may have to rely on the opinions of experts who are asked to score each country for which they possess expertise in terms of the freedoms we have included. Some examples of operationalizing such variables will be discussed later.

INDEX AND SCALE CONSTRUCTION

For attitudinal variables, the operational definition usually is based on a subject's response to one or more questions designed to tap the variable being studied. In a previous example, we determined one's attitude toward abortion using a Likert-type response set.

Statement: Abortion should be illegal.
Response: Strongly Agree
Agree
Unsure
Disagree
Strongly Disagree

We could code each response as an ordinal ranking from (1) Strongly Agree to (2) Agree and so on to (5) Strongly Disagree, thus creating a rank ordering on opposition to abortion. By simply reversing the rankings, (5) Strongly Agree to (1) Strongly Disagree, we would have a rank ordering on support for abortion rather than opposition to abortion as originally ranked.

A variation on this idea is a *ladder question.*

Imagine a ladder on which those most opposed to abortion stand on the top rung and those least opposed stand on the bottom rung. Where on the ladder would you place yourself?	|__| 1 Most Opposed |__| 2 |__| 3 Unsure |__| 4 |__| 5 Least Opposed

The respondent self-selects his or her place on the ladder and the researcher codes that response by indicating the number (rank) of the rung chosen.

A second variation is a *feeling thermometer.* Instead of a ladder, the subject sees a picture of a thermometer ranging, for instance, from 0° to 100°. The accompanying statement asks the respondent to self-assign his or her own "temperature," with 100° most opposed, 50° unsure, and 0° least opposed.

Such questions may suffice to measure attitudes along single issues. A problem arises when what we are measuring is a compound variable made up of many differing attitudes. Suppose we want to measure an individual's *social conservatism.* While in its broadest sense conservatism relates to mistrust of change, in the social context we associate conservatives as taking certain positions on issues. Instead of asking the respondent to simply indicate whether he or she is conservative, we might better tap the issue by asking a series of questions, each designed to tap a separate aspect or dimension of conservatism. The issues chosen must be carefully selected to be meaningful in the current social and political context because attitudes change over time. Forty years ago, many, if not most, conservatives opposed mandatory desegregation of racially separate schools in the U.S. South. Today, few conservatives would be opposed.

Suppose we decided on five **items** (*questions*) that we considered good differentiators of conservatives from liberals in contemporary U.S. politics. The respondent would provide a Likert-type (strongly agree through strongly disagree) response to each of the following items.

1. Abortion should be illegal.
2. "Family Values" should be taught in schools.
3. Full funding for the Defense Department is needed for national security.
4. Educational and welfare issues should be primarily handled by the states or localities, not by the federal government.

5. The controlling or outlawing of handguns by the government is wrong.

As worded above, we would expect a very conservative individual to respond "strongly agree" to most of the five items whereas a strongly liberal, least conservative individual would "strongly disagree" with most of the items. *Thus we could assign points for each item's response and then sum the points for each item, giving us an* **index** *of social conservatism.*

Suppose we score each question's response as follows:

Strongly Agree	20 points
Agree	15 points
Unsure	10 points
Disagree	5 points
Strongly Disagree	0 points

If a respondent gave the most conservative response, strongly agree, to each of the five questions, that respondent would score 100 ($20 \times 5 = 100$) on the index. The person strongly disagreeing with each item would receive a total score of zero ($0 \times 5 = 0$) on the index. One who is completely unsure, assumed to be in the middle on all five items, would score 50 ($10 \times 5 = 50$).

These scores on our social conservatism index (or social conservatism *scale*) are treated, for purposes of data manipulation, as *interval* level of measurement.

We would do several other things to refine our index before utilizing it for actual research purposes. First, we initially set up our questions so that the most conservative response to each item was strongly agree, but in doing so we may have introduced bias into the response set. After several questions, the conservative respondent might automatically answer strongly agree or agree without carefully reading the question. To avoid this, we "reverse" some of the questions so that at times the most conservative response would be strongly disagree instead of strongly agree. In such instances, the most conservative response will still receive 20 points even though it was strongly disagree rather than strongly agree. In the following example, we reword two of the items, show the possible responses, and indicate (in parentheses) the number of points we will assign. In the actual questionnaire, the number of points for each response should not appear in print, but the ones coding the scores later on would use the point values to determine the final index score for each subject.

Directions: Circle the response to each of the following questions that most closely reflects your own opinion.

1. Abortions should continue to be legal.

Strongly Agree	Agree	Unsure	Disagree	Strongly Disagree
(0)	(5)	(10)	(15)	(20)

2. "Family Values" should be taught in schools.

Strongly Agree	Agree	Unsure	Disagree	Strongly Disagree
(20)	(15)	(10)	(5)	(0)

3. Full funding for the Defense Department is needed for national security.

Strongly Agree	Agree	Unsure	Disagree	Strongly Disagree
(20)	(15)	(10)	(5)	(0)

BOX 3.1

Interval Level Scores from Ordinal Level Data

These scores on our social conservatism index (or social conservatism *scale*) are treated as *interval* level of measurement for purposes of data manipulation, even though the Likert response set for each item is really only ordinal. We arbitrarily assigned the point spread and arbitrarily assumed that the difference between each adjacent response would be worth 5 points (strongly agree: 20 points – agree: 15 points equals a differential of 5 points). We have no evidence to verify that these points reflect the true amount of difference between the two responses. We did violate some mathematical assumptions in creating an interval level of measurement index out of ordinal components, but as previously indicated, this is common practice in the social and behavioral sciences. While our index was developed from only five questions, most such indices contain many more items than five. The more items we add, the more possible options of opinion we add to our index, and the closer our index gets to being truly interval level data.

4. Educational and welfare issues should be primarily handled by the federal government, not the states.

Strongly Agree	Agree	Unsure	Disagree	Strongly Disagree
(0)	(5)	(10)	(15)	(20)

5. The controlling or outlawing of handguns by the government is wrong.

Strongly Agree	Agree	Unsure	Disagree	Strongly Disagree
(20)	(15)	(10)	(5)	(0)

The very conservative respondent will answer strongly agree to items 2, 3, and 5 (for a total of 60 points) and answer strongly disagree to items 1 and 4 (for an additional 40 points). The grand total will still be 100 points for that individual.

VALIDITY

Once our questionnaire is reordered, we would *pretest* it on a group of subjects, administering it once and possibly readministering it to the same group several weeks later. During this pretesting phase we would be seeking to refine the scale by determining two things—validity and reliability.

Validity *is the extent to which the concept one wishes to measure is actually being measured by a particular scale or index.* Does the scale measure the concept it claims to measure? Is it congruent to the generally accepted definitions of the concept? For instance, if occupational income alone is being used as a measure of poverty, those with low incomes will be considered to be poor. In most instances, the measure is valid, but what about the millionaire who does not need to work and therefore has no income? This individual is not poor by anyone's definition. Thus work-related income is not necessarily a valid index of poverty.

There are several strategies for determining a measure's validity. The first two—face validity and content validity—rely on the internal logic of the measure. **Face validity** *is the extent to which the measure is subjectively viewed by knowledgeable individuals as covering the concept.* For instance, my conservatism scale developed earlier in this chapter seems valid to me. Each of the five items seems to tap a relevant distinction between more and less conservative people. If I showed the scale to others with knowledge of the subject matter and they confirmed that each item measured conservatism, I could say that the measure had face validity. If there was controversy about some item, say, the abortion question, I would have to ask if in reality the abortion stand was a valid aspect of conservatism.

Content validity is related to face validity, being based on logic and expertise. It *asks whether the measure covers* all *the generally accepted meanings of the concept.* What if I showed the conservatism index to my judges and they responded that each item had face validity but that the scale was incomplete? Several of my experts say, "What about communism? How can you measure conservatism without asking the respondent about communism?" If we concur that this item must be included in the scale to give it content validity, then I would need to add a statement such as this: "Worldwide aid for anticommunist insurgents should be increased."

Two other types of validity are less subjective and more empirical. They are known as criterion validity and construct validity.

Criterion validity *is based on our measure's ability to predict some criterion external to it.* The criterion could be in the present and currently predictable (*concurrent validity*) or it could be in the future (*predictive validity*). For instance, suppose we have designed a scale for determining whether an individual would be good in a management position with a firm. We can look at those who later became managers and compare their performance evaluations with their scale scores. If the index has criterion validity, those scoring high on it would also be expected to perform well as managers. If some aptitude test claims to measure mathematical aptitude, we would expect those receiving high scores to also earn higher grades in math. If the opposite situation should result, high scores and low grades, or if both high and low aptitude scores performed equally well in class, then the aptitude test would be a poor predictor of performance and would lack criterion validity.

Construct validity *has to do with the ability of the scale to measure variables that are theoretically related to the variable that the scale purports to measure.*

Imagine that you have developed a scale to measure overall life satisfaction. The higher the score on the index the greater is the person's life satisfaction. To establish construct validity for the scale, ask what characteristics are likely to be related to overall life satisfaction. For instance, a satisfied individual would be less likely to be a heavy drinker or a spouse or child abuser. Is this the case with those scoring high on your life satisfaction scale? If these or other theoretical attributes are associated with life satisfaction, we should be able to empirically test the relationship between one's score on your scale and alcohol consumption or incidents of abuse. If these associations are indeed found to be the case, then your measure is likely to be a valid index of life satisfaction. You have established construct validity.

Both the criterion and construct validity may be measured using techniques similar to the association and correlation measures presented in later chapters of this text.

RELIABILITY

For a measure to be **reliable,** *it must be free of measurement errors.* That is, (a) the overall score should correspond to the scores of its components, a type of internal consistency; and (b) if the measure is taken over intervals of time, the scores of individuals should remain consistent over time as well.

Split-half reliability is one way to measure internal consistency. To see if the items are all measuring the same concept, *we split our overall scale into two scales, each containing half the original items.* Suppose our original index was a 20-item scale designed to predict whether a teenager was prone to juvenile delinquency. We break the 20-item scale into two 10-item scales either by putting the odd-numbered items in one group and the even-numbered items in the other, or by assigning 10 of the items at random to one group and putting the remaining 10 items in the second. Then we compare scores by subscale. A person appearing prone to delinquency on one subscale should also appear prone to delinquency on the other. If this is the case, we may assume that the original 20-item scale is reliable in terms of internal consistency.

The second kind of reliability is **test-retest reliability,** also known as reliability over time. *Reliability in this context has to do with an individual's consistency in responding the same way to a specific item over time.* Suppose we were to administer the conservatism questionnaire twice to the same group of people and compare each set of responses. If the responses remain about the same over time, that scale is considered reliable. If responses change, the scale may not be reliable. The cause for unreliability may lie in the fact that one or more questions were vague or confusingly worded. As a result, the reader's interpretation at the second reading may have differed from the initial interpretation of the same item. For example, suppose the question on the conservatism measure about regulating handguns showed that many people opposed handgun legislation the first time they filled out the questionnaire but showed changes in their response to support legislation the second time they filled it out. Or the responses may have changed from favoring to supporting the legislation. Under normal circumstances we would consider the item unreliable and delete it from the final questionnaire, concluding that gun control attitude is not a consistent and reliable indicator of political conservatism.

Before concluding unreliability, be sure that no intervening event occurred between the first and second administration of the questionnaire that would cause a consistent one-directional shift of opinion. For instance, every time there is an attempted or successful assassination of a popular public figure, attitudes favoring gun control legislation increase. In such a case, the consistency of the response changes suggests that the item may still be a reliable indicator of social conservatism.

To be a good measure, a scale or index must be both valid and reliable. This is often a difficult order to fill given the fact that many social science concepts are difficult to define and, once defined, are subject to measurement and other human error. In addition, as Babbie (1989)[2] has pointed out, there is a certain tension between validity and reliability. Validity seeks to be inclusive, extending a measure to cover all of the meanings and nuances of the concept in question. Reliability tends to exclude nuances and multiple aspects of a variable so as to focus on what can be specifically scored. One solution is to create several measures for the same concept and see if they produce similar results.

CONCLUSION

Understanding the nature of operational definitions and formulating actual operational definitions are among the hardest tasks for students to master. While in many courses our task is to *broaden* the scope of a definition to include more and more nuances and examples, here our task is to *narrow* that definition, making it ever more specific. In many ways this task parallels the formulation of specific legal definitions. When members of a jury determine facts and thus the guilt or innocence of the accused, they base their determination in part on the judge's instructions, which include the legal definitions of the charged crimes. For example: What constitutes murder? How do first-degree murder, second-degree murder, and manslaughter differ under current state law? What facts must be proven for a jury to conclude a guilty verdict? The legal definitions given to the jury are as specific as possible, and the jury members then determine if the facts presented to them match those required by the definitions.

An expert coder, like a juror, can take the researcher's operational definition and conclude on a case-by-case basis whether the definition is met. For example, based on the operational definition supplied, the coder can determine whether or not country *x* is economically developed. But suppose we have no coder. How do we then decide whether country *x* is an economically developed country or not? One way is to make use of one or more variables as stand-ins for, or indicators of, economic development. We would pick specific variables, each of which may tap only part of the concept of economic development, such as percentage of the population in agriculture, radios per 1000 population, and so on. Finding such data and selecting valid and reliable indicators of the concept we want to measure are often not easy.

Furthermore an improper operational definition will lead to improper statistical results, because the statistics will only be as good as the data. In popular terminology, this is the GIGO Principle: "Garbage in; garbage out." Following are four general situations that lead to misleading operational definitions.

1. *Ideological Assumptions.* An aspect of the operational definition may be a debatable ideological assumption. For instance, there is an organization that rates each country on its adherence to principles of human rights, particularly its treatment of prisoners. In its rating system, capital punishment is considered to be an indicator of *reduced* human rights. Several countries, including the United States, get reduced ratings because they have capital punishment.

2. *Situational Factors.* Situations specific to the subject lead to misleading conclusions. For example, a researcher studying levels of freedom in various countries gives a certain country a low score because it is practicing censorship. The researcher does not take into account the fact that the country was at war at the time of the study. Thus censorship of militarily sensitive subjects had been instituted, whereas in peacetime there would have been no censorship. Another example would be a recent immigrant with a low IQ score. The reason for the score being low was that the IQ test administered was not in that person's native language. Thus it was the testing situation, not the person's intelligence, that led to the low score.

3. *Key Word Inconsistency.* Respondents identify incorrectly with popular terms. For example, students were asked to assign themselves to one of three categories: liberal, moderate, or conservative. Later, they responded to items dealing with policy issues normally thought to differentiate liberals from conservatives. Several who had identified themselves as conservatives responded to the specific policy items with clearly liberal preferences.

4. *Poor Predictability.* The operational definition has a poor track record in predicting what it claims to predict. For instance, many high school students in the United States take standardized aptitude tests to determine their probable performance in college. These test scores are often used as criteria for college admission and for qualification for varsity sports. Yet actual studies of the relationship between aptitude test scores and first year college grade point averages show that only about 6% of the variation in grade point average can be accounted for by such aptitude test scores. (It has been argued, though, that this is because not all who take the tests actually attend college—only those who score high enough. That may have been the case in the past, but today almost everyone can get admitted to some 2- or 4-year college in the United States. It would be interesting to see if the predictability of GPAs from such scores is now going up.)

In all of the above examples, weaknesses in the operational definitions could lead to misleading statistical results. One must guard against such pitfalls. The ancient Greek dictum of "Know thyself!" could well be expanded to say, "Know thyself . . . and thy subject!"

EXERCISES

Exercise 3.1

Assume that you are developing a written questionnaire. Develop questions and categories of response (or scoring instructions) that together form the operational definitions of the following concepts:

1. Age
2. Religion
3. Marital status
4. Party identity

5. Attitude on environmental problems
6. Attitude on rights of homosexuals
7. Attitude on compulsory national service (military or nonmilitary)
8. Attitude on tolerance toward racial, religious, or linguistic minorities
9. Attitude on tolerance of sexually related publications
10. Attitude on tolerance of cigarette smoking by others

Exercise 3.2

Assume that you are developing indices in which countries are the units of analysis. What factors would you consider in developing scales for each of the following? How might you weight these factors?

1. Political tolerance
2. Harshness of criminal penalties
3. Freedom of religion
4. Disability awareness

NOTES

1. As defined in *The American Heritage Dictionary of the English Language*, New College Edition (Boston, MA: Houghton Mifflin, 1981), p. 524.
2. Earl Babbie, *The Practice of Social Research*, 5th ed. (Belmont, CA: Wadsworth, 1989), pp. 125-126.

KEY CONCEPTS

central tendency
mean/arithmetic mean
summation symbol (capital sigma: Σ)
$\bar{x}$, $\bar{y}$, and so on
median (*Md.*)
median position (*Md. Pos.*)
array
cumulative frequency (*cf*)
mode
modal class/modal category
x-axis
f-axis
origin (of a graph)
frequency polygon
histogram
smooth curve
continuous variable
modality
unimodal, bimodal, and trimodal frequency distributions
skewness
symmetric frequency distribution
positively skewed/skewed to the right
negatively skewed/skewed to the left
stem and leaf displays
boxplots/box and whisker plots
fractiles
quartiles
deciles
percentiles

4 Measuring Central Tendency

INTRODUCTION

In this and the next chapter, we examine how to describe a set of scores on one particular variable for some group so that we may compare that group to other groups measured on the same variable. Two types of measures exist for this task: the measures of central tendency and measures of dispersion. In this chapter we discuss *the measure of* **central tendency** (also called *average* or *measure of location*), which *finds a single number that reflects the middle of the distribution of scores—the "average" (meaning typical) score for that group.* We will discuss **measures of dispersion** in Chapter 5 and then, with these topics discussed, we will be able to return to the issue of relationships between variables.

CENTRAL TENDENCY

Suppose you want to study public opinion on the issue of censorship of the arts, specifically, whether governmental agencies funding artists should refuse to fund erotic or other controversial art projects. Your subjects are alumni at a five-year college class reunion. You determine each subject's major field of study in college and ask each subject to self-assign a score on a 0-to-10 scale, where 10 indicates the most support for artistic freedom (or the least amount of censorship). Because this is to be a pretest of a much wider study, other questions will also be asked. You hypothesize that alumni who majored in the liberal arts disciplines would be far more in favor of artistic freedom than those majoring in other fields such as the sciences, business, or health. Let us assume for computational ease that your study includes 9 non-liberal arts majors (Group A) and 10 Liberal Arts majors (Group B).

The traditional way of seeing whether or not the two groups differ is to compare the average artistic freedom score for each of the groups. To find the "average" score (as you probably learned it), we add up all the scores for each group and divide by the number of students in the group. Suppose the scores are as follows:

	Group A		*Group B*
	8		9
	8		9
	8		8
	7		8
	7		8
	7		7
	6		7
	6		7
	6		6
Total =	63		6
		Total =	75

In Group A we note that we have 9 scores. Adding the 9 scores together, we get 63. Dividing 63 by 9, we get an average score of 7.0. In Group B, we have 10 scores, and the sum of those scores is 75. Dividing 75 by 10, we get an average score of 7.5. Thus Group B scored higher than Group A.

Making such a comparison would be simple were it not for the fact that in reality there are several kinds of "averages." The average we found above, called the mean, is only one of many measures of central tendency, each appropriate to certain kinds of data or certain analytical needs. In this chapter we deal with three of these measures: the mean, the median, and the mode.

THE MEAN

What we have called the "average" (a term we will now avoid, since there are several "averages") is actually called the **arithmetic mean.** We will simply call it the **mean** since although there are other kinds of means (the geometric mean and the harmonic mean), only the arithmetic mean will be used in this text.

Let us label our variable, the artistic freedom score, variable *x*. We use *x* simply to distinguish our variable from other variables that could apply to the same group. If we also wanted to know the mean social status for the same group, we could designate status as variable *y*. We might also want to know the mean income and we could designate income as variable *z*. Right now, assume that we are only interested in one variable—artistic freedom—variable *x*. To find the mean, we add up or sum all the artistic freedom scores. We call this *the summation of* x and designate it with *the uppercase Greek letter sigma* (Σ), known as the **summation symbol,** followed by the letter *x*, which designates what variable we are summing up.

$$\Sigma x = \text{the summation of } x$$

We will divide Σx by the total number of people in the group, which we designate as *n*, which stands for number, meaning *the number of cases* (people, places, or things being measured).

$$n = \text{the number of cases}$$

The quotient when we divide Σx by *n* is *the mean score* for our group *along variable* x. We designate that as $\bar{x}$, often read as ***x-bar*** because of the bar over the x. (Logically, then, the mean social status would by $\bar{y}$ and the mean income would be $\bar{z}$.)

$$\bar{x} = \text{the arithmetic mean for variable } x$$

Therefore,

$$\bar{x} = \frac{\sum x}{n}$$

$$\text{for Group A,} \quad \bar{x} = \frac{\sum x}{n} = \frac{63}{9} = 7.0$$

$$\text{for Group B,} \quad \bar{x} = \frac{\sum x}{n} = \frac{75}{10} = 7.5$$

Sometimes the scores for the group are not individually listed, but rather are presented in an *ungrouped frequency distribution.* Then we would have two columns of information: column x, which lists every theoretically possible score that actually was found in that group, and column *f,* which lists the frequency or number of times that score actually occurred in the group. For example, for Group A we note that only the scores of 8, 7, and 6 occurred. Thus the frequency distribution would look like this:

Group A

x =	f =
8	3
7	3
6	3

In this instance, it is *inappropriate* to use the formula for finding the mean that we used above! Do *not* add the x column! Do *not* count up the numbers in the x column and call it *n*! In this case, *n is the summation of the f column* and that becomes the denominator for the mean's formula. Adding the x column gives meaningless information—we must instead count every *x* the actual number of times it occurs in Group A. To do this, we *multiply* each score by the number of times it occurs, that is, by the frequency in the *f* column. In doing this, we generate a new column labeled *fx* (for *f* times *x*). The sum of that new column, $\sum fx$, becomes the numerator in the formula for the mean.

Group A

x =	f =	fx =
8	3	24
7	3	21
6	3	18
	$n = \sum f = 9$	$\sum fx = 63$

$$\bar{x} = \frac{\sum fx}{n} = \frac{\sum fx}{\sum f} = \frac{63}{9} = 7.0$$

Group B

x =	f =	fx =
9	2	18
8	3	24
7	3	21
6	2	12
	$n = \sum f = 10$	$\sum fx = 75$

$$\bar{x} = \frac{\sum fx}{n} = \frac{\sum fx}{\sum f} = \frac{75}{10} = 7.5$$

BOX 4.1

More on the Summation Symbol

Those of you who have taken several mathematics courses may be aware of the fact that we are using Σ to say "add up all the scores." Various notations placed around Σ can be used to exclude certain scores from the addition. Since in this book we will have no need to exclude any scores in a listing, we merely use Σ unadorned by other symbols. Technically, though, the full formula for the mean looks like this:

$$\bar{x} = \frac{\sum_{i=1}^{n} x_i}{n} = \frac{x_1 + x_2 + x_3 + \ldots + x_n}{n}$$

Letting i indicate the particular person whose score is being counted, the numerator is read "the summation of x-sub-i as i ranges from one (the first person) to n (the last person)." Suppose we listed the scores for Group A associating each person with a number (i) as if the i were his or her name.

Group A

$i =$	$x_i =$
9	8
8	8
7	8
6	7
5	7
4	7
3	6
2	6
1	6
	$\sum x_i = 63$

$$\bar{x} = \frac{\sum_{i=1}^{n} x_i}{n} = \frac{x_1 + x_2 + x_3 + \ldots + x_n}{n}$$

$$= \frac{x_1 + x_2 + x_3 + \ldots + x_n}{9}$$

$$= \frac{6+6+6+7+7+7+8+8+8}{9}$$

$$= \frac{63}{9}$$

$$= 7.0$$

For Group A, we simplify the notation.

$$\bar{x} = \frac{\sum x}{n} = \frac{63}{9} = 7.0$$

Note that using this formula with the frequency distribution yields the same mean as using the original formula for individual data. Conceptually the formulas do the same thing, one working from a listing of all scores and the other from the shorter *fx* summary data.

The mean is the most mathematically sophisticated and most commonly used measure of central tendency of those presented in this chapter. Mathematically the arithmetic mean is the value of *x* that satisfies the following algebraic expression:

BOX 4.2

Making Use of the Definition of the Mean

The definition of the mean provides us with one method of checking to see if we correctly calculated the mean. Let us go back to the original scores for Group A since the $\Sigma(x - \bar{x}) = 0$ formula applies to individual ungrouped data not in a frequency distribution. Remembering that the mean for Group A was 7.0, subtract 7.0 from each value of *x* and add those differences algebraically.

Group A

$x =$	$\bar{x} =$	$x - \bar{x} =$	
8	7	+1	
8	7	+1	+3
8	7	+1	
7	7	0	
7	7	0	
7	7	0	
6	7	−1	
6	7	−1	−3
6	7	−1	
$\sum x = 63$		$\sum (x - \bar{x}) = 0$	

$$\bar{x} = \frac{\sum x}{n} = \frac{63}{9} = 7.0$$

continued

$$\sum(x - \bar{x}) = 0$$

If we subtract the mean from each of the original scores and then sum the differences algebraically, then the sum of those differences is 0. Note that this formula takes into consideration not only the value of each score, but also its distance from the mean $(x - \bar{x})$. Other measures of central tendency are less sophisticated in that they do not incorporate such distances.

BOX 4.2

Continued

Had $\Sigma(x - \bar{x})$ not been 0, there would be a great likelihood that we had made a mistake in the original calculation of $\bar{x}$. If $\Sigma(x - \bar{x})$ is not 0 but is quite small, however, that may be due to rounding error.

For data in a frequency distribution, the mean is defined by the following formula.

$$\sum[(x - \bar{x})f] = 0$$

Here we factor in the frequency in which each value of x appears.

Group A					
$x =$	$f =$	$\bar{x} =$	$x - \bar{x} =$	$(x - \bar{x})f =$	
8	3	7	+1	$1 \times 3 =$	+3
7	3	7	0	$0 \times 3 =$	0
6	3	7	−1	$-1 \times 3 =$	−3
				$\sum[(x - \bar{x})f] =$	0

We will encounter such "deviation scores" as $\Sigma(x - \bar{x})$ again in the next chapter.

THE MEDIAN

A second measure of central tendency is the *median.* In a list where values of x are arranged from highest to lowest score, the score that falls in the middle is the median score. Thus the **median** *is that value of x such that there are as many scores greater than the median as there are scores less than the median.*

To find the median, *we begin by finding the person, place, or thing that possesses the median score. This middle position is known as the* **median position.** Whatever the score possessed by the person, place, or thing at the median position is the median itself. Note that the median is *not* the median position! The median is the *value* of the variable that is associated with the person, place, or thing in the median position.

Let us take it step by step. First, place the scores in an **array,** *a listing from highest to lowest* (or lowest to highest). Second, find the median position (*Md. Pos.*) by using the following formula:

$$Md.\ Pos. = \frac{n+1}{2}$$

Third, find the score associated with the median position. That score is the median (*Md.*).

Now let us look at a specific example, the list for Group A.

$x =$
8
8
8
7
7
7
6
6
6

The scores are listed in an array from highest to lowest. Since there are nine scores, $n = 9$, the median position is

$$Md.\ Pos. = \frac{n+1}{2} = \frac{9+1}{2} = \frac{10}{2} = 5$$

Thus the fifth person in the array is the one possessing the median score.

	$i =$	$x =$	
	9	8	
	8	8	
	7	8	
	6	7	
Md. Pos.—	5	7	— *Md.*
	4	7	
	3	6	
	2	6	
	1	6	

The $i =$ column is a convention for identifying the person possessing the adjacent value of x. Instead of using a name, we identify the person with the lowest score as $i = 1$, the person with the second lowest score as $i = 2$, and work up to the person with the highest score, $i = n$. The person at the median position here is therefore $i = 5$.

Note that we could have counted down from the top, the highest value of x, and arrived at the *same* conclusion.

	$i =$	$x =$	
	1	8	
	2	8	
	3	8	
	4	7	
Md. Pos.—	5	7	— *Md.*
	6	7	
	7	6	
	8	6	
	9	6	

Counting down to the 5th person ($i = 5$), we see that the adjacent value of x is 7. The median is 7.

When we calculate the median for Group B, we encounter a new complication.

$$Md.\ Pos. = \frac{n+1}{2} = \frac{10+1}{2} = \frac{11}{2} = 5.5$$

	$i =$	$x =$
	10	9
	9	9
	8	8
	7	8
	6	8
Md. Pos. = 5.5 —		
	5	7
	4	7
	3	7
	2	6

Whenever the original n is an odd number, as in Group A, $n + 1$ is even and the median position is a whole number.

Whenever the original n is an even number, as in Group B, $n + 1$ is odd and the median position is a number with a .5 decimal (in this case, 5.5).

1	6

There is no 5.5th person in the array, so we take the score of the person just below the hypothetical 5.5th and the score of the person just above that 5.5th score. In other words, we take the score of $i = 5$ and the score of $i = 6$, which are 7 and 8, respectively. The median is the midpoint of those two values and is calculated in the same way that a mean is calculated, by adding the two scores and dividing by 2. Therefore

Group B

$i =$	$x =$
10	9
9	9
8	8
7	8
6	8
5	7
4	7
3	7
2	6
1	6

Md.Pos. = 5.5 (between i = 6 and i = 5)

$$Md. = \frac{7+8}{2} = \frac{15}{3} = 7.5$$

Note that in both groups, the means equalled their respective medians.

For Group A, $\bar{x} = Md. = 7$, and

for Group B, $\bar{x} = Md. = 7.5$.

However, this is often *not* the case. Consider the following problem.

$i =$	$x =$
5	500
4	50
Md. Pos.—3	30—*Md.*
2	20
1	10
	$\sum x = 610$

There is one large score ($x = 500$) plus four much smaller scores.

$$\bar{x} = \frac{\sum x}{n} = \frac{610}{5} = 122$$

$$Md.\ Pos. = \frac{n+1}{2} = \frac{5+1}{2} = \frac{6}{2} = 3$$

Where $i = 3$ (the median position), the adjacent value of x is 30.
Thus, $Md. = 30$.
$Md. \neq \bar{x}$. The median is not equal to the mean.

Thus we see an important difference between the two measures of central tendency: the calculation of the mean takes into consideration the distance between the mean and each score; the calculation of the median does not. The one extreme value of 500 in our problem causes the mean to increase in magnitude toward that extreme value, but it does not increase the median. To illustrate, let us change the previous problem by replacing 500 with 60.

	$i =$	$x =$	
	5	60	
	4	50	
Md. Pos.——	*3*	*30*——	*Md.*
	2	20	
	1	10	
		$\sum x = 170$	

$$Md.\ Pos. = \frac{n+1}{2} = \frac{5+1}{2} = \frac{6}{2} = 3$$

Thus *Md.* = 30 and is unchanged from the previous problem.

The mean, however, is reduced considerably:

$$\bar{x} = \frac{\sum x}{n} = \frac{170}{5} = 34$$

The mean falls from 122 to 34, whereas the median—not affected by the actual values of *x*—remains the same.

GROUPED DATA

In the past, large data sets were often grouped first to ease the job of calculating by hand or by mechanical calculator, and then from the grouped data means and medians were estimated.

Many statistics books present the techniques for doing so, but we will not cover those techniques for a variety of reasons. First, they are only *estimates* of the true mean and median, which lessens their value to us. Second, in the case of the median, the technique is complex and time-consuming. Third, with today's calculators and computers, it is possible to find these measures even for very large data sets.

Why not just use computers to calculate means and medians *all* the time and instead of learning the previous techniques? Because without understanding the logic of the formulas used, a researcher may select an inappropriate measure for his or her data. Also, the risk of incorrectly interpreting the findings would increase. Finally, with small data sets, unless a personal computer with statistical software is readily at hand,

text continues on page 91

BOX 4.3

Finding the Median in a Frequency Distribution

In the case of a frequency distribution, the procedure for finding the median parallels the procedure for ungrouped data. We find the median position remembering that in a frequency distribution, $n = \Sigma f$. Therefore

$$Md.\ Pos. = \frac{n+1}{2} = \frac{\sum f + 1}{2}$$

We count up or down the frequency column until we encounter the median position and see what value of x is associated with this frequency.

Group A

$x =$	$f =$
8	3
7	3
6	3
	$n = \sum f = 9$

$$Md.\ Pos. = \frac{n+1}{2} = \frac{\sum f + 1}{2} = \frac{9+1}{2} = \frac{10}{2} = 5$$

The 5th person in the array is at the median position. Looking at the f column, we see by counting up that the first three people have the score of 6, and persons number 4, 5, and 6 all fall in the category adjacent to $x = 7$. Thus, 7 is the median.

This process is sometimes eased, particularly when there are many scores or n is large, by generating a **cumulative frequency** (cf) column. We can do this either from the top working down or from the bottom working up, although we generally work from the bottom up. Working from the bottom, we begin with the smallest score and enter its frequency in the cf column.

$x =$	$f =$	$cf =$
6	3	3

The cf column tells us that in counting our lowest score, $x = 6$, we have accounted for the first three people ($i = 1$, $i = 2$, and $i = 3$).

We move to the next highest score, $x = 7$, note its frequency, and *add* that to the number in the cf column below it. This number tells us the total accumulated number of people we have accounted for after passing a score of $x = 7$ or below.

continued

BOX 4.3

Continued

$x =$	$f =$	$cf =$
7	3	6
6	3	3

We have now accounted for the first six people ($i = 1, i = 2, i = 3, i = 4, i = 5$, and $i = 6$).

Finally, we add to our cumulative frequency column the additional three people possessing a score of $x = 8$.

$x =$	$f =$	$cf =$
8	3	9
7	3	6
6	3	3

We do this until we run out of scores possessing a frequency other than 0. At that point cf would be n, and it would remain at that number for any subsequent values of x.

Listing all possible scores:

$x =$	$f =$	$cf =$
8	3	9
7	3	6
6	3	3
5	0	0
4	0	0
3	0	0
2	0	0
1	0	0
0	0	0

We know already that our median position is 5. Since cf is 3 where $x = 6$ and cf is 6 where $x = 7$, we observe that the person possessing the median position has entered where $x = 7$. (Persons $i = 4$, $i = 5$, and $i = 6$ enter where $x = 7$.)

Since person $i = 5$, possessing the median position, must have entered where $x = 7$, 7 is the median.

Looking at Group B:

$$Md.\ Pos. = \frac{n+1}{2} = \frac{\sum f + 1}{2} = \frac{10+1}{2} = \frac{11}{2} = 5.5$$

$x =$	$f =$	$cf =$
9	2	10
8	3	8
7	3	5
6	2	2

$n = \sum f = 10$

We need to find the scores of persons $i = 5$ and $i = 6$. Where $cf = 5$, $x = 7$. Person $i = 5$ is included among those where $x = 7$.

continued

BOX 4.3

Continued

We see that since *cf* for $x = 8$ is 8, persons $i = 6$, $i = 7$, and $i = 8$ are entered where $x = 8$. Thus, person $i = 6$ possesses a score of $x = 8$.

Since the score for $i = 5$ is 7 and the score for $i = 6$ is 8, we take the midpoint of the two to find the median.

$$Md. = \frac{7+8}{2} = \frac{15}{2} = 7.5$$

The same result is found as when the median was found from the individual scores in Group B.

Let us find the mean and median for one more example of a frequency distribution.

$x =$	$f =$
10	10
9	20
8	40
7	20
6	10
5	5
4	5
3	10
2	30
1	10
0	5

To find the mean, we will need to determine n, which is Σf. We will need Σfx, so we construct an fx column. Finally, we will need a cf column to help us locate the median position and subsequently the median itself.

$x =$	$f =$	$fx =$	$cf =$
10	10	100	165
9	20	180	155
8	40	320	135
7	20	140	95
6	10	60	75
5	5	25	65
4	5	20	60
3	10	30	55
2	30	60	45
1	10	10	15
0	5	0	5
	$n = \Sigma f = 165$	$\Sigma fx = 945$	

continued

BOX 4.3

Continued

Finding the mean:

$$\bar{X} = \frac{\sum fx}{n} = \frac{\sum fx}{\sum f} = \frac{945}{165} = 5.727272 = 5.727$$

For the median position:

$$Md.\ Pos. = \frac{n+1}{2} = \frac{\sum f + 1}{2} = \frac{165+1}{2} = \frac{166}{2} = 83$$

Looking up the *cf* column, we see that when we account for the scores through $x = 6$, 75 subjects are accounted for; when we account for the scores through $x = 7$, 95 subjects are accounted for. Thus the subject at the median position, $i = 83$, enters where $x = 7$. Therefore, $Md. = 7$.

it is faster to grind out these statistics using a calculator than it is to go to a computer center, input the data, and wait for a printout.

USING CENTRAL TENDENCY

Remember that the primary purpose in calculating a measure of central tendency is to compare it to similar measures. Which class did better on a standard quiz, section one, section two, or section three? What region of the United States has the highest voter turnout, the Northeast, the South, or the Midwest? Are families in Asia larger than families in Latin America? What European country has the largest per capita GNP? In making such comparisons, we always compare means with means or medians with medians; never is the mean of one group compared to the median of another. The same point holds true for the mode, which will be discussed shortly.

Also, each measure of central tendency assumes a different level of measurement. The mean requires interval level data; the median requires at least ordinal level data; and the mode is the only one of the three which may have some limited applicability to nominal data. Accordingly, we may find and use all three measures on interval level data; we may use the median and mode (but not the mean) on ordinal level data. Only the mode can be used at the nominal level. Further discussion of this point will come a bit later in this chapter.

Before going on to the mode, the last of the measures of central tendency, it would be good to get some practice in both calculating and comparing these measures by completing Exercises 4.1 to 4.4 at the end of this chapter. When your calculations are done, compare them to the answers given at the back of the book.

THE MODE

The **mode** is a third measure of central tendency. *It is a category of a variable that contains more cases than can be found in either category adjacent to it.* Generally, it simply is the category with the largest frequency. Consider the following age distribution:

Age	*f* =
40 - 59	15
20 - 39	30
0 - 19	10
Total	55

We would call the 20 - 39 age group the **modal class** or **modal category** since it has a higher frequency than either adjacent category. Note that like the median, but unlike the mean, extreme values of the variable have no impact on the value of the mode.

One unusual characteristic of the mode is that there may be more than one mode in a particular frequency distribution. For example,

Age	*f* =
50 - 59	15
40 - 49	45
30 - 39	20
20 - 29	10
10 - 19	35
0 - 9	5
Total	130

Both the 40 - 49 and 10 - 19 age groups have more cases than the adjacent classes (above or below them), thus both are modal classes. Note that they need not each have the *same* frequency.

Another characteristic of the mode is that it requires a fairly large *n* for all modes to be correctly identified. Suppose for a sample of 26 respondents the following age distribution had been observed.

Age	$f =$
50 - 59	3
40 - 49	9
30 - 39	4
20 - 29	2
10 - 19	7
0 - 9	1
Total	26

Modal classes are 40 - 49 and 10 - 19.

If the sample size were increased from 26 to 130, the pattern with two modes might appear as before:

Age	$f =$
50 - 59	15
40 - 49	45
30 - 39	20
20 - 29	10
10 - 19	35
0 - 9	5
Total	130

The same two modal classes observed before, 40 - 49 and 10 - 19, appear.

But it is also possible that the identity of one of the modes could disappear with increased sample size.

Age	$f =$
50 - 59	15
40 - 49	45
30 - 39	20
20 - 29	20
10 - 19	15
0 - 9	15
Total	130

When the sample size was 26, the 10 - 19 class with $f = 7$ appeared to be a mode, but when the sample size increases to 130, it becomes clear that the 10 - 19 "modal class" was really due to the small size of the original sample.

For individual, interval level data, we can most clearly understand the mode by changing the data format to an ungrouped frequency distribution and then generating a graph of that distribution. To graph the frequency distribution, we lay out *two perpendicular lines, a horizontal line labeled x and a vertical line labeled f,* referred to as the ***x*-axis** and the ***f*-axis,** respectively.[1] These axes each have numerical scales much like those on rulers.

Where the x-axis intercepts the f-axis—the **origin** of the graph—both *x* and *f* are zero. The numbers on the *x*-axis increase as one moves to the right of the origin. The distance between each unit and the unit that follows is always the same; that is, the distance from zero to one is the same as the distance from one to two, and so on. On the *f*-axis, the units increase in size as one moves above the origin. Distances between the units on the *f*-axis are also equal, though these distances need not be the same as the equivalent distances on the *x*-axis.

Let us graph the ungrouped frequency distribution from Box 4.3 for which we found the mean of 5.727 and the median of 7.

x =	f =
10	10
9	20
8	40
7	20
6	10
5	5
4	5
3	10
2	30
1	10
0	5

The numbers on the axes reflect the ranges of scores. Since in our problem *x* ranges from 0 to 10, we lay out units of 0, 1, 2, 3, and so on, up to 10. Since *f* ranges from 0 to 40, we lay out distances on the *f*-axis in units of 5: 0, 5, 10, 15, . . . 40. This procedure is illustrated in Figure 4.1. For each value of *x*, we find its corresponding value of *f* and move up the graph above the value of *x*, placing a dot at the point where we are adjacent to the appropriate *f* value. Thus since where $x = 0, f = 5$, we move up the *f*-axis to $f = 5$ and place a dot. Since where $x = 1, f = 10$, we move up directly above $x = 1$ until we are on the same level as $f = 10$ and place a dot. We do this until we exhaust all values of *x* in our range of scores.

The connection of the dots may be done in three general forms. The first of these, a **frequency polygon,** is *formed by drawing a straight line from each dot to the next dot, as x increases.* The end points would be on the *x*-axis one-half unit above the highest appearing frequency and one-half unit below the lowest appearing frequency. The second form

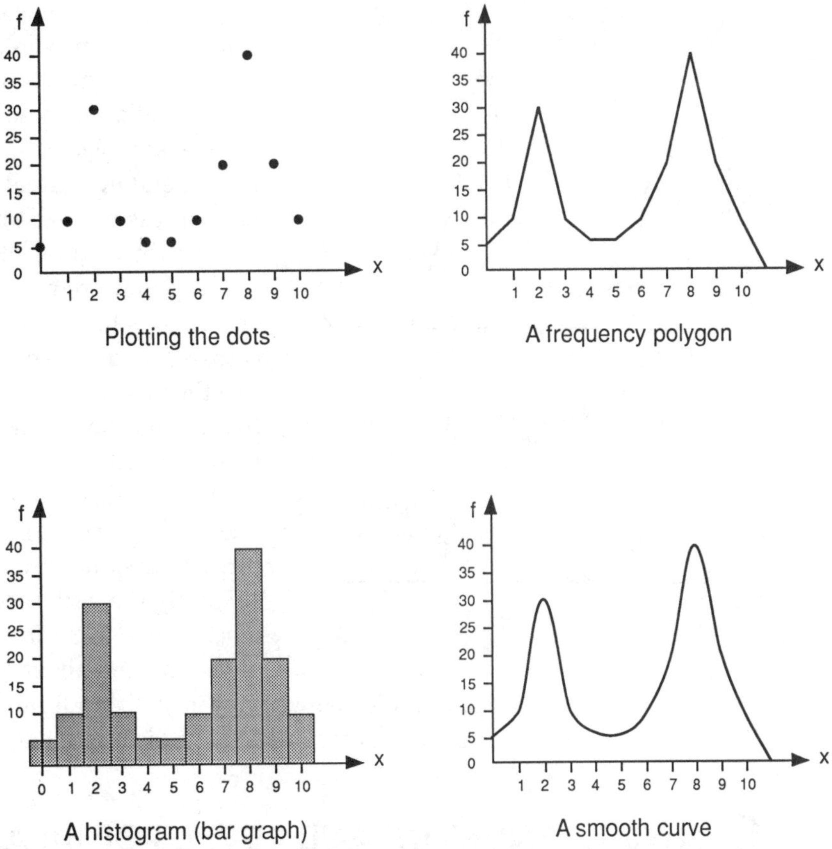

Figure 4.1 Graphs for the Ungrouped Frequency Distribution of Box 4.3

is a type of bar graph known as a **histogram.** *Bars are created from one-half unit below each value of x to one-half unit above that value.* The third form is made by joining the points in a **smooth curve.**

Technically a smooth curve is most appropriate where n is large and x is a continuous variable. By a **continuous variable** we mean *one that is not limited to whole number scores.* For example, suppose 1000 people were measured on a quiz containing 100 questions worth 1/10 of a point each. While scores range as before from 0 to 10, scores such as 9.7 or 6.9 could also appear. The smooth curve would give an accurate portrayal of the pattern produced by the many resulting dots. In the case of a smaller n or where x is a discrete, rather than a continuous, variable (the scores are only whole numbers from 0 to 10), the use of a smooth curve seems less logical than the other forms of graphing. Nevertheless the smooth curve is often utilized, no matter what the n or the nature of the variable. We use it in

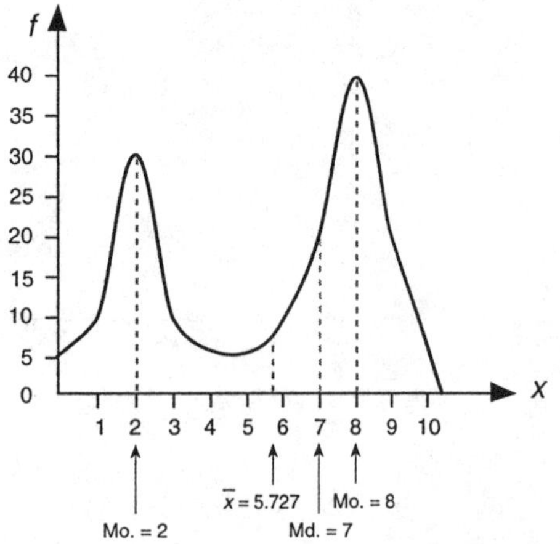

Figure 4.2
Measures of Central Tendency

our example because although the scores only range from 0 to 10 in whole numbers, they represent an underlying continuous variable. It is similar to the age variable. Although we generally only express adult ages in years, underlying the years are months, weeks, and even smaller units of time, which we do not express but exist nonetheless.

Our obtained results for all three measures of central tendency are presented in Figure 4.2. The modes are those values of x where the curve peaks, in this case, where $x = 2$ and again where $x = 8$. (Remember that the modes are the *values* of x and *not* their respective frequencies. It would be wrong to say that the modes are 30 and 40.) Since there are two modes, we say that the distribution is **bimodal.** If there were only one mode, it would be called **unimodal**; if there were three modes, the distribution would be **trimodal,** and so on. By the term **modality,** *we mean the number of modes found in the frequency distribution.* A great deal of information about a frequency distribution can be communicated verbally just by indicating its modality and *skewness,* another characteristic to be discussed shortly.

CENTRAL TENDENCY AND LEVELS OF MEASUREMENT

Now that all three measures of central tendency have been presented, we note again that the usage of a measure of central tendency is determined in part by the level of measurement of the data. The mode is the only measure that may be used on data of all measurement levels. We have already seen it applied to interval level data, both ungrouped (where we found modes of 2 and 8) and grouped (where 40 - 49 was the modal class). Similarly we could apply the mode to ordinal level data.

Feelings of Verbal Efficacy	$f =$
Very High	15
Somewhat High	30
Moderate	50
Somewhat Low	15
Very Low	10
Total	120

The modal category of feelings of verbal efficacy (effective verbal communication) is the moderate category.

We could also apply it to nominal level data, although care must be taken to remember that *all* we learn is which category had the largest frequency.

Region of Canada	$f =$
Atlantic Canada	10
Quebec	30
Ontario	35
Prairie Provinces	20
British Columbia	15
Northwest and Yukon Territories	5
Total	115

So, for our 115 respondents, the modal Canadian was from Ontario. (You could say that the modal Canadian was from either Quebec and Ontario; however, with no ordering to our categories—this is nominal data—how big a frequency do you need to determine that it is a mode?)

By contrast the median assumes either ordinal or interval level data, so we can find no median region of Canada. We have calculated the median for interval level data. To calculate it for ordinal level data, let us reexamine our Feelings of Verbal Efficacy example.

Feelings of Verbal Efficacy	$f =$	$cf =$
Very High	15	120
Somewhat High	30	105
Moderate	50	75
Somewhat Low	15	25
Very Low	10	10
Total	120	

$$Md.\ Pos. = \frac{n+1}{2} = \frac{120+1}{2} = \frac{121}{2} = 60.5$$

Examining the *cf* column, we see that both the 60th and 61st respondents enter at the moderate category. Thus moderate is the median level of Feelings of Verbal Efficacy.

Of the three measures of central tendency, the most sophisticated measure—the mean—is reserved for the most sophisticated levels of measurement—interval level and ratio level. The mean assumes that one is working with scores whose true distances from the mean can be ascertained. Since rankings do not reveal actual distances, they should not be utilized for finding a mean.

Skewness

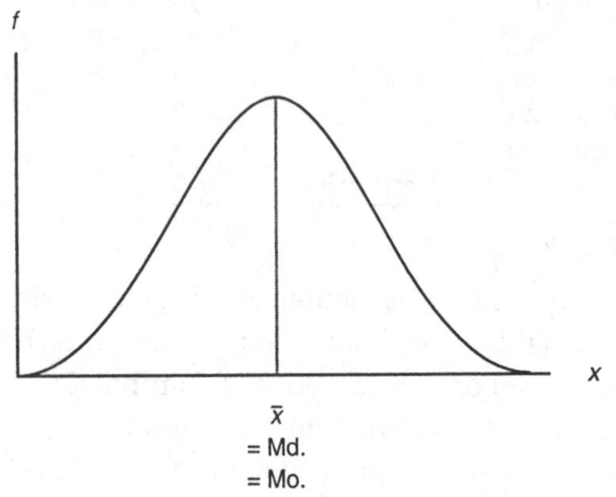

Figure 4.3
A Unimodal, Symmetric Frequency Distribution

We can also describe the nature of a frequency distribution by making reference to **skewness,** *the extent to which the distribution deviates from* **symmetry** or *the balance between the right and left halves of the curve.* In fact, we will see that we determine skewness from two measures of central tendency—the mean and median. From skewness plus the number of modes, we can visualize a very close replica of the actual appearance of a given frequency distribution. *A distribution with no skewness (skewness equals zero) is said to be a* **symmetric distribution.** In a symmetric distribution the curve to the left of the mean is a mirror image of the curve to the right of the mean.

An illustration is given in Figure 4.3. When a curve is unimodal and symmetric, the mean, median, and mode are all of the same value, as illustrated by the vertical line in the center of the figure. Imagine Figure 4.3 lying flat on a page of tracing paper folded down the vertical line at the point of central tendency. If we folded the page as we would close a book, the curve on the side to the left of the fold would fall *exactly* on the curve to the right, as shown in Figure 4.4. The left side is a mirror image of the right side, thus we have symmetry.

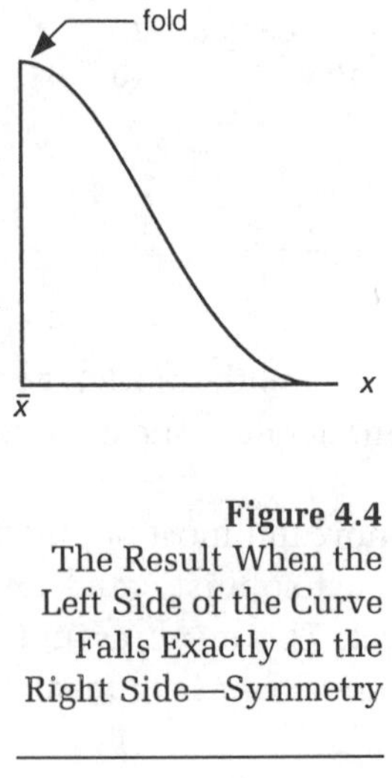

Figure 4.4
The Result When the Left Side of the Curve Falls Exactly on the Right Side—Symmetry

In a bimodal, symmetric distribution as shown in Figure 4.5, the modes are different from the mean or median, but the mean and median are equal.

Symmetry is illustrated in Figure 4.6 where the page is folded along the vertical line where $\bar{x} = Md.$ and the curve to the left of the fold falls directly over the curve to the right.

If a distribution is *not* symmetric, we say it is *asymmetric* or *skewed.* For example, look at Figure 4.7. Here the far right (called the *right tail*) of the curve extends much farther than the tail on the left.[2] If we fold the page at the mean, as in Figure 4.8, the left side of the curve does *not* fall directly on the right side. There is *no* mirror image.

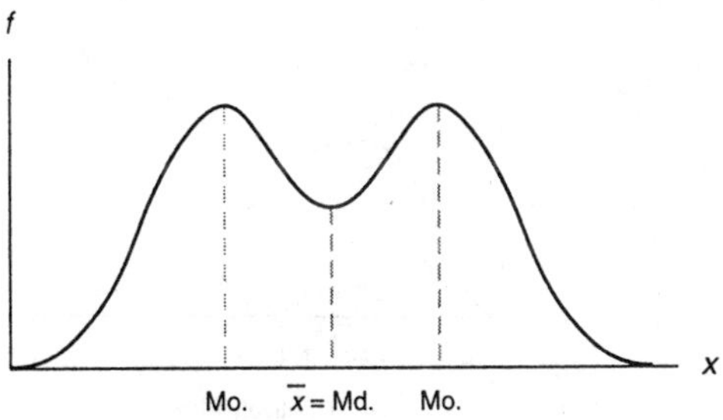

Figure 4.5
A Bimodal, Symmetric Frequency Distribution

When a curve is skewed, we indicate the direction of skewness with reference to the longer tail. Thus in Figure 4.7 the distribution is **skewed to the right** or, more formally, **positively skewed,** *since skewness is in the direction of increasing positive values on the x-axis.* If the left tail were longer than the right tail, then the curve would be **skewed to the left** or, more formally, **negatively skewed,** since in that case, *skewness would run in the direction of increasing negative values* on the *x*-axis. Both cases are illustrated in Figure 4.9.

In both examples in Figure 4.9, note that in comparing the mean to the median, the mean is always the measure of central tendency pulled most in the direction of skewness, the direction of the more extreme values of *x*. We already alluded to this phenomenon in our discussion of the median, and it points to a situation where we must choose between the mean and median as the most appropriate measure of central tendency for describing a particular distribution. (Ignore the mode for the moment.) Generally, the arithmetic mean, being the more mathematically sophisticated measure of the two, is preferable to the median. In fact, the mean is so widely used as a measure of central tendency that, as we know, most people call it the average. Nevertheless, if a distribution is highly skewed, the median may be more appropriate or honest than the mean for explaining central tendency.

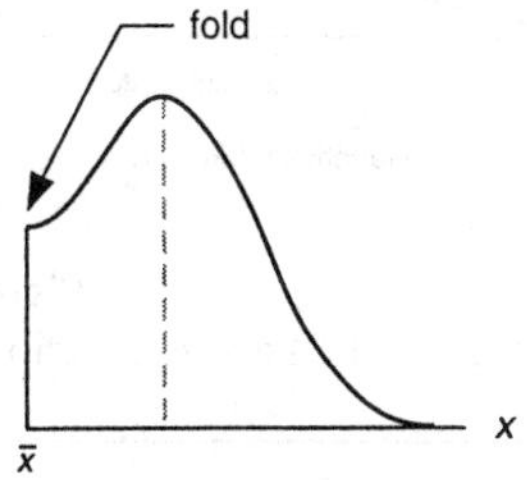

Figure 4.6
Symmetry Illustrated by Folding the Left Side of the Distribution in Figure 4.5 Over the Right Side

A clear example is found in reporting information pertaining to the variable income. In 1987, for instance, the median income for the United States was around $16,000. Here the median is preferable to the mean because although the dollar range below the median is $16,000, the same range above the median ends at $32,000. Many people earn incomes above $32,000 and have the impact of pulling the mean higher in the direction of skewness. Thus the mean figure is much higher than the $16,000 median income. Think of it this way: In calculating the median, a person earning $1 million per year has the same impact or weight as the person making $1.00 per year. In calculating the mean, however, it could take one million individuals earning $1.00 per year to counterbalance the impact of one person earning $1 million. So, although the mean is usually

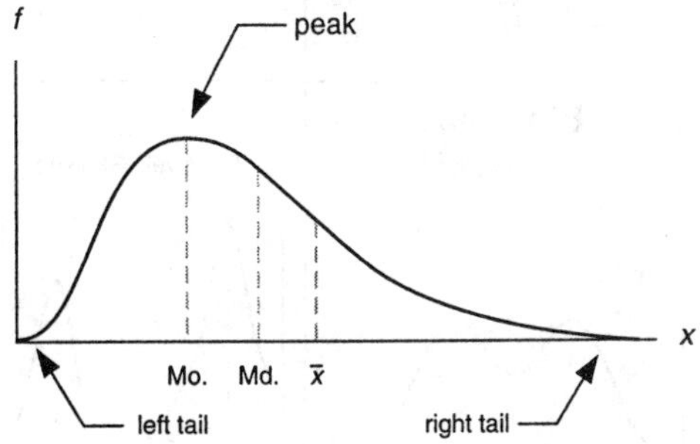

Figure 4.7
An Asymmetric or Skewed Distribution

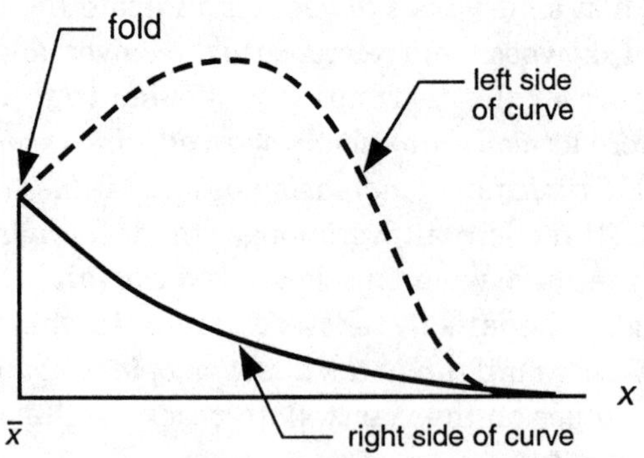

Figure 4.8
Result of Folding Figure 4.7 Over Right at the Mean

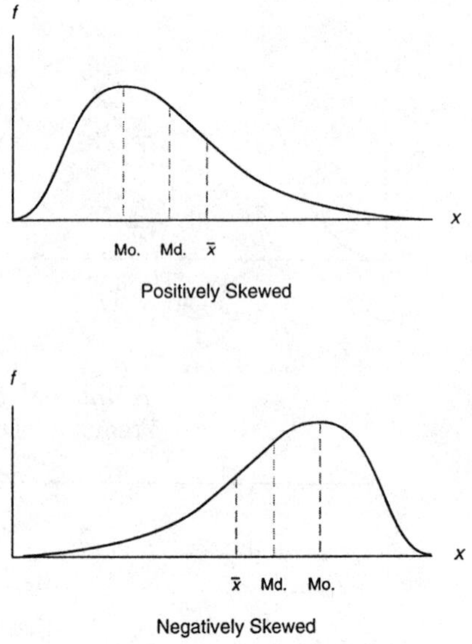

Figure 4.9
Examples of Skewed Distributions

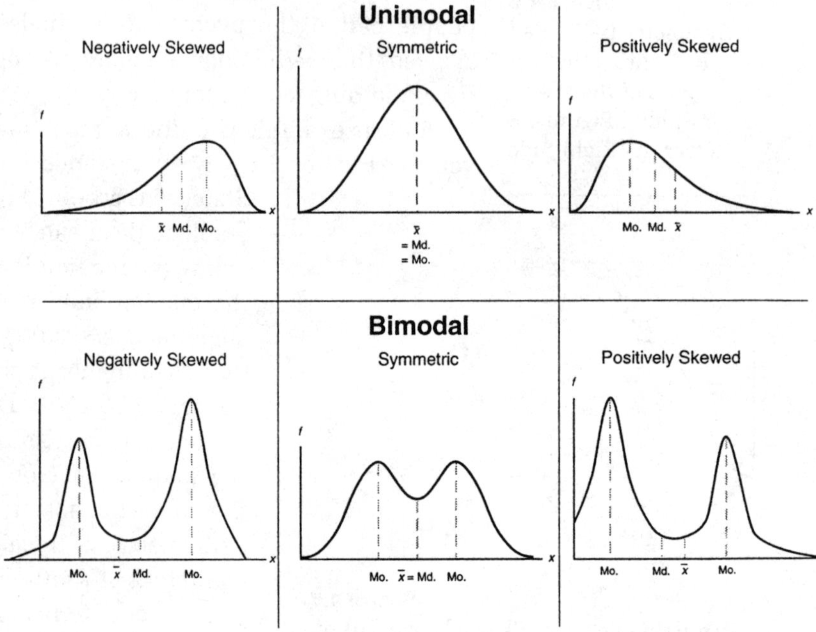

Figure 4.10
Modality and Skewness of Frequency Distributions With Reference to Central Tendency

the measure of choice, we calculate the median along with the mean to add a degree of protection to our procedure by screening out those instances where the mean may mislead.

In Figure 4.10, symmetry and skewness are compared for both unimodal and bimodal distributions and points of central tendency are noted for each example. Be aware that the two skewed bimodal distributions are drawn so that the lower peak lies in the direction of skewness, but this does not have to be the case. It is sometimes very difficult to ascertain skewness on bimodal distributions; the examples in Figure 4.10 are for illustrative purposes.

OTHER GRAPHIC REPRESENTATIONS

STEM AND LEAF DISPLAYS

In the past several years a graphic relative of the histogram, known as the stem and leaf display, has enjoyed growing popularity as a means of summarizing social data. A **stem and leaf display** *combines the visual effect of a histogram, but preserves the actual scores in small- to medium-sized data sets.*

Note the following frequency distribution, which appeared earlier in this chapter.

Age	$f =$
50 - 59	3
40 - 49	9
30 - 39	4
20 - 29	2
10 - 19	7
0 - 9	1
Total	26

Suppose we list, in ascending order, the actual ages of the 26 people whose ages comprise the frequency distribution: 8, 10, 12, 13, 16, 17, 17, 18, 21, 26, 32, 33, 35, 38, 40, 41, 42, 43, 44, 45, 47, 47, 48, 52, 53, 59.

To get our stem and leaf display, we list our class intervals in *descending order* from the top. Each class interval becomes a "stem" on the stem and leaf display. We then go through the 26 scores just presented, find the score's appropriate class interval, and list the last digit of the score to the right of the appropriate class interval. For instance, the first score in the listing is 8. We place it beside its class interval, the 0 to 9 stem, and it becomes the "leaf." The second score on our listing, 10, is in the 10 to 19 class interval. Thus the last digit of that score, 0, becomes the first leaf on the 10 to 19 stem. We get the following results for our 26 scores:

Age	
0 - 9	8
10 - 19	0 2 3 6 7 7 8
20 - 29	1 6
30 - 39	2 3 5 8
40 - 49	0 1 2 3 4 5 7 7 8
50 - 59	2 3 9

Sometimes, instead of listing the whole class interval as a stem, we use only the first digit of the lowest end of the class interval, as follows:

0	8
1	0 2 3 6 7 7 8
2	1 6
3	2 3 5 8
4	0 1 2 3 4 5 7 7 8
5	2 3 9

Also, while the scores were ordered from lowest to highest in our listing, that is not necessarily a requirement for our display. Suppose we had the following ages, listed in no particular order: 47, 25, 42, 56, 58, 37, 38, 21, 42, 42, 45, 51, 32, 48, 36. The stem and leaf display for these 15 ages would be as follows:

Age	
20 - 29	5 1
30 - 39	7 8 2 6
40 - 49	7 2 2 2 5 8
50 - 59	6 8 1

In Figure 4.11, we see how this last stem and leaf display resembles a histogram. In that figure, the histogram is drawn with horizontal rather than vertical bars. The bars resemble the display of leaves on each stem but, of course, the bars do not tell you the original scores, whereas the stem and leaf display does.

Age						
20–29	5	1				
30–39	7	8	2	6		
40–49	7	2	2	2	5	8
50–59	6	8	1			

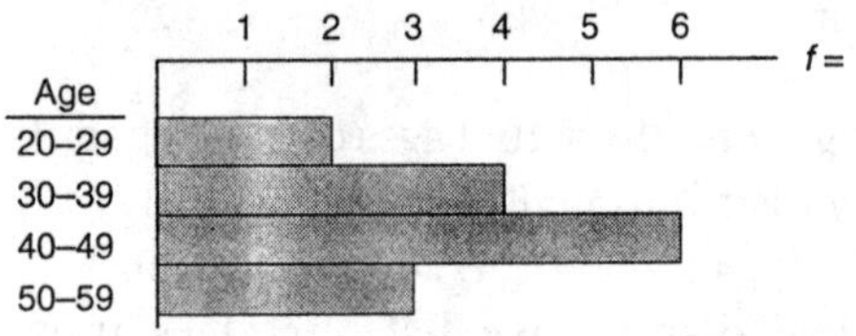

Figure 4.11
A Stem and Leaf Display and Equivalent Histogram

BOXPLOTS

Boxplots, or as they are often called, **box and whisker plots,** *are useful representations of small data sets for which the kinds of graphs presented previously would not generally provide useful information.* For example, suppose we took the 10 youngest ages from the first example used in the previous section. These are: 8, 10, 12, 13, 16, 17, 17, 18, 21, 26.

To understand boxplots, we must first introduce the concept of fractiles. **Fractiles** *divide scores into smaller groups of scores of approximately equal size.* We have already learned one such fractile, the median, which divides the scores into two subsets: the half of the scores lower than the median and the half greater than the median. We may also divide our scores into *4 sets of scores, called* **quartiles**; *10 sets of scores, called* **deciles**; or *100 sets of scores, called* **percentiles.** You may already be familiar with these terms in relation to standardized tests where, in addition to raw scores being reported, scores are often also reported as fractiles. Thus if you took such a test and scored in the 87th percentile, 87% of the scores were below or equal to your own.

Let's first find the median for our data set. Since $n = 10$, the median position is 11/2 or 5.5. Using the procedures already learned, we take the midpoint of the fifth and sixth scores in the sequence to find the median. Here, the fifth score is 16 and the sixth score is 17. So the median is 16.5 and there are 5 scores below 16.5 and 5 scores above it. To find the first quartile—the lowest one fourth of our scores—we in effect find the "median" of the given scores below the median for all the data. For these scores—8, 10, 12, 13, 16—the median position is 3 and the median is 12. Thus we say that the first quartile is 12, implying that the lowest one fourth of the scores is below 12.

The second quartile for our full data set is really the median of 16.5. This is because the two lowest quarters of the scores lie below 16.5. (Note that $1/4 + 1/4 = 1/2$, so the lowest half of all the scores is the same as the two lowest quarters of the scores.)

Finally, the third quartile divides the highest one fourth from the second highest one fourth of the scores. To find it, take the scores greater than the median of the original data set, 16.5, and find the median of that subset. The scores are 17, 17, 18, 21, and 26, the median position is 3, and the median is therefore 18.

To summarize:

26 is the highest score
18 is the third quartile
16.5 is the median and second quartile

12 is the first quartile
8 is the lowest score

A simple boxplot of these scores consists of a rectangle (our box) beginning at the first quartile and extending to the third quartile. The box is subdivided into two parts by a vertical line at the median or second quartile. From either end of the box there is a straight line (a whisker) running to the highest value and the lowest value, respectively. Our boxplot is depicted in Figure 4.12.

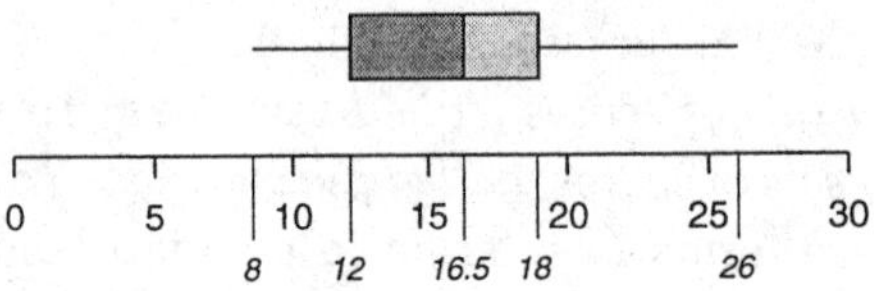

Figure 4.12
A Boxplot

The location of the median line in the box may also tell us the skewness of the data. If the left "sub-box" is the same size as the right sub-box, the distribution of scores is symmetric. If the right sub-box is much larger than the left one, the data are positively skewed and if the left sub-box is much larger than the right, the data are negatively skewed. In Figure 4.12, the left sub-box is slightly larger than the right one, suggesting a negative skew. If we calculate the mean for the data, we find that it is 15.8, less than the median, and it confirms our suspicion of negative skewness. In some cases, a skewness decision may also be made based on the length of the whiskers, with the longer whisker indicating the direction of skewness. However, in Figure 4.12, though the right (positive whisker) is longer than the left one, we already know that the skewness is negative, not positive. So we really cannot use whiskers to determine skewness unless the bigger sub-box and the longer whisker are both on the same side.

Often extremely large or extremely small scores known as *outliers*, which are far from the range of the other scores, are left out of a boxplot. This further confounds our use of sub-boxes and longer whiskers to determine skewness, so always calculate the mean to make sure. One convention used to eliminate outliers is to run the whiskers out not to the low and high scores, but only to the 10th percentile on the left and the 90th percentile on the right.

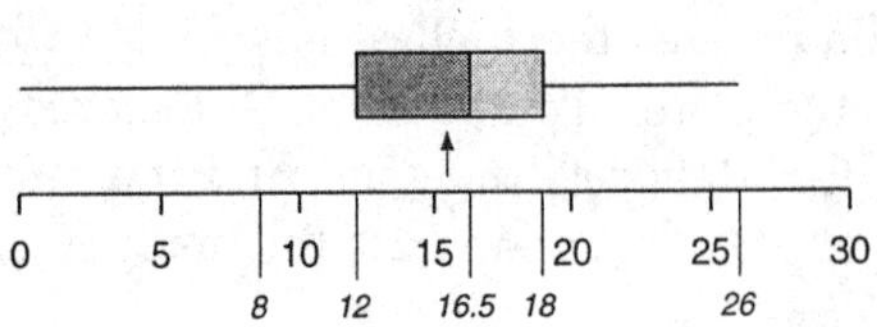

Figure 4.13
A Boxplot Consistent With Negative Skewness

Suppose that the lowest score in our data set had not been 8 but 1. The new boxplot is redrawn in Figure 4.13, where both the left whisker and left sub-box are greater than their right-sided equivalents. Negative skewness is clearly indicated. The mean for our data (indicated by an arrow) is recalculated to be 15.1, is less than the median, and confirms negative skewness. (This figure also suggests that a better name for this kind of plot might be a "box, whisker, and toothpick" plot!)

CONCLUSION

Measures of central tendency enable us to compare different groups to determine the relative relationship of their "middle" values. However, we must be aware that each of the measures discussed has its own idiosyncrasies. First, each is keyed to certain levels of measurement. Also, skewness of the data may make the use of the mean problematic. Finally, if our group's size is too small, false modes may appear in a graph.

Yet, knowing central tendency, modality, and skewness enables us to describe each group we are studying very accurately. We have covered some very powerful descriptive tools. Another useful set of descriptive tools, measures of dispersion, will be presented in the next chapter. There we will be looking for measures that give us an idea of the extent to which scores are clustered about the central tendency.

EXERCISES

Exercise 4.1

Suppose you are interested in studying the impact of oil wealth on those Middle Eastern countries fortunate enough to be petroleum exporters. To keep calculations simple, you limit your study to the Arab nations situated on the Asian continent, leaving out all non-Arab states as well as the Arab states of North Africa. Compare oil-exporting countries with nonexporting countries in terms of male life expectancies at birth. (In line with patterns elsewhere, female life expectancy is greater than male life expectancy. In the case of the nations listed below, female life expectancy is anywhere from 1 to 5 years greater than for males.)

Oil Exporters[3]		*Nonexporters*	
Country	*Male Life Expectancy*	*Country*	*Male Life Expectancy*
Bahrain	68	Jordan	60
Iraq	56	Lebanon	63
Kuwait	68	Syria	65
Oman	51	South Yemen	41
Qatar	72	North Yemen	37
Saudia Arabia	54		
United Arab Emirates	41		

Calculate the means for each group and compare them. What are your conclusions?

Exercise 4.2

Calculate the medians for the data in Exercise 4.1 and compare them. What are your conclusions? How do you reconcile the conclusions here with the ones in Exercise 4.1?

Here is where knowledge of facts and quantitative skills interface. Although there could be several others, one possible explanation is the effect of the United Arab Emirates, which is an oil exporter that resembles a nonexporting state in terms of life expectancy and several other health and economic variables. This observation might send you into the literature on the U.A.E., where you would discover that this country is really a federation of seven states, only three of which—Abu Dhabi, Dubai, and Sharjah—export oil. If we had data at hand for these seven states, we could treat them as three oil-exporting states and four nonexporting states, respectively. Barring that, we might consider excluding the U.A.E. from our study altogether due to the inconsistent nature of its constituent states.

Exercise 4.3

With apologies to the U.A.E., exclude it from the data of Exercise 4.1 and make a new comparison of both means and medians. What are your conclusions now?

Exercise 4.4

Let us now compare the two groups, oil-exporters and nonexporters, along another variable: each country's resident's average caloric intake as a percentage of the minimum caloric intake recommended by the U.N. Food and Agricultural Organization (FAO). Since we have excluded the U.A.E. and have no caloric intake data for three other countries, we have expanded our data set to include the Arab North African countries. Compare the means and medians of the two groups. Note that the scores presented below are placed in frequency distributions; be sure to use the formulas appropriate for such cases. What are your conclusions?

x = Average Citizen's Caloric Intake as a Percentage of the FAO Minimum

Oil Exporters		*Nonexporters*	
$x =$	$f =$	$x =$	$f =$
147	1	120	1
127	1	116	1
116	2	115	1
96	1	102	1
		99	2
		86	1
		76	1

SOURCE: Spencer, 1988, pp. 30-40.

Exercise 4.5

In the graphs to the left, the modes are indicated by an *M* and the other central tendencies by the letters a through h.

Match each of the six distributions to the right to the appropriate statement: Each of I through VI is used once.

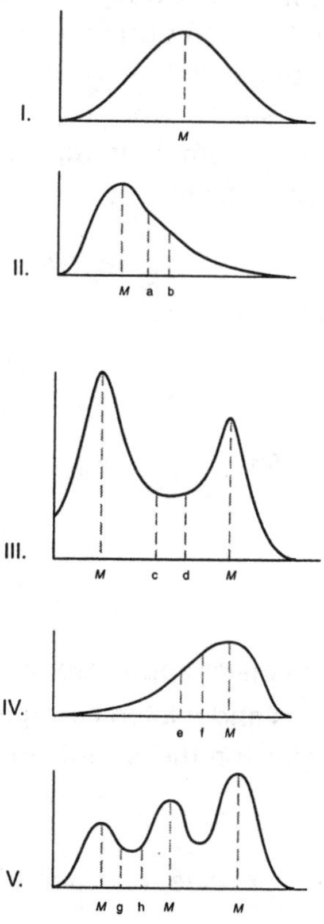

VI. None of the above

I. ______ Bimodal, positively skewed
II. ______ $\bar{x}$ = *Md.* ≠ Mode(s)
III. ______ Unimodal, symmetric
IV. ______ Trimodal, negatively skewed
V. ______ $\bar{x}$ < *Md.* < mode
VI. ______ Unimodal, positively skewed

Exercise 4.6

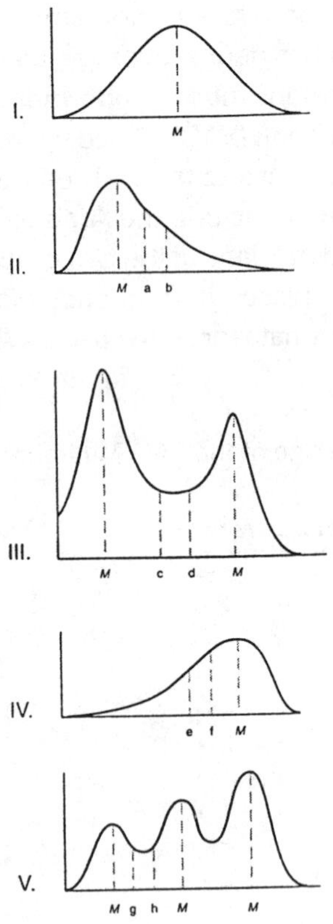

VI. None of the above

Using the graphs on the left, indicate for letters a through h whether each is a mean or a median.

a. ______
b. ______
c. ______
d. ______
e. ______
f. ______
g. ______
h. ______

Exercises 4.7–4.9

In Exercises 4.7 to 4.9, you will be working with the following simulated data. A sociologist is looking into the relative importance of selected social and economic problems as perceived by a variety of groups in a particular county. Respondents are asked to rate the importance of each issue on a scale ranging from 0 (unimportant) to 100 (most important). The issues are as follows:

HEALTH: Health care costs
SUICIDE: Suicides among senior citizens
GROWTH: Economic growth in the county
ABUSE: Child abuse
INSURANCE: Availability of health insurance
AIDS: AIDS and HIV cases in the county

Data were collected from two groups:

Officials: All elected county officials and the major appointed officials
Social Workers: A sample of social workers and related care-giving professionals working either directly for the county or for private agencies in the county

Following are the mean importance ratings, by issue, for each of the two groups.

Issue	*Officials*	*Social Workers*
HEALTH	74.95	78.22
SUICIDE	48.59	60.20
GROWTH	63.52	53.52
ABUSE	62.47	71.30
INSURANCE	73.87	73.82
AIDS	70.00	72.50

To reach conclusions about a given issue, we compare the means of the officials with those of the social workers. If we subtract the mean of the social workers from the mean of the officials, we will see the amount of difference between the two groups. Also if the difference is positive, then officials consider the issue more important than do the social workers. Negative differences mean the social workers consider the issue more important than do officials. In the case of AIDS, for instance, we see that social workers consider that issue to be slightly more important than do the officials.

	Officials	*Social Workers*	*Difference*
AIDS	70.00	72.50	–2.50

Exercise 4.7

1. How are health care costs viewed by the two groups?
2. Which group is more concerned about economic growth?
3. Compare the "social" to the "economic" variables. Which group (predictably) considers the social issues most important?
4. Do the groups really differ on the issue of health insurance coverage?

EXERCISES

Exercise 4.8

Suppose we subdivide the officials into two groups: elected and appointed.

Issue	*Elected*	*Appointed*
HEALTH	78.64	70.20
SUICIDE	42.10	50.00
GROWTH	68.75	62.75
ABUSE	60.00	65.50
INSURANCE	65.25	76.66
AIDS	68.00	72.00

Compare the two groups in terms of the importance of each issue. What overall conclusions do you reach?

Exercise 4.9

Suppose we subdivide the social worker group into two groups: managers and case workers.

Issue	*Managers*	*Case Workers*
HEALTH	84.20	69.95
SUICIDE	59.00	62.25
GROWTH	57.52	51.22
ABUSE	72.00	70.50
INSURANCE	73.82	73.82
AIDS	70.00	74.50

Compare the two groups in terms of the importance of each issue. What do you conclude?

Exercises 4.10–4.12

In Exercises 4.10 to 4.12, you will be working with data from a hypothetical study of a major U.S. automobile manufacturer. The variables, all on scales ranging from 0 to 100, are as follows:

ATTEND: Employee attendance/participation. The percentage of workdays (excluding paid vacation) that the employee was *not* absent in the 12 months prior to the study.

BOARD: The employee's support for policies supported by the corporation's board of directors in the 12 months prior to the study.

DIV: The employee's support for positions taken by the director of the employee's own division in the 12 months prior to the study.

SECUR: The employee's feelings of security in keeping his or her present job during the coming year.

PARTIC: The employee's support for a participatory decision-making approach similar to the one used by several Japanese competitors.

OPPOR: The employee's self-assessment of opportunity for promotion/advancement in the coming year.

UNION: The employee's support for expanded trade union activities in the firm.

SALARY: The employee's perception of the likelihood of receiving a major pay raise in his or her current job sometime in the next year.

From this study, two data sets were established.

MGTPOP: All 89 of the highest-level managers in the corporation. (Top and upper-middle levels.)

EMPLOY: A randomly selected sample of 50 employees working under the top and middle managers at the firm.

Following are the mean scores for the management population and the employee sample.

VARIABLE	*MGTPOP*	*EMPLOY*
ATTEND	92.35	92.92
BOARD	57.11	37.38
DIV	74.94	78.22
SECUR	56.16	50.70
PARTIC	48.59	60.20
OPPOR	42.52	36.64
UNION	62.47	71.30
SALARY	53.87	53.82

EXERCISES

Exercise 4.10

1. Do managers and employees differ in terms of attendance?
2. Which group supports the board of director's decisions more?
3. Which group has the greater sense of both security and opportunity?
4. Assuming that support for union activity expansion is a measure of employee discontent, is your conclusion in comparing the means for UNION consistent with your conclusion in part 3?

Exercise 4.11

The management group was broken down into Top and Upper-Middle Management subgroups. Mean scores for these subgroups were as follows:

	MGTPOP	
	Top	*Upper-Middle*
Variable	*Management*	*Management*
ATTEND	93.07	91.80
BOARD	69.79	47.22
DIV	70.20	78.64
SECUR	77.15	39.80
PARTIC	18.46	72.10
OPPOR	75.92	16.48
UNION	32.74	85.66
SALARY	76.66	36.10

1. Compare the two groups of managers in terms of:
   a. Attendance
   b. Support for the board of directors
   c. Support for division directors' decisions
2. Do you sense a certain level of discontent in upper-middle management?

Exercise 4.12

The employee group was divided into white collar and blue collar subgroups. Mean scores for these subgroups were as follows:

	EMPLOY	
Variable	*White Collar*	*Blue Collar*
ATTEND	93.31	92.38
BOARD	23.89	56.00
DIV	84.20	69.95
SECUR	31.31	77.47
PARTIC	81.89	30.23
OPPOR	13.17	69.04
UNION	92.44	42.09
SALARY	35.00	79.80

Compare the two groups along the eight variables. What can you conclude about them? What conclusion, if any, was unexpected to you?

Exercise 4.13

A social psychologist has developed an index to measure extroversion. It ranges from 0 to 59 (59 = most extroverted). She administers the index to one of her classes and obtains the following scores: 7, 8, 9, 10, 12, 14, 17, 20, 21, 22, 24, 28, 28, 29, 30, 30, 33, 34, 36, 37, 38, 39, 41, 43, 45, 47, 48, 52, 57, and 59. Prepare a stem and leaf display of the scores.

Exercise 4.14

For the data in Exercise 4.13, find the median, first quartile, and third quartile. Prepare a boxplot.

Exercise 4.15

A math anxiety scale ranges from a low score of 0 to a high of 100. This scale was administered to 35 students in a college orientation program. Following are the scores: 32, 20, 19, 89, 38, 39, 12, 65, 75, 21, 29, 27, 27, 93, 43, 54, 21, 33, 10, 9, 92, 18, 20, 77, 88, 47, 35, 87, 16, 87, 25, 23, 76, 22, 88. Prepare a stem and leaf display. What can you conclude about the distribution?

Exercise 4.16

Prepare a boxplot for scores above 50 in the math anxiety scale data of Exercise 4.15. How are the scores skewed?

NOTES

1. In later chapters, what we call the *f*-axis here will be used not for a frequency but for a second variable, *y*. At that time we will refer to that axis as the *y*-axis.

2. In Figures 4.7, 4.9, and 4.10, where skewness is represented, the positioning of the means and medians has been somewhat altered to aid visualization. If drawn to exact scale, the median would divide the area under each curve into two halves of equal size. As for the mean, if the curve were a solid object placed on a fulcrum, the object would be exactly balanced at the mean, much like two people on a seesaw balancing one another. In Figure 4.7, for instance, the actual locations of the mean and median are farther to the left than they appear.

3. These figures originate from several sources and are summarized in Spencer, W. (1988), *Global studies: The middle east* (pp. 37-144). Guilford, CT: The Dushkin Publishing Group. Note that subsequent to the printing of that book, the two Yemens merged into a single state.

KEY CONCEPTS

dispersion
range

mean deviation/average deviation/mean absolute deviation
absolute value

variance
standard deviation
definitional formula
computational formula

5 Measuring Dispersion

INTRODUCTION

In addition to finding measures of central tendency for a set of scores, we also calculate measures of dispersion to aid us in describing the data. **Measures of dispersion,** also called *measures of variability,* address the degree of clustering of the scores about the mean. Are most scores relatively close to the mean or are they scattered over a wider interval and thus farther from the mean? *The extent of clustering or spread of the scores about the mean determines the amount of* **dispersion.** In the instance where all scores are exactly at the mean, there is no dispersion at all; dispersion increases from zero as the spread of scores widens about the mean. In this chapter we will cover four measures of dispersion: the range, the mean deviation, the variance, and the standard deviation.

VISUALIZING DISPERSION

To begin our discussion, let us suppose that in a penology class, three teaching assistants—Tom, Dick, and Harriet—had their respective discussion groups role-play court-employed social case workers who read the files of convicted criminals and recommended to the judge the penalty to be imposed for each criminal. The teaching assistants then compared each student's recommended sentence to the one actually imposed by the real judge. The teaching assistants then rated each student on a 0-to-10 scale, with 10 being a totally accurate reproduction of the sentences that were actually handed down. There were four students in each discussion group. The results were as follows:

Tom's Group	*Dick's Group*	*Harriet's Group*
$x =$	$x =$	$x =$
8	9	10
8	8	10
8	8	6
8	7	6
$\sum x = 32$	$\sum x = 32$	$\sum x = 32$

$$\bar{x}_{\text{Tom}} = \frac{32}{4} = 8 \qquad \bar{x}_{\text{Dick}} = \frac{32}{4} = 8 \qquad \bar{x}_{\text{Harriet}} = \frac{32}{4} = 8$$

The three groups share the same mean, but the dispersion of the scores varies from none in Tom's group to some in Dick's group to even more in Harriet's group. This is illustrated in the histograms to the left. Because the distribution of individual scores clearly differed from each other in terms of their dispersion, we need to measure that dispersion in addition to measuring central tendency.

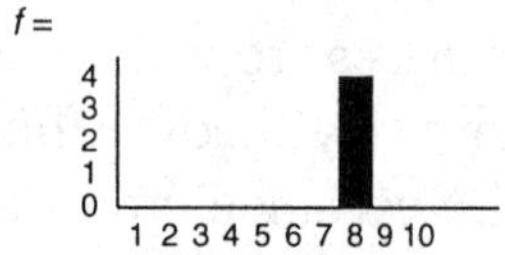

Tom's Group

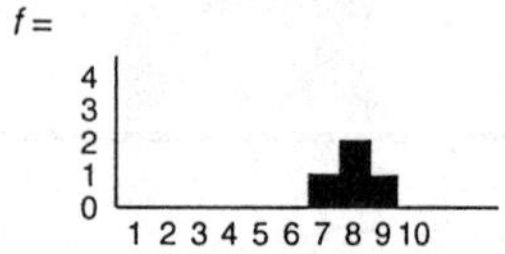

Dick's Group

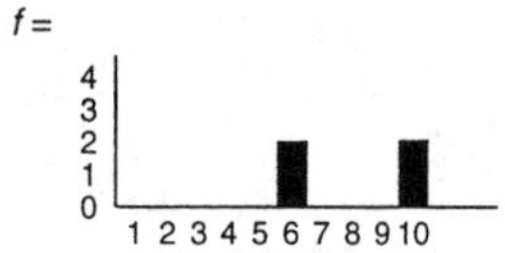

Harriet's Group

In this chapter, we will discuss measures of dispersion in an order that will ultimately bring us to the two measures used to the virtual exclusion of the others, the *variance* and its positive square root, the *standard deviation.* The first two measures we will discuss, the *range* and the *mean deviation,* may be thought of as building blocks for understanding the variance and standard deviation. Since such measures are rarely used with data having a level of measurement less sophisticated than interval level, they are usually calculated along with the calculation of the mean. With the mean as our measure of central tendency, we then calculate a measure of dispersion, most often the standard deviation.

THE RANGE

The **range** *is the simplest measure of dispersion.* It *compares the highest score and the lowest score achieved for a given set of scores.* The range can be expressed in two ways: (a) with a statement such as, "the scores ranged from _____ (the lowest score) to _____ (the highest score)," or (b) with a single number representing the difference between the highest and lowest score.

In the case of Harriet's group, whose scores were 6, 6, 10, and 10, we would say: "the scores ranged from 6 to 10." Or we could express the range as the difference between 6 and 10 (10 – 6) or 4. "The scores in Harriet's group had a mean of 8 and range of 4." Now we can compare the ranges of the three group.

- Harriet's Group:
  Scores ranged from 6 to 10. Range = 10 – 6 = 4.
- Dick's Group:
  Scores ranged from 7 to 9. Range = 9 – 7 = 2.
- Tom's Group:
  Scores ranged from 8 to 8. Range = 8 – 8 = 0.

These ranges correspond to the spread on the histograms for the three groups, with Harriet's group's scores being most dispersed about the mean, Dick's being less dispersed, and Tom's having no dispersion at all.

Although we commonly make use of the range in our day-to-day discourse, it really is not a very meaningful measure of dispersion. Because only the highest and lowest scores are taken into consideration in finding the range, the other scores have no impact. Just as in the case of the mean where an extreme value of x can distort the mean and lessen its usefulness, the use of only the extreme values can render the range less useful. Our next measure, the mean deviation, rectifies this situation.

THE MEAN DEVIATION

The **mean deviation** (*M.D.*) (also called the *average deviation* or the *mean absolute deviation*) is sensitive to every score in the set. It is based on a strategy of first finding out how far each score deviated from the mean of the scores (the distance from each score to the mean), summing these distances to find the total amount of deviation from the mean in the entire set of scores, and dividing by the number of scores in the set. The result is a mean, or "average," distance that a score deviates from the mean.

To get the mean deviation, we first find the distance between each score and the mean by subtracting the mean from each score. Let us use Harriet's group as an example.

Harriet's Group

$x =$	$\bar{x} =$	$x - \bar{x} =$
10	8	2
10	8	2
6	8	−2
6	8	−2
$\sum x = 32$		

$$\bar{x} = \frac{32}{4} = 8$$

At this juncture we encounter a problem: We cannot add up the $x - \bar{x}$ column to get the total amount of deviation in the system. Recalling that the mean is the value of x that satisfies the expression $\Sigma(x - \bar{x}) = 0$, we can see that if $\bar{x} = 8$, adding algebraically, the $x - \bar{x}$s for each student in Harriet's group produce a sum of zero:

$$\sum(x - \bar{x}) = 2 + 2 - 2 - 2 = 4 - 4 = 0$$

This is because the positive deviations (where x is greater than the mean) exactly balance the negative deviations (where x is less than the mean).

Recall that we currently are seeking the distance from each score to the mean, without regard to direction, that is, we do not care whether x is greater or less than $\bar{x}$. Like a car's odometer, we want to count the distances traveled, disregarding the direction or directions in which we drove. We do this by taking the **absolute value** of each $x - \bar{x}$, *the distance disregarding its sign* (in effect treating all $x - \bar{x}$'s as if they were positive numbers). We symbolize the absolute value of a deviation as $|x - \bar{x}|$. When we add up all these absolute values, $\Sigma|x - \bar{x}|$, we get the total amount of deviation of the scores from the mean. When we divide that sum by the total number of scores, we get the "average" amount (the mean amount) that a score deviated from the mean of all of the scores: the mean deviation.

Thus:

$$M.D. = \frac{\sum |x - \bar{x}|}{n}$$

For Harriet's Group:

	$x =$	$\bar{x} =$	$x - \bar{x} =$	$\lvert x - \bar{x} \rvert =$
	10	8	2	2
	10	8	2	2
	6	8	−2	2
$n = 4$	6	8	−2	2
	$\sum x = 32$			$\sum \lvert x - \bar{x} \rvert = 8$

$$\bar{x} = \frac{32}{4} = 8 \qquad M.D. = \frac{\sum |x - \bar{x}|}{n} = \frac{8}{4} = 2.0$$

For Dick's Group:

	$x =$	$\bar{x} =$	$x - \bar{x} =$	$\lvert x - \bar{x} \rvert =$
	9	8	1	1
	8	8	0	0
	8	8	0	0
$n = 4$	7	8	−1	1
	$\sum x = 32$			$\sum \lvert x - \bar{x} \rvert = 2$

$$\bar{x} = \frac{32}{4} = 8 \qquad M.D. = \frac{\sum |x - \bar{x}|}{n} = \frac{2}{4} = 0.5$$

For Tom's Group:

	$x =$	$\bar{x} =$	$x - \bar{x} =$	$\lvert x - \bar{x} \rvert =$
	8	8	0	0
	8	8	0	0
	8	8	0	0
$n = 4$	8	8	0	0
	$\sum x = 32$			$\sum \lvert x - \bar{x} \rvert = 0$

$$\bar{x} = \frac{32}{4} = 8 \qquad M.D. = \frac{\sum |x - \bar{x}|}{n} = \frac{0}{4} = 0$$

These results are in keeping with our expectations: Harriet's group has the largest mean deviation, Dick's has a smaller one, and Tom's has the smallest (a value of zero).

THE VARIANCE AND STANDARD DEVIATION

The formula for the **variance** *resembles that of the mean deviation except that* $\Sigma|x-\bar{x}|$ *is replaced by the expression* $\Sigma(x-\bar{x})^2$. Instead of taking the absolute value of each deviation, we square it to get rid of negative numbers. (Remember that a negative number times itself is a positive number, just as a positive number times itself is a positive number.) Since the squares of the deviations greater than one unit will be much larger than their respective absolute values, $\Sigma(x-\bar{x})^2$ will usually be larger than $\Sigma|x-\bar{x}|$ and the final variance will usually be larger than the mean deviation. *To adjust for this and produce a result more comparable to the mean deviation (more like an "average" amount of deviation), we often take the positive square root of the variance, thus producing the* **standard deviation,** indicated for now by the letter *s*. Thus

$$\text{Variance} = s^2 = \frac{\sum (x-\bar{x})^2}{n}$$

$$\text{Standard Deviation} = s = \sqrt{\frac{\sum (x-\bar{x})^2}{n}}$$

Let us calculate s^2 and s for our three groups—Tom's, Dick's, and Harriet's—whose mean deviations were 0, 0.5, and 2.0, respectively.

Tom's Group

$x =$	$\bar{x} =$	$x - \bar{x} =$	$(x-\bar{x})^2 =$
8	8	0	0
8	8	0	0
8	8	0	0
8	8	0	0
			$\Sigma(x-\bar{x})^2 = 0$

Thus,

$$s^2 = \frac{\sum (x-\bar{x})^2}{n} = \frac{0}{4} = 0$$

$$s = \sqrt{\frac{\sum (x-\bar{x})^2}{n}} = \sqrt{\frac{0}{4}} = \sqrt{0} = 0$$

The variance and standard deviation both equal zero, as does the mean deviation, for this group in which there is *no* dispersion at all.

Dick's Group

$x =$	$\bar{x} =$	$x - \bar{x} =$	$(x - \bar{x})^2 =$
9	8	1	1
8	8	0	0
8	8	0	0
7	8	−1	1
			$\sum(x - \bar{x})^2 = 2$

Thus,

$$s^2 = \frac{\sum (x - \bar{x})^2}{n} = \frac{2}{4} = \frac{1}{2} = 0.5$$

$$s = \sqrt{\frac{\sum (x - \bar{x})^2}{n}} = \sqrt{\frac{2}{4}} = \sqrt{\frac{1}{2}} = 0.707 = 0.7$$

Remember that it is the standard deviation (0.7), not the variance, which substitutes for the mean deviation (0.5).

Harriet's Group

$x =$	$\bar{x} =$	$x - \bar{x} =$	$(x - \bar{x})^2 =$
10	8	2	4
10	8	2	4
6	8	−2	4
6	8	−2	4
			$\sum(x - \bar{x})^2 = 16$

Thus,

$$s^2 = \frac{\sum (x - \bar{x})^2}{n} = \frac{16}{4} = 4.0$$

$$s = \sqrt{\frac{\sum (x - \bar{x})^2}{n}} = \sqrt{\frac{16}{4}} = \sqrt{4} = 2.0$$

Let us compare our measures. See histograms at top of next page.

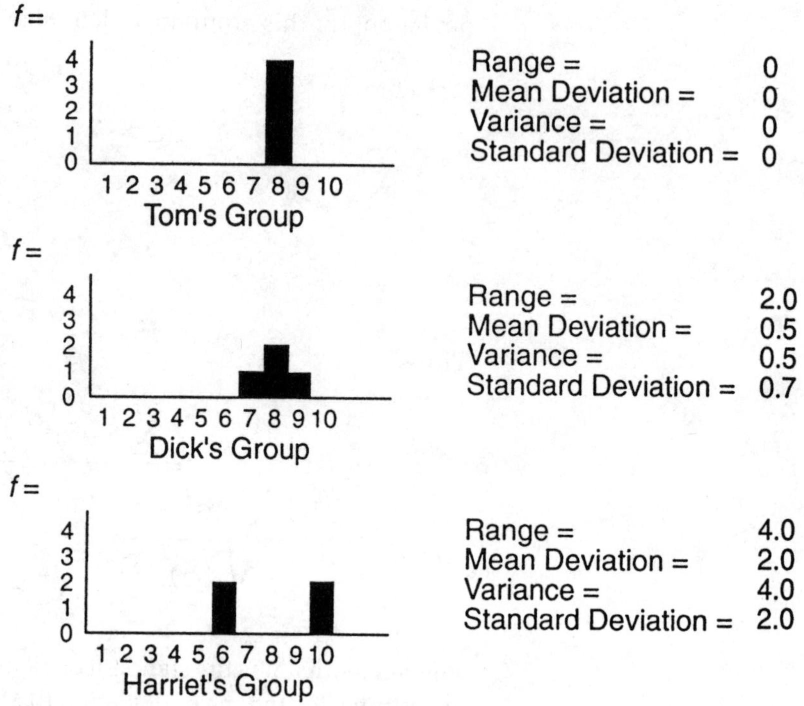

Below are the dispersion measures for the artistic freedom for the non-Liberal Arts majors, Group A, presented in Chapter 4.

Group A

	$x =$	$\bar{x} =$	$x - \bar{x} =$	$\|x - \bar{x}\| =$	$(x - \bar{x})^2 =$
	8	7	1	1	1
	8	7	1	1	1
	8	7	1	1	1
	7	7	0	0	0
	7	7	0	0	0
	7	7	0	0	0
	6	7	−1	1	1
	6	7	−1	1	1
$n = 9$	6	7	−1	1	1
	$\Sigma x = 63$			$\Sigma \|x - \bar{x}\| = 6$	$\Sigma (x - \bar{x})^2 = 6$

The scores range from 6 to 8. Range $= 8 - 6 = 2$

$$\bar{x} = \frac{\sum x}{n} = \frac{63}{9} = 7.0$$

$$M.D. = \frac{\sum |x - \bar{x}|}{n} = \frac{6}{9} = \frac{2}{3} \cong 0.67$$

$$\text{Variance} = s^2 = \frac{\sum (x - \bar{x})^2}{n} = \frac{6}{9} = \frac{2}{3} \cong 0.67$$

$$\text{Standard Deviation} = s = \sqrt{0.67} = 0.82$$

Summary	*Group A*
Range =	2.00
Mean Deviation =	0.67
Variance =	0.67
Standard Deviation =	0.82

As mentioned, the variance and standard deviation are the most widely used measures of dispersion in statistics, even though on the face of it the mean deviation would appear to be the most logical measure (and easiest to calculate) of the three. The reason is that the standard deviation has meaning in terms of a common frequency distribution known as the *normal curve,* which we will encounter later in this text.

THE COMPUTATIONAL FORMULAS FOR VARIANCE AND STANDARD DEVIATION

The variance formula $s^2 = \Sigma(x - \bar{x})^2/n$ is often referred to as the **definitional formula** since it *not only calculates the variance but also defines or explains what the variance is: the mean amount of the squared deviations of the scores from the mean.* (It is often quite difficult for those long away from algebraic formulas to "see" that definition, but it is there.)

For computational purposes, however, it is often easier to use one of several alternative formulas, known as **computational formulas,** particularly if a calculator is available. One such computational formula is the following.

$$\text{Variance} = s^2 = \frac{\sum x^2 - \frac{\left(\sum x\right)^2}{n}}{n}$$

$$\text{Standard Deviation} = s = \sqrt{\frac{\sum x^2 - \frac{\left(\sum x\right)^2}{n}}{n}}$$

Before we apply these formulas, we should make note of the difference between two parts of the formula; Σx^2 and $(\Sigma x)^2$, which are *not* the same. The first, Σx^2, read "summation of x squared," tells us to square each x and then add up all of the x^2's. The second, $(\Sigma x)^2$, read "summation of x, quantity squared" tells us to first add up all the x's to get Σx and then square Σx to get $(\Sigma x)^2$. (This follows the convention of first doing what is *inside* a set of parentheses before doing what is outside of the parentheses.) Thus we must add the original scores and square the sum and we must also square each original score and add up the squared values.

Group A

$x =$	$x^2 =$
8	64
8	64
8	64
7	49
7	49
7	49
6	36
6	36
6	36
$\Sigma x = 63$	$\Sigma x^2 = 447$

$n = 9$

$$(\Sigma x)^2 = (63)^2 = 63 \times 63 = 3969$$

$$s^2 = \frac{\sum x^2 - \frac{\left(\sum x\right)^2}{n}}{n} = \frac{447 - \frac{(63)^2}{9}}{9}$$

$$= \frac{447 - \frac{3969}{9}}{9} = \frac{447 - 441}{9}$$

$$= \frac{6}{9} = \frac{2}{3} = 0.67$$

and

$$s = \sqrt{0.67} = 0.82$$

The answers are obviously the same as when we use the definitional formula. Often the two results will differ slightly due to rounding error, particularly if the mean used in the definitional formulas is not a whole number (such as 7, in this case) but possesses several decimals (such as 7.2, 7.23, 7.234, and so on). Notice that the computational formula requires the calculation of several large intermediate figures, such as the $(\sum x)^2 = 3969$. Since such large numbers are not needed when using the definitional formula, we may question the need for a computational formula. If, however, there are many scores (even as few as the 9 scores in Group A), it is faster and easier to use the computational formulas. It is even easier to use the computational formulas with today's advanced scientific, business, and statistical calculators, which usually store $\sum x$ and $\sum x^2$ in their memories for easy retrieval.

VARIANCE AND STANDARD DEVIATION FOR DATA IN FREQUENCY DISTRIBUTIONS

If the data are in frequency distributions, the formulas given above will not find the correct variance or standard deviation. In a frequency distribution we must account not only for each possible value of x but also for the number of times, or frequency, that value occurs. This is the same reason we modified the formula for finding the mean of a frequency distribution in the previous chapter. Recall that in calculating the mean for the liberal arts majors, Group B, we first established an fx column and added it up to get $\sum fx$. We then divided $\sum fx$ by $\sum f$ (our n) to get the mean. For frequency distribution data, the *definitional formula* for the *variance* is also adjusted so that before adding the squared deviations, we multiply each squared deviation by the frequency of that particular value of x.

$$s^2 = \frac{\sum[(x-\bar{x})^2 f]}{n} = \frac{\sum[(x-\bar{x})^2 f]}{\sum f}$$

Therefore,

Group B

$x =$	$f =$	$fx =$	$\bar{x} =$	$x - \bar{x} =$	$(x-\bar{x})^2 =$	$(x-\bar{x})^2 f =$	
9	2	18	7.5	1.5	2.25	$2.25 \times 2 =$	4.50
8	3	24	7.5	0.5	0.25	$0.25 \times 3 =$	0.75
7	3	21	7.5	−0.5	0.25	$0.25 \times 3 =$	0.75
6	2	12	7.5	−1.5	2.25	$2.25 \times 2 =$	4.50
$x =$	$\sum f = 10$	$\sum fx = 75$				$\sum[(x-\bar{x})^2 f] =$	10.50

$$\bar{x} = \frac{\sum fx}{n} = \frac{\sum fx}{\sum f} = \frac{75}{10} = 7.5$$

Thus the variance is

$$s^2 = \frac{\sum [(x-\bar{x})^2 f]}{n} = \frac{\sum [(x-\bar{x})^2 f]}{\sum f} = \frac{10.50}{10} = 1.05$$

and the standard deviation is

$$s = \sqrt{1.05} = 1.0246 = 1.03$$

For data in frequency distributions there is also an adjusted *computational formula.*

$$s^2 = \frac{\sum x^2 f - \frac{\left(\sum fx\right)^2}{n}}{n} = \frac{\sum x^2 f - \frac{\left(\sum fx\right)^2}{\sum f}}{\sum f}$$

To apply this to Group B, we must generate columns for x^2 in order to find $\sum x^2$ and $x^2 f$ in order to find $\sum x^2 f$. We have already generated an fx column, but we need to square its summation.

$x =$	$f =$	$fx =$	$x^2 =$	$x^2 f$
9	2	18	81	$81 \times 2 = 162$
8	3	24	64	$64 \times 3 = 192$
7	3	21	49	$49 \times 3 = 147$
6	2	12	36	$36 \times 2 = 72$
	$n = \sum f = 10$	$\sum fx = 75$		$\sum x^2 f = 573$

$$\left(\sum fx\right)^2 = (75)^2$$

$$= 75 \times 75$$

$$= 5625$$

Thus the variance is

$$s^2 = \frac{\sum x^2 f - \frac{\left(\sum fx\right)^2}{\sum f}}{\sum f} = \frac{573 - \frac{(75)^2}{10}}{10} = \frac{573 - \frac{5625}{10}}{10}$$

$$= \frac{573 - 562.5}{10} = \frac{10.5}{10} = 1.05$$

and the standard deviation is

$$s = \sqrt{1.05} = 1.03$$

The results are identical to those found using the definitional formulas.

We now know the primary measures for describing a single interval or ratio level variable: the mean for central tendency and the standard deviation or variance for dispersion. With the latter two, we generally use the standard deviation for descriptive purposes but retain the variance for use in procedures that will be discussed later in this text.

With the exception of the range, the measures of dispersion presented in this chapter all assume interval level of measurement. (The range may be applied also to ordinal data: "The guests at the $100-a-plate charity fund raiser ranged from middle class to affluent.") While measures of dispersion are widely used with interval level data, they are only rarely used with lower levels of measurement. Accordingly, such usage will not be covered here.

CONCLUSION

We have now covered the last of the basic tools of descriptive data analysis. With the introduction of dispersion measures, particularly the variance and the standard deviation, we can begin the study of several statistical techniques widely applied in many disciplines. We will see that in addition to their the role as useful descriptive tools, the mean and the variance often plug into other formulas. Thus they do double duty. Armed with the tools introduced so far, we will eventually return to the task of finding and describing relationships between two variables.

EXERCISES

Note: For the following exercises, refer to the exercises at the end of Chapter 4 for definitions of the variables.

Exercise 5.1

In the social worker sample (Exercises 4.7 to 4.9), a group of 9 private agency employees was compared to a group of 16 public employees. Following are the health care cost ratings for the private agency employees. Remember that the higher rating indicates more concern about the issue.

Private Agency Employees

Health
70
55
15
10
5
5
5
0
0

1. Find the mean Health score.
2. Find the median.
3. Find the mean deviation.
4. Find the variance using the definitional formula.
5. Find the variance using the computational formula.
6. Find the standard deviation.

Exercise 5.2

Following are the health care cost ratings for the public employees:

Public Employees

Health
95
95
95
90

90
90
90
90
90
80
80
75
75
60
40
35

Form a frequency distribution from the above, and using the appropriate formulas:

1. Find the mean Health score.
2. Find the median.
3. Find the variance using the definitional formula.
4. Find the variance using the computational formula.
5. Find the standard deviation.
6. Compare the mean and standard deviation of the public employees to those of the private agency employees found in Exercise 5.1. Which group's scores cluster more closely about its mean?

Exercise 5.3

Management personnel have been scored on a scale measuring assertiveness of leadership style, where more assertiveness indicates less accommodativeness. Are financial and banking managers more assertive than their colleagues in other service industries? Following are scores for 7 managers in finance or banking related firms.

Assertiveness
24
49
92
92
11
68
97

1. Find the mean Assertiveness score.

2. Find the median. (Note that you must first array the data from high to low scores.)
3. Find the mean deviation.
4. Find the variance using the definitional formula.
5. Find the variance using the computational formula.
6. Find the standard deviation.

Exercise 5.4

Following are assertiveness scores for 18 managers from nonfinancial service industries listed in an ungrouped frequency distribution.

x = Assertiveness	*f =*
100	1
97	1
92	1
86	3
54	1
30	1
27	3
24	2
22	1
5	1
3	1
0	2

1. Find the mean Assertiveness score.
2. Find the median.
3. Find the variance using the definitional formula.
4. Find the variance using the computational formula.
5. Find the standard deviation.
6. Compare the means and standard deviations of the nonfinancial institution managers to those found in Exercise 5.3. Which group is more assertive? Which group's scores are more spread out about the mean?

Exercise 5.5

Below are the results, in printout format, for the employee sample of Exercise 4.10 (refer to Exercise 4.10 for a definition of the variables).

Please note that this was run using SAS, one of several statistical packages available. Like most such packages, data are presented with far more decimal places than social scientists need. While suitable for engineers and some scientists, this level of precision is not suitable for the less exact measures that we use. Thus when discussing the results we will round to one or two decimal places.

In this exercise, workers have been broken down by region, Midwest versus all other regions combined. Suppose it had been rumored that the corporation was planning to close several plants and move those jobs to plants in other countries with lower wage scales. Suppose it had also been rumored that only plants in the Midwest would be exempt; in all other regions, some plants would be shut down. Let us compare the attitudes of the employees.

			Reg = Midwest
Variable	*N*	*Mean*	*S.D.*
ATTEND	13	90.6153846	12.2782902
BOARD	13	44.7692308	19.2663517
DIV	13	76.6153846	16.8302231
SECUR	13	67.7692308	28.4580213
PARTIC	13	39.6153846	35.0868885
OPPOR	13	55.4615385	38.5413531
UNION	13	55.3846154	35.5844968
SALARY	13	65.6923077	25.9466909

			Reg ≠ Midwest
Variable	*N*	*Mean*	*S.D.*
ATTEND	37	93.7297297	5.8720082
BOARD	37	34.7837838	18.1615209
DIV	37	78.7837838	16.1832134
SECUR	37	44.7027027	32.6129361
PARTIC	37	67.4324324	30.5646315
OPPOR	37	30.0270270	32.4349661
UNION	37	76.8918919	29.4380553
SALARY	37	49.6486486	27.4764710

1. Compare the means for each variable. What do you conclude?
2. Which region usually has the greater diversity on these dimensions as determined by comparing the standard deviations? In which two scales is that tendency reversed?

Exercise 5.6

Following is a comparison of the managerial group to the employee group.

MGTPOP

Variable	*N*	*Mean*	*S.D.*
ATTEND	89	92.3595506	9.9307228
BOARD	89	57.1123596	15.1840513
DIV	89	74.9438202	16.4305422
SECUR	89	56.1685393	32.4479429
PARTIC	89	48.5955056	34.6737126
OPPOR	89	42.5280899	36.3509065
UNION	89	62.4719101	31.6307136
SALARY	89	53.8764045	24.2575294

EMPLOY

Variable	*N*	*Mean*	*S.D.*
ATTEND	50	92.9200000	8.0997899
BOARD	50	37.3800000	18.7832861
DIV	50	78.2200000	16.2081989
SECUR	50	50.7000000	32.9274093
PARTIC	50	60.2000000	33.7602592
OPPOR	50	36.6400000	35.5486214
UNION	50	71.3000000	32.2118307
SALARY	50	53.8200000	27.7501167

You have already compared the means in Exercise 4.10.

Now compare the standard deviations for each variable. What can you conclude? For which variables are the managers more diverse (have larger standard deviations)? For which variables are the employees more diverse?

Exercise 5.7

The two discontented groups, upper-middle management and white collar employees, are compared in the following sets of data.

UPPER-MIDDLE MANAGEMENT

Variable	*N*	*Mean*	*S.D.*
ATTEND	50	91.8000000	12.8364914
BOARD	50	47.2200000	10.9195388
DIV	50	78.6400000	15.1600442
SECUR	50	39.8000000	31.0227040
PARTIC	50	72.1000000	22.0178499
OPPOR	50	16.4800000	18.2043278
UNION	50	85.6600000	10.8130873
SALARY	50	36.1000000	11.1158097

WHITE COLLAR EMPLOYEES

Variable	*N*	*Mean*	*S.D.*
ATTEND	29	93.3103448	5.1137311
BOARD	29	23.8965517	6.9710873
DIV	29	84.2068965	9.4354169
SECUR	29	31.3103448	26.7956598
PARTIC	29	81.8965517	17.8992115
OPPOR	29	13.1724137	16.7333477
UNION	29	92.4482758	10.9628796
SALARY	29	35.0000000	14.7672417

Compare the means and then the standard deviations for each variable. What do you conclude?

KEY CONCEPTS

contingency table	spurious relationships	antecedent variable
control variable	causal models	intervening variable

6 Constructing and Interpreting Contingency Tables

INTRODUCTION

In this chapter we further develop the subject of tables. We begin by discussing how to further recode the variables in a data set to create useful tables that relate two of the variables from that data set. We will also present procedures for changing the table's entries from frequencies to percentages, which makes most tables easier to interpret. We will look at examples of both the construction and the interpretation of cross-tabulations.

In the latter part of the chapter, we will discuss the impact that a third variable might have on the relationship between two other variables, in particular, the impact of the third variable on tables relating the first two. We will also discuss how to imply a direction of causation among the variables and will look at the variety of causal relationships that may be found within a set of three variables.

CONTINGENCY TABLES

In Chapters 1 and 2 we introduced the logic behind the interpretation of tables. Contingency tables (or cross-tabulations or just tables) continue to be tools for both interpretation and presentation of data. Consequently, contingency tables are often presented in scholarly research even if far more complex statistical techniques of analysis were also used by the researcher. Thus table construction and interpretation remain essential components of data analysis.

The **contingency table** *depicts a relationship between the independent variable and the dependent variable*. Each variable is in grouped data format and is generally nominal or ordinal level of measurement, even if the original data were individual interval level (as explained in Chapter 2). If this is to be the case, once the variables have been selected the researcher returns to the original data and creates grouped categories, often with new labels such as high, medium, or low. In effect these newly created categories reflect a new operational definition on the part of the researcher, since the researcher determines the cutoff points for high, medium, and low. Since tables with nominal and ordinal variables were presented earlier, we will now emphasize such ordinal variables resulting from grouping individual interval data.

For example, suppose we select a sample of 20 countries and select data from the *World Handbook of Political and Social Indicators* mentioned earlier.[1] Let us examine the relationship between two variables: per capita Gross National Product (GNP/capita) and Political Rights as shown in Table 6.1.

We wish to test the hypothesis that the level of wealth enjoyed by a nation's citizens is positively related to that nation's level of political rights. We know that historically those countries with competitive democratic political systems and high levels of political and social rights and freedoms have been among the world's wealthier industrialized countries, though that is not exclusively true. To see how much of a relationship exists, we use per capita GNP as the independent variable.

The dependent variable is an index of political rights developed by Gastil and reproduced in the *World Handbook*.[2] Gastil's index originally ranged from 1 to 7 with the score of 1 going to the countries with the highest levels of political rights. To simplify this example, countries are grouped into a three-category ordinal scale with categories of High, Medium, or Low (H, M, and L). The independent variable, GNP/capita, is the total value of all final goods and services produced by a nation's

Table 6.1 Partial Data List from *World Handbook of Political and Social Indicators*

Country	*Per Capita GNP c. 1978 $ U.S.*	*Political Rights*
U.S.A.	9,770	H
U.K.	5,720	H
U.S.S.R.	3,710	L
Canada	8,670	H
France	8,880	H
Poland	3,650	L
Israel	3,730	M
Lebanon	1,070	L
Sweden	10,540	H
New Zealand	5,530	H
Argentina	2,030	L
South Africa	1,580	L
Ireland	3,810	H
Zimbabwe	480	L
Brazil	1,510	L
India	180	M
Australia	8,060	H
El Salvador	640	L
Chad	150	L
Japan	7,700	M

SOURCE: C. L. Taylor and D. Jodice, *World Handbook of Political and Social Indicators* (1983). New Haven, CT: Yale University Press. Used by permission.

economy divided by the size of the population—a widely applied index of relative wealth.

REGROUPING VARIABLES

Although the political rights index has already been grouped, we will need to group per capita GNP. There are two basic ways to approach this problem. One is to see if there is already some external standard in use. For instance, the International Monetary Fund or the United Nations may have already created operational definitions for high, medium, and low per capita GNP. If this is the case, we use those definitions.

If no such precedents exist, the way we group the data will be based solely on our own preferences and priorities. Although there are no hard and fast rules for doing this, the following procedure is helpful. Look at the range of scores (lowest to highest) and create class intervals along this range. Do a frequency distribution of scores and examine that distribution. Then combine some of the class intervals with two objectives in mind: (a) reducing the number of class intervals to a manageable number (3 or 4 categories) and (b) having in each category at least a few cases and if possible a roughly equal number of cases in each class interval. Rarely will the final breakdown meet the criteria for grouped interval level data as discussed in Chapter 2. In this case we end up with grouped ordinal data.

To ease our task, we change the scale to reflect units of $1,000. (That is, we divide each GNP by 1000. Thus, for example, the U.S.A. score of $9,770 becomes a score of 9.77). We do not have to do this, but data presentation is easier if we do. Now, we note that the scores range from a low of .150 (Chad) to a high of 10.54 (Sweden) and we have a spread of about 10 units. Let us set up class intervals covering 2 units each and count the number of countries in each category.[3]

GNP/Capita (in thousands of dollars)	*f* =	*f* =
8.0 - 9.9	|||||	5
6.0 - 7.9			1
4.0 - 5.9	||	2
2.0 - 3.9	|||||	5
0 - 1.9	||||| ||	7

We now combine categories to attain a smaller number of roughly equal-sized categories. For example, there is only one country in the 6.0 to 7.9 category—Japan (7.7). We can add Japan to the top category by setting the lower limit of the category below Japan's score, say at 7.5. The new category can be labeled 7.5 to 9.9 or, more simply, 7.5 and above. (We now have unequal-sized class intervals; the data are no longer interval level of measurement, so we might as well make the top class interval open-ended for ease of presentation.)

Using the same logic, we can ultimately establish two other new categories, 2.0-7.4 and 1.9 and below, to get the following distribution.

	f =	*f* =
7.5 and above	|||||		6
2.0 - 7.4	||||| ||	7
1.9 and below	||||| ||	7

Finally, we label the three categories High, Medium, and Low.

GNP/Capita (in thousands of dollars)	*Range*	*f =*
High	7.5 and above	6
Medium	2.0 - 7.4	7
Low	1.9 and below	7

Throughout this process, the researcher chose class interval sizes, cutoff points, and category names. You might easily have chosen other alternatives to the ones selected here.

We used a similar procedure for the Political Rights index to produce a four-category scale.

Political Rights	*Rank*	*f =*
High	1	8
Medium	2 - 3	3
Low	4 - 5	5
Very Low	6 - 7	4

Then, to simplify the table to be generated, we combine low and very low into one category. These are the values in Table 6.1.

	Rank	*f =*
High	1	8
Medium	2 - 3	3
Low	4 - 7	9

We are now ready to construct our table. To do so, we follow a series of conventions used for table construction. (You will find in the research literature many tables that deviate from some of these conventions. What follows is the way tables are generally constructed.) By convention, we generally place the independent variable so that its categories become the columns of the table, and the categories of the dependent variable become the table's rows (see Table 6.2).

Once we have laid out the table, we examine each country in Table 6.1, go to the column of the table within which that country's GNP/capita falls, and go down the column to the row within which that country's Political Rights score can be found. In the cell where the appropriate column and appropriate row intersect, we put a tally mark. We continue with the process until all 20 countries have been tallied.

For example, from our original data we see that the U.S.A. has GNP/capita of $9,770 which, after dividing by 1000, gives a score of 9.7, which falls in the High category (7.5 and above). We also see that the

Table 6.2

	GNP/Capita		
	High	*Medium*	*Low*
Political Rights	*7.5 & above*	*2 - 7.4*	*Below 1.9*
High	|||||	|||	
Medium						
Low		|||	|||||	

U.S.A. ranks High for Political Rights. We find the cell where the High GNP column intersects the High Political Rights row and place a tally mark in that cell.

When we have finished all 20 countries, we will have a total of 5 tally marks in that (High-High) cell, corresponding to the U.S.A., Canada, France, Sweden, and Australia. In the lowest right-hand cell (Low GNP/Capita and Low Political Rights), we will have 6 marks, corresponding to Lebanon, South Africa, Zimbabwe, Brazil, El Salvador, and Chad.

As we discussed in Chapter 1, a clear clustering appears along the main diagonal of the table, indicating the presence of a positive relationship between the variables.

We replace the tally marks in each cell with the sum of the tally marks and put in marginal totals and the grand total for reference, as shown in Table 6.3.

GENERATING PERCENTAGES

Since the category marginal totals of the independent variable are very close to one another (6, 7, 7), it is possible to interpret Table 6.3 from the cell entries, as we did in Chapter 1. Often, however, this is not the case. The category totals are often larger numbers and they differ, sometimes considerably, from one another. Accordingly, it is almost always advantageous to change the table from frequencies to percentages before interpreting the data.

Percentaging in this manner has the effect of creating an equal number of cases in each category of the independent variable, so that trends in the table may be more readily observed. It is as if there were 100 high GNP/capita countries, 100 medium, and 100 low. With equal numbers of cases, comparisons across the categories are easier to make.

Table 6.3

	GNP/Capita			
	High	Medium	Low	
Political Rights	7.5 & above	2 - 7.4	Below 1.9	Total
High	5	3	0	8
Medium	1	1	1	3
Low	0	3	6	9
Total	6	7	7	20

Consider the following table with frequencies as its cell entries. Both the top and bottom rows increase as one scans from left to right.

8	15
2	85

If the table is percentaged so that each column adds up to 100%, however, the relationship is far more apparent.

80%	15
20	85
100%	100%

Percent means "per 100" and the idea behind percentaging is to make the individual cell entries add up to a common base, a total of 100%. There are three ways to base our percentages (and most computer programs do all three for us, whether we need them or not).

1. Treat each cell entry as a percentage of the *grand total*. For instance, we have 5 cell entries in the High-High cell of Table 6.3 and a grand total of 20.

$$5 \div 20 = .25 = 25\%$$

   If we percentaged similarly for each of the 9 cells in the body of the table and then added all of these percentages together, they would add up to 100%. In the analysis of tables, however, this kind of percentaging is rarely useful. We really are not interested in the fact that 25% of the countries in our study had high GNPs/capita and also high levels of political rights.

2. Treat each cell entry as a percentage of its *row total*. In the High-High cell there are 5 countries out of a total of 8 countries having high Political Rights levels.

$$5 \div 8 = .625 = 62.50\%$$

Thus 62.5% of the countries with high Political Rights scores also have high per capita GNPs. We repeat the process for the other GNP categories in the same row and add up the three percentages for the row.

Countries with a high Political Rights score having:	
High GNPs/Capita	62.5%
Medium GNPs/Capita	37.5%
Low GNPs/Capita	0
Total of all countries with a high level of Political Rights	100.0%

We then do the same for the other rows.

3. Treat each cell entry as a percentage of its *column total*. There are a total of 6 countries with high per capita GNPs, 5 of which have high political rights levels.

$$5 \div 6 = .8333 = 83.3\%$$

One country is in the Medium Political Rights category.

$$1 \div 6 = .1666 = 16.7\%$$

No country is in the Low Political Rights category. Adding these up, we get

Countries with high (7.5 and above) GNPs per capita having:	
High Political Rights Scores	83.3%
Medium Political Rights Scores	16.7%
Low Political Rights Scores	0
Total of all countries with high GNPs per capita	100.0%

Table 6.4 Political Rights, by Per Capita GNP (in thousands of dollars) (in percentages)

	GNP/Capita		
	High	*Medium*	*Low*
Political Rights	*7.5 & above*	*2.0 - 7.4*	*Below 2.0*
High	83.3%	42.8	0
Medium	16.7	14.3	14.3
Low	0	42.8	85.7
Total	100.0%	99.9%	100.0%
(*n* =)	(6)	(7)	(7)

SOURCE: C. L. Taylor and D. Jodice, *World Handbook of Political and Social Indicators* (1983). New Haven, CT: Yale University Press; and R. Gastil, *Freedom in the World* (1980). New York: Freedom House. Used by permission.

The convention generally followed is to percentage so that each category of the *independent variable* sums to 100%. The independent variable is usually the one whose categories are the columns of the table, thus percentaging is usually such that each column of the table sums to 100%.

Once our percentaging is done, we will be able to compare across categories of per capita GNP. Without percentaging, differences among the scores in each GNP group could merely reflect the different overall sizes of these groups. With percentaging, all GNP categories have the same overall size, namely 100. Thus if within a single category of the dependent variable the percentage in each GNP group differs, we know that the differences are likely due to a relationship between the variables.

We need one more thing to complete our table—a title. There are many possible formats for the title, but the format often encountered in the literature will be utilized here. First name the dependent variable, followed by a comma and the word *by*, and then name the independent variable. Other supporting information, such as year and location of the study, may also be inserted parenthetically where appropriate. In simple form the title of Table 6.4 would be:

Political Rights, by Per Capita GNP

We could expand this as follows:

Political Rights Levels, by Per Capita GNP (in thousands of U.S. dollars), c. 1978, for Selected Countries (in percentages)

The title for Table 6.4 is a compromise between the two.

Note three other things about Table 6.4. First, a minor point: It is only necessary to put one percentage sign (%) in the body of the table, in the upper left-hand cell. (Some people include the percentage sign in each of the entries of the top row to indicate that percentages add to 100% for each column.) Also, include percentage signs by the 100.0% totals at the bottom. Second, a major point: We always report the number of cases ($n =$) for each category of the independent variable. This is because most tests and measures that we will learn to calculate from tables will be based on numbers of cases, not percentages. By reporting the n's we can always retrieve the data found in Table 6.3, the original cell entries. For instance, in the High-High cell we find 83.3% of the high GNP countries and 6 high GNP countries indicated at the bottom of the table. Thus 83.3% of 6 = $.833 \times 6 = 4.998$ (due to rounding error) = 5 = the cell entry in Table 6.3.

The final point to note about Table 6.4—one of great importance—is the footnote citing the sources of the data and (where possible) page references. Here, a reference is given to a bibliography at the end of the article where the complete citation would be provided. Note: Failure to cite sources is plagiarism.

INTERPRETING

In the following discussion, we will review some of the points presented in Chapter 1. A positive relationship will be indicated by a clustering on the *main diagonal* (upper left to lower right). An inverse relationship will be indicated by clustering on the *off diagonal* (lower left to upper right). Now, however, we have percentages in our table instead of the original frequencies.

We interpret our table by examining increases or decreases in percentages of the dependent variable as we move from one category of the independent variable to the other. In the case of a 3×3 table set up so that the upper left-hand cell represents the largest quantity on both variables (High-High) and the lower right-hand cell is the least quantity for both variables (Low-Low), if the relationship is *positive* we expect approximately the following.

	GNP/Capita		
Political Rights	*High*	*Medium*	*Low*
High	Highest % in this row	2nd Highest % in this row	Lowest % in this row
Medium	________	________	________
Low	Lowest % in this row	2nd Highest % in this row	Highest % in this row

The pattern in the Medium Rights category is harder to interpret. Generally, we would expect less change across this row than across any other. If there had been more than three categories of Political Rights, the changes in intermediate categories may be vague. For instance, if a category existed between High and Medium, call it "Medium-High Political Rights," changes across that row could be similar to those in the High row but less severe, or the changes could be quite small, as in the Medium row.

This sort of pattern in a table is not the only pattern that indicates a relationship, but it is the most commonly found pattern and is called a *linear relationship*. We will discuss linear relationships again later.

Beginning at the top row (High Rights) of Table 6.4, we look at the percentages from left to right: 83.3% to 42.8% to 0%. Clearly this goes from the highest to the second highest to the lowest percentage. In the Medium Rights row, the percentages (from left to right) read: 16.7% to 14.3% to 14.3%; relatively little change. For the Low Rights row, we expect a turnaround, with the percentages increasing from left to right, and that row does follow the expected pattern. The percentage starts at 0% for High GNP and increases dramatically to 42.8% and finally to 85.7%. With few inconsistencies, therefore, Table 6.4 suggests a positive relationship between the variables.

If the relationship were *inverse*, we would expect the upper row to increase from left to right and the lower row to decrease as shown below.

High	Lowest %	2nd Highest %	Highest %
Medium			
Low	Highest %	2nd Highest %	Lowest %

The percentages in such a table might look like this:

	High	*Medium*	*Low*
High	2.5%	42.8	70.5
Medium	20.5	20.0	19.5
Low	77.0	37.2	10.0
	100.0%	100.0%	100.0%

Suppose there is *no* relationship between the variables. Then for each row we would expect the percentages to be the same (or nearly the same). The following is an example of no relationship.

	High	*Medium*	*Low*
High	60.0%	60.0	60.0
Medium	25.0	25.0	25.0
Low	15.0	15.0	15.0
	100.0%	100.0%	100.0%

Here, 60% of all countries have high levels of political rights *regardless* of GNP/capita; 25% have medium political rights, again regardless of GNP; and 15% have low political rights for all three categories of GNP.

AN EXAMPLE

Nine variables discussed in the *World Handbook of Political and Social Indicators* are presented below. Operational definitions for these variables have been arbitrarily established by assigning scores to class intervals labeled High, Medium, Low, and Very Low. A data list for a sample of 20 countries (Table 6.5) follows the operational definitions.

Per Capita GNP (in $1000 U.S.)		*f =*
High	7.5 and above	6
Medium	2.0 - 7.4	7
Low	0 - 1.9	7

Telephones per 1000 Population		*f =*
High	400 and above	5
Medium	150 - 399	4
Low	50 - 149	5
Very Low	49 and Below	5

Percentage of Labor Force Engaged in Agriculture		*f =*
High	55% and over	3
Medium	20 - 54%	5
Low	10 - 19%	6
Very Low	0 - 9%	6

Political Rights		*f =*
High	1	8
Medium	2 - 3	3
Low	4 - 5	5
Very Low	6 - 7	4

Civil Rights		*f =*
High	1	8
Medium	2 - 3	3
Low	4 - 5	6
Very Low	6 - 7	3

Protest Demonstrations per 1 Million Population		$f =$
High	10 and above	6
Medium	2 - 9.9	9
Low	0 - 1.9	5

Deaths From Political Violence per 1 Million Population		$f =$
High	70 and above	4
Medium	5 - 69	7
Low	1 - 4	4
Very Low	Below 1	5

Imposition of Political Sanctions (by the Government) per 1 Million Population		$f =$
High	25 and above	6
Medium	5 - 24.9	8
Low	0 - 4.9	6

Executive Adjustments (Peaceful Administrative and Ministerial Personnel Changes)		$f =$
High	80 and above	5
Medium	45 - 79	7
Low	0 - 44	8

Experts in economic and political development argue that the percentage of the labor force engaged in agriculture in a country's economy is a good index of industrialization, modernization, and thus, of wealth. The idea is that the less economically developed a nation, the greater its agricultural workforce. To test this, assume Percentage of Labor Force engaged in Agriculture to be the independent variable and Per Capita GNP to be the dependent variable, since it is an estimate of the nation's wealth. We set up a table as a tally sheet and fill it in with information from the data list of Table 6.5. On the data list, the 20 countries are listed alphabetically down the left-hand side and the variables are listed along the top. Argentina, for instance, has been assigned to the medium category for Per Capita GNP. It is low for Telephones per 1000 Population, low for Percentage of Labor Force in Agriculture, and very low for Political Rights (these data were collected prior to the collapse of military rule in Argentina and the subsequent restoration of democracy). Thus in Table 6.6, Argentina would fall into the cell delineated by the row labeled Medium for Per Capita GNP and the column labeled Low for Percentage of Labor Force in Agriculture.

Table 6.5 Data List from *World Handbook of Political and Social Indicators*

	Per Capita GNP	Telephones per 1000 Population	Percentage of Labor Force in Agriculture	Political Rights Score	Civil Rights Score	Protest Demon-stration per 1 Million Population	Deaths From Political Violence per 1 Million Population	Imposition of Political Sanctions	Executive Adjustments
Argentina	M	L	L	VL	L	M	H	H	H
Australia	H	M	VL	H	H	M	VL	L	L
Brazil	L	VL	M	L	L	L	L	L	M
Canada	H	H	VL	H	H	M	VL	M	M
Chad	L	VL	H	VL	VL	L	H	M	L
El Salvador	L	VL	M	L	L	L	M	M	L
France	H	M	L	H	M	M	L	M	M
India	L	VL	H	M	M	L	M	L	H
Ireland	M	L	M	H	H	H	M	H	L
Israel	M	M	VL	M	M	H	M	H	M
Japan	H	H	L	M	H	M	VL	L	M
Lebanon	L	No Data	L	L	L	H	H	H	M
New Zealand	M	H	L	H	H	M	VL	L	L
Poland	M	L	M	VL	L	M	M	M	H
South Africa	L	L	M	L	VL	M	M	H	L
Sweden	H	H	VL	H	H	M	VL	M	L
UK	M	M	VL	H	H	H	M	M	H
USA	H	H	VL	H	H	H	L	M	M
USSR	M	L	L	VL	VL	L	L	L	H
Zimbabwe	L	VL	H	L	L	H	H	H	L

SOURCE: Recoded from C. L. Taylor and D. Jodice, *World Handbook of Political and Social Indicators* (1983). New Haven, CT: Yale University Press. Used by permission.

Table 6.6

Per Capita GNP	High 55% and Above	*Percentage of Labor Force in Agriculture* Medium 20 - 54%	Low 10 - 19%	Very Low 0 - 9%
High (7.5 and above)			||	||||
Medium (2.0 - 7.4)		||	|||	||
Low (0 - 1.9)	|||	|||	|	

Table 6.7

Per Capita GNP	High 55% and above	*Percentage of Labor Force in Agriculture* Medium 20 - 54%	Low 10 - 19%	Very Low 0 - 9%
High (7.5 and above)	0	0	2	4
Medium (2.0 - 7.4)	0	2	3	2
Low (0 - 1.9)	3	3	1	0
Total	3	5	6	6

Table 6.8 Per Capita Gross National Product for Selected Nations, by Percentage of the Labor Force in Agriculture

Per Capita GNP *(in $1000 U.S.)*	High 55% and above	*Percentage of Labor Force in Agriculture* Medium 20 - 54%	Low 10 - 19%	Very Low 0 - 9%
High (7.5 and above)	0%	0	33.3	66.7
Medium (2.0 - 7.4)	0	40.0	50.0	33.3
Low (0 - 1.9)	100.0%	60.0	16.7	0
Total (*n* =)	100.0% (3)	100.0% (5)	100.0% (6)	100.0% (6)

A clustering appears in the off diagonal, indicating the anticipated inverse relationship: as percentage in agriculture increases, GNP/capita decreases. In Table 6.7, we replace the tallies with the actual frequencies.

From the frequencies in Table 6.7, we calculate the percentages shown in the final table, Table 6.8. The percentage changes, particularly in the high and low Per Capita GNP rows, indicate the existence of a relatively strong inverse relationship between the two variables.

One small technical note: The low agriculture column could have been rounded as follows:

Low
10-19%
33.3
50.0
16.6
99.9%

with a footnote in Table 6.8 stating that rounding error is why percentages do not sum to 100%. You will see such footnotes from time to time in journal articles and papers, particularly those done some years ago. Notice that earlier, in Table 6.4, the footnote was omitted.

CONTROLLING FOR A THIRD VARIABLE

We have been dealing so far with the relationship between only two variables at a time. At this point, we explore the influence of a third variable, called the **control variable**, on the relationship between the first two variables. Later, we will discuss the exact role played by that third variable; that is, is it also an independent variable or is it another dependent variable? For now, however, let us merely examine some of the influences of a third variable.

What we are doing is making use of statistical analysis in a manner analogous to a scientific experiment in a laboratory setting. In many experiments, one group of subjects, known as the *control group*, is given *no* experimental treatment, but is compared to others who do receive treatment. If we are studying the effects of time constraints on decision making, one experimental group may have 10 minutes in which to make a decision. A second experimental group may have 20 minutes to do so. But a third group, the control group, would have no time constraints placed upon it. If all three groups reach similar decisions, the researcher could argue that time constraints have no influence on decision making. If one or both experimental groups reach different decisions than the control group, the researcher could argue that time constraints do have an influence.

At the conclusion of this chapter, Exercise 6.2 is a table in which a marked inverse relationship between percentage of the labor force in agriculture and the number of telephones per 1000 people. Why is this the case? Probably since telephones are more likely to be found in wealthier industrialized societies, which by their nature have fewer people employed in agriculture than do preindustrial societies. We might look for a control variable such as per capita GNP (wealth) or proportion of people in metropolitan areas (urbanization) and explore what happens to the original relationship when the effects of the control variable are taken into account. If the original relationship turns out to be the result of the presence of wealth or income, we say that the original relationship between agriculture and telephones is an *indirect* relationship and specifically in this case, a **spurious relationship**, caused by the presence of the control variable. We use the term *spurious* when *the relationship between two variables is the product of a common independent variable*. Thus agriculture and telephones are functions of a single independent variable—wealth. If in time order wealth did not come first but followed agriculture and preceded telephones, we would say wealth provides *interpretation* of the relationship between agriculture and telephones. By the same token, suppose that among low per capita GNP countries the same relationship between agriculture and telephones exists as can be found among high per capita GNP countries. In this case wealth would appear to have *no* influence on the original relationship.

Let us use an example in which the control variable functions as a second independent variable. We want to look at what happens to the relationship between the dependent and independent variable when we "control for" the influence of this new variable. By "control for" we mean to explore the relationship between the initial two variables for each of the categories of our new control variable. In doing so, we can see the impact of the control variable on the initial relationship.

Suppose we conduct a study for a sample of 200 newly commissioned military officers. One of the variables we examine has to do with the extent that these subjects favor armed intervention abroad under circumstances where the country's vital interests are in jeopardy. We think that the gender of the respondents will influence their attitudes on intervention. The following table is generated.

Attitude on Military Intervention	*Gender*	
	Male	*Female*
Favor	70%	60
Oppose	30	40
Total	100%	100%
($n =$)	(100)	(100)

Table 6.9

Attitude on Intervention	Gender		Race			
			White		Nonwhite	
	Male	Female	Male	Female	Male	Female
Favor	70	60	42	36	28	24
Oppose	30	40	18	24	12	16
Total (*n* =)	100	100	60	60	40	40

There appears to be some relationship between gender and attitude with the males a bit more inclined toward favoring intervention. We now wish to see to what extent, if any, this relationship is changed with the introduction of a control variable. Since our study is being done in the United States, we choose race (white, nonwhite) as the control variable. In other countries different variables might impact the relationship. For example, if these were Canadian Armed Forces personnel, we might use language (Anglophone, Francophone) as the control variable. In the United Kingdom we might use region (England, Scotland, Ulster, Wales). We use race in this example in order to work with an easy-to-represent dichotomy.

PARTIAL TABLES

There are two general outcomes that may ensue when the control variable is entered.

1. The control variable has no impact on the initial relationship.
2. The presence of the control variable changes the initial relationship or is necessary for there to be a relationship between the independent and dependent variables.

We look first at the case where the control variable has no impact. To illustrate this, we generate two *partial tables*. Each one shows the relationship between attitude and gender for a specific category of the control variable. Assuming 120 white and 80 nonwhite respondents, we will have three tables: the initial table, the table for white respondents, and the table for nonwhite respondents. The first set of tables will be in actual frequencies (see Table 6.9).

Table 6.10

Attitude on Intervention	Gender		Race			
			White		Nonwhite	
	Male	Female	Male	Female	Male	Female
Favor	70%	60	70%	60	70%	60
Oppose	30	40	30	40	30	40
Total	100%	100%	100%	100%	100%	100%
(*n* =)	(100)	(100)	(60)	(60)	(40)	(40)

Table 6.11

Attitude on Intervention	Gender		Race			
			White		Nonwhite	
	Male	Female	Male	Female	Male	Female
Favor	70%	60	80%	70	60%	50
Oppose	30	40	20	30	40	50
Total	100%	100%	100%	100%	100%	100%
(*n* =)	(100)	(100)	(60)	(60)	(40)	(40)

Note that if you add the corresponding frequencies in the partial tables together, you retrieve the original data. Thus adding the 42 white males who favor intervention to the 28 similarly inclined nonwhite males yields 70, the frequency in the initial table on the left.

To interpret the tables, we now generate the percentages, as shown in Table 6.10. The percentage is the same for each corresponding cell: 70% of all males favor intervention as do 70% of white males and 70% of nonwhite males. In this example, race has *no* effect on the original relationship.

When the presence of a control variable changes the initial relationship, there are several possible outcomes, creating partial tables from the same initial table that we have been using. First is a case where the partial table percentages differ from the original ones, but the strength of the relationship remains about the same within each category of the control variable (see Table 6.11).

Second, we have a case where the relationship is stronger in one category of the control variable (here, among whites) than in the other category (see Table 6.12.).

Table 6.12

Attitude on Intervention	Gender		Race			
			White		Nonwhite	
	Male	Female	Male	Female	Male	Female
Favor	70%	60	80%	67	55%	50
Oppose	30	40	20	33	45	50
Total	100%	100%	100%	100%	100%	100%
(*n* =)	(100)	(100)	(60)	(60)	(40)	(40)

Table 6.13

Attitude on Intervention	Gender		Race			
			White		Nonwhite	
	Male	Female	Male	Female	Male	Female
Favor	70%	60	83%	67	50%	50
Oppose	30	40	17	33	50	50
Total	100%	100%	100%	100%	100%	100%
(*n* =)	(100)	(100)	(60)	(60)	(40)	(40)

Table 6.14

Attitude on Intervention	Gender		Race			
			White		Nonwhite	
	Male	Female	Male	Female	Male	Female
Favor	70%	60	93%	92%	38%	39%
Oppose	30	40	7	8	62	61
Total	100%	100%	100%	100%	100%	100%
(*n* =)	(100)	(100)	(58)	(39)	(42)	(61)

Third, we have a case where the relationship exists for one category of the control variable but not for the other. Here, the relationship disappears for nonwhites (see Table 6.13).

Finally, a very strange pattern emerges in Table 6.14.

In Table 6.14 the original relationship all but disappears when we control for race. To understand this phenomenon, examine the frequencies shown in Table 6.15, from which percentages were generated in the partial tables.

Table 6.15

Attitude on Intervention	Race			
	White		Nonwhite	
	Male	Female	Male	Female
Favor	54	36	16	24
Oppose	4	3	26	37
Total	58	39	42	61

Table 6.16

Attitude on Intervention	Race	
	White	Nonwhite
Favor	90	40
Oppose	7	63
Total	97	103
Percentaging:		
Favor	93%	39
Oppose	7	61
Total	100%	100%
(*n* =)	(97)	(103)

Note the small number of whites, regardless of gender, who oppose intervention. The original relationship is not really between gender and attitude, but between race and attitude. Whites overwhelmingly favor intervention while nonwhites overall tend to oppose intervention. To see more clearly that race, not gender, is what causes attitude to vary, let us reconstruct the table showing the relationship between race and attitude. We add the 54 white males who favor intervention to the 36 white females who favor it, giving a total of 90 whites in favor. We repeat this for those who oppose get the frequencies shown in Table 6.16.

In this situation, where the original relationship between attitude and gender was based on race, we call the original association indirect: the relationship between attitude and gender was really due to the racial factor—not gender at all. We use partial tables such as those above, as well as partial correlation coefficients (see Chapter 14), to help us discover spurious and other relationships.

Another possible outcome of a control variable may be a case whereby a relationship between two variables may not be apparent except in the presence of the control variable. This is illustrated in Table 6.17. Note

Table 6.17

Attitude on Intervention	Gender		Race			
			White		Nonwhite	
	Male	Female	Male	Female	Male	Female
Favor	50%	50	100%	0	0%	100
Oppose	50	50	0	100	100	0
Total	100%	100%	100%	100%	100%	100%
(*n* =)	(100)	(100)	(50)	(50)	(50)	(50)

Table 6.18

Attitude on Intervention	Gender		Race			
			White		Nonwhite	
	Male	Female	Male	Female	Male	Female
Favor	50%	50	80%	20	20%	80
Oppose	50	50	20	80	80	20
Total	100%	100%	100%	100%	100%	100%
(*n* =)	(100)	(100)	(50)	(50)	(50)	(50)

that unlike in previous examples, the initial table before controlling shows no relationship.

The original lack of relationship was brought about by two offsetting relationships. Among whites, all males favor and all females oppose intervention. Among nonwhites, all females support intervention and all males oppose it. This corresponds to the one instance mentioned in Chapter 1 where association would not be necessary for causation. The offsetting relationships must balance one another, but they need not be perfect relationships. Refer to Table 6.18.

There are other occurrences that we might encounter with partial tables. It is possible under certain circumstances that there is no original association, but upon partialing, each partial table shows a small relationship. Unlike the above relationships, however, the partial tables' relationships do not offset one another, but actually run in the same direction. Also, we may occasionally encounter an initial table with some amount of association in it, and, upon partialing, find that the association in both partial tables runs in the direction opposite the association in the initial table. These two cases are rare enough that we need not illustrate them here.

CAUSAL MODELS

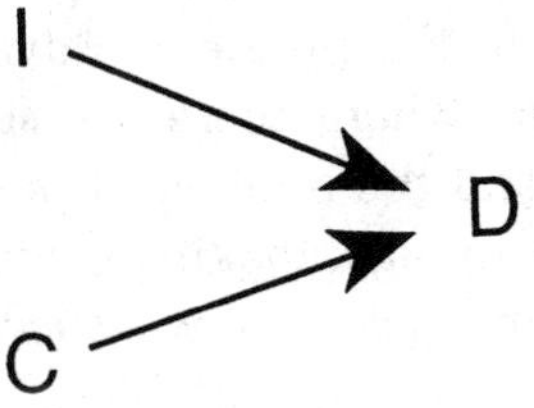

Figure 6.1

The findings in the partial tables are combined with other assumptions to produce **causal models** of these relationships. Often we portray these models by use of schematic diagrams indicating the independent, dependent, and control variables—*I, D,* and *C,* respectively. Lines are drawn between each pair of variables having association. (Of course, if there is no initial association, no line is drawn.) If the association is later proven to be indirect, it is replaced by a dotted line. Finally, an arrowhead is placed on the line to indicate the probable direction of causality. Since *D* is the variable being explained, at least one arrow must point to that variable. Thus our diagrams are based on observation and logic. We observe the relationship between each pair of variables to determine whether a line should be drawn between them and whether the line should be solid or dotted. Logic determines the direction of each arrowhead: if D is dependent, what is the logical flow of causation?

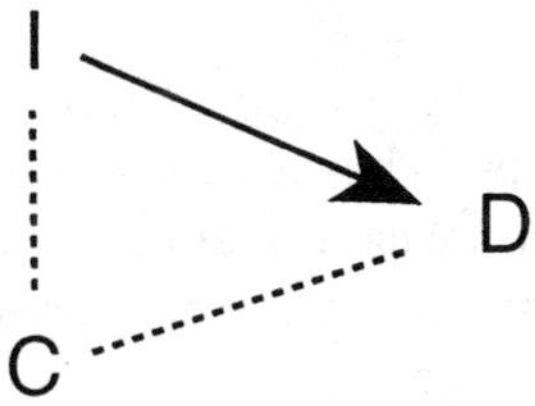

Figure 6.2

In the example we have been using, *D* is attitude on military intervention, *I* is gender, and *C* is race. Since gender and race develop at about the same time biologically and attitudes are shaped by both factors roughly concurrently, it is logical to assume that both gender and race are independent variables acting on attitude. Accordingly, a schematic of the relationships in Tables 6.11, 6.12, and 6.13 could look like the one in Figure 6.1.

For Table 6.10, where race had no impact, the schematic might look like the one in Figure 6.2.

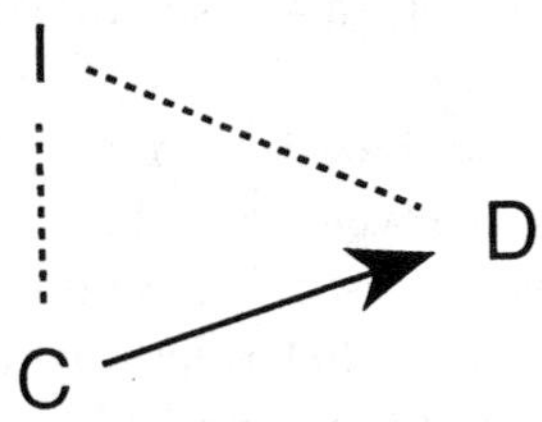

Figure 6.3

For Table 6.14, where the I - D relationship was indirect, we could use the schematic in Figure 6.3.

A double-headed arrow between two variables (or two parallel arrows pointing in opposite directions) could indicate reciprocal causation between two variables (see Figure 6.4). However, be aware that certain techniques using such models exclude the option of reciprocal causation and require the selection of a single direction for each arrow.

Figure 6.4

One last point: In our example, race *(C)* and gender *(I)* develop concurrently, what if that were not the case? Suppose *C* were not race but socialization, the process whereby attitudes (including those pertaining to armed intervention) are learned and internalized. Our model then might be similar to the one in Figure 6.5.

Gender "determines" the type of values to which one is socialized, and socialization shapes attitudes toward a specific military intervention. (Assumption: Boys are led to believe that it is best to fight back; girls learn that peaceful solutions are preferable.) Here we say that gender, *I*, is **antecedent** and socialization, *C*, is **intervening**. That is, *I* leads to *D* by way of *C*.

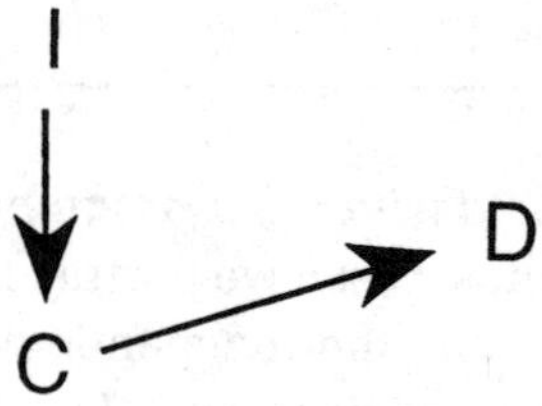

Figure 6.5

Assuming no double arrows, Figure 6.6 demonstrates several possible models for a three-variable situation with D being dependent. However, this only scratches the surface of the problem. Advanced techniques enable us to explore situations where there are several control variables working simultaneously and the number of possible models increases rapidly. For our purposes, though, looking at just one control variable is sufficient to appreciate the variety of ways external factors can influence a relationship between any two variables.

CONCLUSION

Tables traditionally have been handy tools for presenting findings and demonstrating relationships between variables. They are relatively easy to learn to interpret, and data at any level of measurement may be put into tabular form. They still represent a common form of data presentation, although perhaps less so than a decade ago.

As we have seen, partial tables enable us to investigate the impact of a third variable on the relationship between two other variables. Moreover, partial tables provide a very thorough way of studying the impact of a control variable. However, partial correlations—which will be covered in Chapter 14—and other techniques are also used for similar purposes today.

The major weakness of tables is that unless there is either a perfect relationship or a total lack of a relationship between variables, tables are vague. As we have seen, we can spot a less-than-perfect relationship in a table, but we cannot specify the extent of that relationship. In Chapter 11 that weakness will be addressed and the concept known as a *measure of association* will be introduced. The measure of association provides a number that seeks to reflect the actual degree of association of the variables in a table.

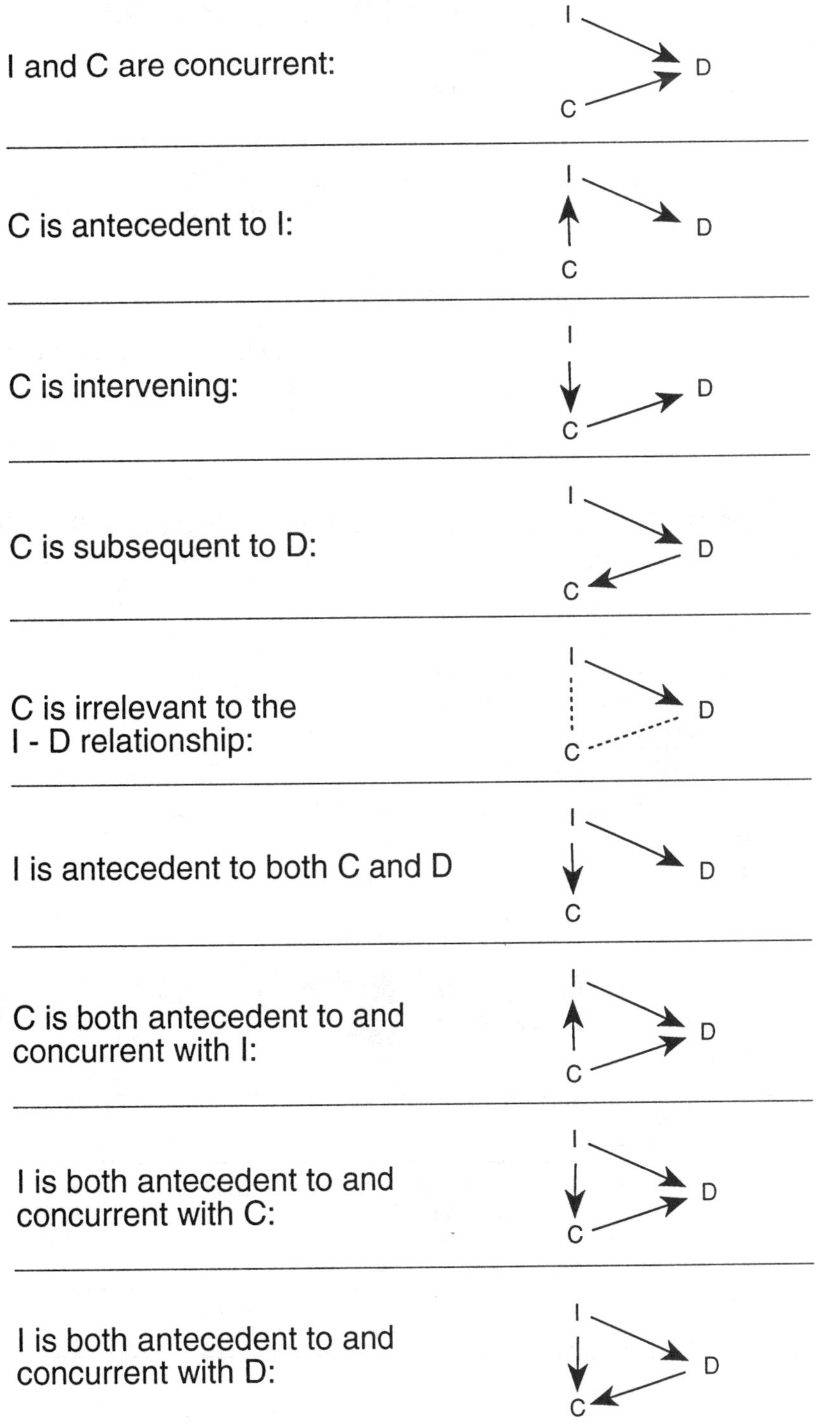
I and C are concurrent:
C is antecedent to I:
C is intervening:
C is subsequent to D:
C is irrelevant to the I - D relationship:
I is antecedent to both C and D
C is both antecedent to and concurrent with I:
I is both antecedent to and concurrent with C:
I is both antecedent to and concurrent with D:

Figure 6.6

EXERCISES

Exercise 6.1

Formulate one or two hypotheses from the variables in the data list from Table 6.5 (other than those used below), create your own tables, and interpret the results.

Exercise 6.2

Two percentage tables are presented below. Write a short paragraph interpreting each table.

Table E6.2.1 Civil Rights Scores by Protest Demonstrations per 1 Million Population

	Protest Demonstrations per 1 Million Population		
Civil Rights Scores	*High* *10 and above*	*Medium* *2 - 9.9*	*Low* *0 - 1.9*
High and medium 1 - 3	66.7%	66.7	20.0
Low and very low 4 - 7	33.3	33.3	80.0
Total	100.0%	100.0%	100.0%
(*n* =)	(6)	(6)	(9)

Table E6.2.2 Telephones per 1000 Population by Percentage of the Labor Force in Agriculture

		Percentage of Labor Force in Agriculture			
Telephones per 1000 Population		*High* *55 and above*	*Medium* *20 - 54*	*Low* *10 - 19*	*Very Low* *0 - 9*
High	400 and above	0%	0	40.0	50.0
Medium	150 - 399	0	0	20.0	50.0
Low	50 - 149	0	60.0	40.0	0
Very Low	49 and below	100.0	40.0	0	0
Total		100.0%	100.0%	100.0%	100.0%
(*n* =)		(3)	(5)	(5)	(6)

Table E6.3 Table of GNP by AGRIC

GNP / Frequency / Percent / Row Pct / Col Pct	AGRIC: H	M	O	VL	Total
H	0 0.00 0.00 0.00	0 0.00 0.00 0.00	2 10.00 33.33 33.33	4 20.00 66.67 66.67	6 30.00
M	0 0.00 0.00 0.00	2 10.00 28.57 40.00	3 15.00 42.86 50.00	2 10.00 28.57 33.33	7 35.00
O	3 15.00 42.86 100.00	3 15.00 42.86 60.00	1 5.00 14.29 16.67	0 0.00 0.00 0.00	7 35.00
Total	3 15.00	5 25.00	6 30.00	6 30.00	20 100.00

NOTE: Low is indicated by the letter O, not LO.

Exercise 6.3

Tables 6.7 and 6.8 could have been produced by a computer, such as the accompanying table from the SAS system. In each cell you will see the frequency and under it three percentages based on the grand total, the row total, and the column total, respectively.

Suppose percentage of the labor force in agriculture was the dependent variable and per capita GNP was the independent variable. How would you interpret this table?

EXERCISES

Table E6.4 Table of CRIGHTS by PRIGHTS

CRIGHTS / *Frequency* / *Percent* / *Row Pct* / *Col Pct*	*PRIGHTS* *H*	*M*	*O*	*VL*	*Total*
H	7 35.00 87.50 87.50	1 5.00 12.50 33.33	0 0.00 0.00 0.00	0 0.00 0.00 0.00	8 40.00
M	1 5.00 33.33 12.50	2 10.00 66.67 66.67	0 0.00 0.00 0.00	0 0.00 0.00 0.00	3 15.00
O	0 0.00 0.00 0.00	0 0.00 0.00 0.00	4 20.00 66.67 80.00	2 10.00 33.33 50.00	6 30.00
VL	0 0.00 0.00 0.00	0 0.00 0.00 0.00	1 5.00 33.33 20.00	2 10.00 66.67 50.00	3 15.00
Total	8 40.00	3 15.00	5 25.00	4 20.00	20 100.00

Exercise 6.4

In the table above, a nation's civil rights score is dependent on the independent variable Political Rights. Interpret the table.

Table E6.5 Table of PHONES by GNP

PHONES *FREQUENCY* *PERCENT* *ROW PCT* *COL PCT*	*GNP* *H*	*M*	*O*	*TOTAL*
H	4 21.05 80.00 66.67	1 5.26 20.00 14.29	0 0.00 0.00 0.00	5 26.31
M	2 10.53 50.00 33.33	2 10.53 50.00 28.57	0 0.00 0.00 0.00	4 21.06
O	0 0.00 0.00 0.00	4 21.05 80.00 57.14	1 5.26 20.00 16.67	5 26.31
VL	0 0.00 0.00 0.00	0 0.00 0.00 0.00	5 26.32 100.00 83.33	5 26.32
TOTAL	6 31.58	7 36.84	6 31.58	19 100.00

Exercise 6.5

Indicate the dependent and independent variables and interpret the table above.

EXERCISES

Table E6.6 Table of DEATHS by PDEMS

DEATHS / Frequency / Percent / Row Pct / Col Pct	PDEMS: H	M	O	Total
H	2 10.00 50.00 33.33	1 5.00 25.00 11.11	1 5.00 25.00 20.00	4 20.00
M	3 15.00 42.86 50.00	2 10.00 28.57 22.22	2 10.00 28.57 40.00	7 35.00
O	1 5.00 25.00 16.67	1 5.00 25.00 11.11	2 10.00 50.00 40.00	4 20.00
VL	0 0.00 0.00 0.00	5 25.00 100.00 55.56	0 0.00 0.00 0.00	5 25.00
Total	6 30.00	9 45.00	5 25.00	20 100.00

Exercise 6.6

In the table above, Deaths from Political Violence is the dependent variable and Number of Protest Demonstrations is the independent variable. Is there a relationship? If so, what kind?

Table E6.7 Table of CRIGHTS by DEATHS

CRIGHTS / FREQUENCY / PERCENT / ROW PCT / COL PCT	DEATHS: H	O	TOTAL
H	4 21.05 36.36 40.00	7 36.84 63.64 77.78	11 57.89
O	6 31.58 75.00 60.00	2 10.53 25.00 22.22	8 42.11
TOTAL	10 52.63	9 47.37	19 100.00

Exercise 6.7

A country's civil rights score is being predicted or explained by the variable Deaths from Political Violence (see table above). Is there a relationship? If so, what kind? (Note: In this and the following problems, some variables have been dichotomized to ease interpretation. In these cases, H contains both the old high and the medium category, and O is both the old low and very low.)

Table E6.8.1 Table 1 of CRIGHTS by DEATHS Controlling for ADJUST = H

CRIGHTS FREQUENCY PERCENT ROW PCT COL PCT	DEATHS H	O	TOTAL
H	2 40.00 100.00 50.00	0 0.00 0.00 0.00	2 40.00
O	2 40.00 66.67 50.00	1 20.00 33.33 100.00	3 60.00
TOTAL	4 80.00	1 20.00	5 100.00

Table E6.8.2 Table 2 of CRIGHTS by DEATHS Controlling for ADJUST = M

CRIGHTS FREQUENCY PERCENT ROW PCT COL PCT	DEATHS H	O	TOTAL
H	1 16.67 20.00 100.00	4 66.67 80.00 80.00	5 83.33
O	0 0.00 0.00 0.00	1 16.67 100.00 20.00	1 16.67
TOTAL	1 16.67	5 83.33	6 100.00

Table E6.8.3 Table 3 of CRIGHTS by DEATHS Controlling for ADJUST = O

CRIGHTS *FREQUENCY* *PERCENT* *ROW PCT* *COL PCT*	*DEATHS* *H*	*O*	*TOTAL*
H	1 12.50 25.00 20.00	3 37.50 75.00 100.00	4 50.00
O	4 50.00 100.00 80.00	0 0.00 0.00 0.00	4 50.00
TOTAL	5 62.50	3 37.50	8 100.00

Exercise 6.8

The relationship in Exercise 6.7 is controlled by the variable Executive Adjustments. What happens to the initial relationship?

Table E6.9.1 Table 1 of CRIGHTS by DEATHS Controlling for GNP = M

CRIGHTS FREQUENCY PERCENT ROW PCT COL PCT	DEATHS H	O	TOTAL
H	3 42.86 75.00 60.00	1 14.29 25.00 50.00	4 57.14
O	2 28.57 66.67 40.00	1 14.29 33.33 50.00	3 42.86
TOTAL	5 71.43	2 28.57	7 100.00

Table E6.9.2 Table 2 of CRIGHTS by DEATHS Controlling for GNP = O

CRIGHTS FREQUENCY PERCENT ROW PCT COL PCT	DEATHS H	O	TOTAL
H	1 16.67 100.00 20.00	0 0.00 0.00 0.00	1 16.67
O	4 66.67 80.00 80.00	1 16.67 20.00 100.00	5 83.33
TOTAL	5 83.33	1 16.67	6 100.00

Exercise 6.9

Here the Civil Rights by Deaths from Political Violence relationship has another control variable, Per Capita GNP. Interpret the results. Note that no table is shown for the high GNP category, since all six countries in that category had low deaths and high civil rights. In effect, such a table would be meaningless.

Table E6.10.1 Table 1 of CRIGHTS by DEATHS Controlling for PRIGHTS = H

CRIGHTS / *FREQUENCY* / *PERCENT* / *ROW PCT* / *COL PCT*	*DEATHS* *H*	*O*	*TOTAL*
H	2 25.00 25.00 100.00	6 75.00 75.00 100.00	8 100.00
O	0 0.00 - 0.00	0 0.00 - 0.00	0 0.00
TOTAL	2 25.00	6 75.00	8 100.00

Table E6.10.2 Table 2 of CRIGHTS by DEATHS Controlling for PRIGHTS = M

CRIGHTS / *FREQUENCY* / *PERCENT* / *ROW PCT* / *COL PCT*	*DEATHS* *H*	*O*	*TOTAL*
H	2 66.67 66.67 100.00	1 33.33 33.33 100.00	3 100.00
O	0 0.00 - 0.00	0 0.00 - 0.00	0 0.00
TOTAL	2 66.67	1 33.33	3 100.00

Exercise 6.10

Here the control variable is the Political Rights score. Note what happens when the control variable is highly associated with one of the other variables in the table. In this case, civil rights is associated with political rights. Interpret the results.

Table E6.10.3 Table 3 of CRIGHTS by DEATHS Controlling for PRIGHTS = O

CRIGHTS FREQUENCY PERCENT ROW PCT COL PCT	*DEATHS* *H*	*O*	*TOTAL*
H	0 0.00 – 0.00	0 0.00 – 0.00	0 0.00
O	3 75.00 75.00 100.00	1 25.00 25.00 100.00	4 100.00
TOTAL	3 75.00	1 25.00	4 100.00

Table E6.10.4 Table 4 of CRIGHTS by DEATHS Controlling for PRIGHTS = VL

CRIGHTS FREQUENCY PERCENT ROW PCT COL PCT	*DEATHS* *H*	*O*	*TOTAL*
H	0 0.00 – 0.00	0 0.00 – 0.00	0 0.00
O	3 75.00 75.00 100.00	1 25.00 25.00 100.00	4 100.00
TOTAL	3 75.00	1 25.00	4 100.00

Table E6.11 Table of PDEMS by CRIGHTS

PDEMS FREQUENCY PERCENT ROW PCT COL PCT	CRIGHTS H	O	TOTAL
H	10 52.63 71.43 90.91	4 21.05 28.57 50.00	14 73.68
O	1 5.26 20.00 9.09	4 21.05 80.00 50.00	5 26.32
TOTAL	11 57.89	8 42.11	19 100.00

Exercise 6.11

Protest Demonstrations (dependent) and Civil Rights (independent) are the variables in Table E6.11. What are your conclusions?

Table E6.12.1 Table 1 of PDEMS by CRIGHTS Controlling for WORLD = F

PDEMS *FREQUENCY* *PERCENT* *ROW PCT* *COL PCT*	*CRIGHTS* *H*	*O*	*TOTAL*
H	10 76.92 83.33 100.00	2 15.38 16.67 66.67	12 92.31
O	0 0.00 0.00 0.00	1 7.69 100.00 33.33	1 7.69
TOTAL	10 76.92	3 23.08	13 100.00

Table E6.12.2 Table 2 of PDEMS by CRIGHTS Controlling for WORLD = T

PDEMS *FREQUENCY* *PERCENT* *ROW PCT* *COL PCT*	*CRIGHTS* *H*	*O*	*TOTAL*
H	0 0.00 0.00 0.00	2 33.33 100.00 40.00	2 33.33
O	1 16.67 25.00 100.00	3 50.00 75.00 60.00	4 66.67
TOTAL	1 16.67	5 83.33	6 100.00

Exercise 6.12

To the relationship in Exercise 6.11, we have added a control variable World. Countries in the data set were assigned to the categories First World (developed, democratic), Second World (communist), and Third World (developing nations). Then, in line with recent world events, the variable was dichotomized by bringing the two communist countries into the First World. Interpret the results.

Notes

1. C. L. Taylor and D. Jodice, *World Handbook of Political and Social Indicators* (New Haven, CT: Yale University Press, 1983). Note that for computational ease, only a small number of countries is being used in this data set. These countries may not be representative of the entire world of nations.

2. R. Gastil, *Freedom in the World* (New York: Freedom House, 1980).

3. Suppose a country possesses a score "outside" a class interval, such as 3.93 or 5.95. In this case our rule will be to round off to the nearest tenth. If the number in the hundredths place is 5 or more, round up; if below 5, round down. Thus 3.93 rounds down to 3.9 and is counted in the 2.0 to 3.9 class interval. Likewise, 5.95 rounds up to 6.0 and is entered in the 6.0 to 7.9 interval.

KEY CONCEPTS

tests of statistical significance
inferential statistics/inductive statistics
descriptive statistics
population/sampling universe
sample
sampling bias/biased sample
random sample
simple random sample
sampling error
null hypothesis/H_0
alternative hypothesis/research hypothesis/H_1
fallacy of affirming the consequent
chi-square test
sample statistics
population parameters
population mean/mu/μ
population standard deviation/ lowercase sigma/σ
one-sample tests
one-sample z test
one-sample t test
two-sample t test
one-way analysis of variance/one-way ANOVA
.05 level of significance
probability
type I error/alpha error
nondirectional alternative hypothesis/two-tailed alternative hypothesis/two-tailed test
directional alternative hypothesis/one-tailed alternative hypothesis/one-tailed test
degrees of freedom
type II error/beta error

7 Statistical Inference and Tests of Significance

INTRODUCTION

We now turn our attention to the issue of sampling and the procedures used to determine the likelihood that data obtained from a sample will reflect the group from which the sample was selected. In other words, to generalize about a very large group—a reference group, all autistic children, or a large legislative body, for example—we must study a subset, or sample, of the whole group. How certain can we be in concluding that what we found in studying the sample really applies to the whole group? To deal with this problem, we use inferential statistics and tests of statistical significance.

In this chapter we will provide a road map to four chapters that follow. We will discuss the nature of sampling and the logic of testing for statistical significance. The simplest of these tests, the one-sample z test, is used as a model for presenting the steps performed in all such tests. Theoretical considerations are postponed until the next and subsequent chapters. We present major problem formats in this chapter, and the tests that match each format are mentioned and referenced to the specific chapters in which they are presented in detail.

A word of caution: Many new ideas are presented here, and careful reading is required. The logical process that underlies a test of significance is likely to seem strange to a newcomer. With new ideas to grasp and a new way of thinking to be mastered, it may take you a while to be comfortable with this new material. But take heart. Once you know this chapter, most of what comes in the next four chapters should be fairly easy to grasp. The "number crunching" will become more complex, but the type of reasoning learned here will remain the same.

WHAT IS STATISTICAL INFERENCE?

To begin our discussion, let us look at Table 7.1. If the hypothesis is that there is a relationship between attitude on the death penalty for top drug dealers and attitude on intervention in Latin America to halt drug production, then it appears that the hypothesis has been verified. Supporters of the death penalty predominate 3:1 among supporters of intervention, whereas the ratio is reversed (3:1 death penalty opponents) among those against intervention. Up to this point in this text, that is the conclusion we would be expected to make.

We now add a new question. About whom are we generalizing? Certainly for 30 of the 40 people studied here the hypothesis is verified, but are we trying to generalize only about that group of 40 people? Assume they are college students. Perhaps we want to generalize about all students at that particular college. Or perhaps we want to generalize about all college students in the country, or even all college students in the world. Often we want to reach a conclusion for more than the people actually studied. But how safe are we in concluding that what is true for the 40 students of Table 7.1 is also true for all students at that college? How safe are we in generalizing beyond that college to the country? The world?

The techniques that help us to answer this question about generalizing to a larger group are known as **tests of statistical significance** and *the body of knowledge that deals with such tests of significance is called* **inferential statistics** or sometimes **inductive statistics.** Up to this point we have only been dealing with **descriptive statistics,** *where frequency distributions or relationships between variables are described.* Now we turn our attention to inferring that what is true for a group we have actually studied is true for the larger group about which we want to generalize. We are inducing—going from the specific to the general—that what is true for the subjects studied, the 40 students, is true for all students in the college, the country, or possibly the planet.

Table 7.1

	Attitude on Drug Intervention		
Attitude on Death Penalty	*For*	*Against*	*Total*
Supports	15	5	20
Opposes	5	15	20
Total	20	20	40

Let us begin by introducing some terms used in studies such as the preceding where a smaller group is selected specifically to reflect a larger group. *The group about which we want to generalize is called the* **population** or, less often, the **sampling universe.** (We call this group the population even though it may not necessarily be a population in a demographic sense, such as all American citizens or all residents of the Bronx.) The population consists of *all* the members about whom we wish to generalize, for example, all felons, all attorneys, all divorced women, or all Methodists. The key word is *all.*

From the population, we select the smaller group that we will study—the **sample.** We call *the selection of the subjects to be in that sample* **drawing a sample,** in the same sense as drawing cards in a card game such as poker. How we go about drawing the sample determines whether or not we can use a test of significance to help us in our generalization to the population.

One way a sample may *not* accurately reflect the population from which that sample was drawn is known as **sampling bias**; that is, a **biased sample** could have been drawn. *In sampling bias, it is the mechanism for selecting the sample that causes the sample to be not representative of the population as a whole.* For example, for the study shown in Table 7.1 suppose we had recruited students by giving priority to recovered drug addicts. This could possibly result in a group with a disproportionate number of "hard liners" on both issues. Thus the resulting sample would be not representative of the distribution of opinion in the college as a whole.

Sometimes a biased sample is intentionally selected since the researcher seeks a particular outcome. For example, suppose I want to prove that "Paingo" is the best-selling headache medicine in town. I interview shoppers leaving the drug store, ask them if they had purchased headache medicine and if so, what brand. Of the 20 people who said they had bought headache medicine, suppose 15 said they bought Paingo. I could then prepare a television commercial claiming that 75% (15 out of 20) of purchasers prefer Paingo. However, suppose I knew in advance that the only headache remedies sold by that particular pharmacy were Paingo and a new medicine for which there was not yet any name brand recognition by the consumer. My sample would be intentionally biased.

Usually there is not such an ill intention by the researcher, but something not perceived causes the sample to be biased. Imagine the same scenario and the same results—three out of four customers prefer Paingo—but unknown to me, the drug store had Paingo on sale that day, at half price! No wonder so many people preferred Paingo.

In American politics, a classic example of sampling bias took place in 1936 when the *Literary Digest* polled an unusually large sample and then predicted that Republican Alf Landon would win the election by nearly 60%. In the election the landslide went not to Landon, who only won 38% of the vote, but to his opponent, Franklin D. Roosevelt. It turned out that the *Literary Digest* had selected its sample from lists of automobile owners and from the phone book. This was during the Great Depression, when only the relatively well-heeled could afford automobiles or telephones, and these groups at the time were staunchly pro-Republican. The F.D.R. supporters came from the far larger group of those less well-off and less likely to own what at that time were luxuries.

We must bear in mind that if a sample is biased, there is *no* statistical technique that can make the sample representative of the population. Accordingly, every technique we will discuss assumes, among other things, that the sample is representative, not biased. If there is reason to believe that our sample is biased, no test of significance exists for turning our sow's ear into a silk purse.

RANDOM SAMPLES

If a sample is not biased, it is a **random sample.** In fact, statistical tests assume a **simple random sample,** *one in which each member of the population has an equal chance or probability of being included in the sample.* Drawing a simple random sample assumes a selection process not unlike an honest lottery. Suppose we list all the students in our college alphabetically. We then assign each student a number beginning with 1 and continuing until all, say, 3000 students have been included. We might then get a rotating drum containing 3000 marbles of equal size and weight, numbered consecutively from 1 to 3000. We turn the drum many times to thoroughly mix the marbles and have a volunteer select a marble. The volunteer takes out the marble, writes down its number on a list, and returns the marble to the drum. The drum is turned again, and the process is repeated until our list contains as many numbers as the size sample we want to select. (Assume for simplicity that no marble is selected twice and that the drum is sturdy, so that nobody loses their marbles.) To get a sample of 40 people, we would take the list of 40 numbers back to the alphabetized list of all students in our population, write down the name adjacent to each number on our list, and use this list as our simple random sample. Every student in the college would have had an equal likelihood of being included in the sample.

This tedious technique is unnecessary today, as we may instead use a pregenerated table of random numbers or a random number-generating computer program. All honest sampling techniques used by legitimate researchers and polling organizations strive to produce unbiased samples. These techniques rarely generate perfect random samples, but they come close enough that we may use them and assume randomization.

Suppose we have used all means to ensure that our sample of 40 students was randomly drawn. We no longer have to worry about sampling bias, but another nemesis awaits: **sampling error.** The term *error* as used here does not mean that a mistake was made in selecting the sample. Rather, error here means *deviation from what actually exists in the population.* This deviation can be accounted for by the laws of chance or probability. For example, imagine a population that is 50% female. Using a proper randomization method, we draw a sample that just by chance is 60% female. We mistakenly conclude that the population is 60% female. In a few samples this can happen by chance. In a somewhat smaller number of samples we might get 70% women. In an again smaller number of samples we might get 80% women. There is even some likelihood ("once in a blue moon") of getting all women in our sample and mistakenly concluding that the college population is entirely female. Similarly, any other characteristic of a sample may deviate from the population due to sampling error—religion, income, party preference, fraternity membership, or in the case of our original problem, attitudes on the two variables.

Suppose in the college of our study, there is *no* relationship between attitudes on the death penalty and intervention, and yet as a result of sampling error we drew a random sample in which there *was* such a relationship. A test of significance would tell us how likely it is that this could have occurred by chance, given random selection of the sample. It would tell us the odds of concluding, based on our sample, that in the population our two variables are related when in fact they are not. Note that such tests do not tell us for any given sample whether that sample accurately reflects the population. The tests only tell us the probability that this may be the case. Thus we will be living with uncertainty from here on.

While the following is rarely seen in actual scholarly research, we begin by stating a **null hypothesis,** *symbolized by* H_0, with the H meaning hypothesis and the subscript meaning zero, for null. The nature of a null hypothesis varies from problem to problem. *In the case* *of our problem in Table 7.1, it is a statement that in the population of all students at this college there is* no *relationship between attitudes toward the death penalty and drug intervention.* It states that the two variables are independent of one another and that the relationship appearing in the table is solely the result of sampling error (chance) and does not reflect a relationship in the population.

H_0: in the population there is no relationship between the variables

If H_0 is true, we would be wrong to conclude that the variables are related in the population. What if H_0 is probably false? Then we might conclude one of several possible **alternative hypotheses** (sometimes called **research hypotheses**), the simplest of which is that *in the population the two variables* are *related.* We call the alternative hypothesis H_1 (we usually only formulate one alternative hypothesis, but we could have others—H_2, H_3, etc.).

If H_1 is true, then the relationship in the table is not due to sampling error but is actually reflective of a relationship in the population. Note that as a researcher, you are really trying to prove H_1, but *a principle of logic known as the* **fallacy of affirming the consequent** *suggests that we can only "prove"* H_1 indirectly by showing that H_0 is probably untrue. If H_0 is false, then H_1 is the only logical conclusion. If H_0: in the population there is no relationship is false, then the only other general conclusion is H_1: in the population there is a relationship. We set up H_0 hoping to disprove it; it is a "straw man" which we hope to knock down. We will return to this problem in Chapter 12 and make use of *a test of significance known as the* **chi-square test** to see if we can reject the null hypothesis. That test will be appropriate whenever two variables are presented in a cross-tabulation such as Table 7.1, regardless of the level of measurement of those variables.

COMPARING MEANS

Let us now consider a situation where we are comparing two groups' means, such as the means of two classes that have taken a common examination. If we treat both groups as *populations,* our conclusions are made simply by examining and comparing the two means. Thus either the mean for class 1 = the mean for class 2, or the mean for class 1 ≠ the mean for class 2. If the latter case is correct, either the mean for class

1 > the mean for class 2, or the mean for class 1 < the mean for class 2. We simply compare the two numbers to reach our conclusion. However, when one or both of the means comes from random samples rather than populations, the possibility of sampling error emerges and we need to perform a test of significance on the data.

Before discussing this further, we need to introduce some new terms. To differentiate data from a sample from data from the population, *we call the statistics computed from sample data* **sample statistics** *and those from the population data* **population parameters.** We designate sample statistics with the same notation we have been using all along; that is,

$\bar{x}$ is a sample's mean.
s is a sample's standard deviation.
s^2 is a sample's variance.
n is a sample's size.

For sample statistics we also continue to use the formulas we have learned thus far.

$$\bar{x} = \frac{\sum x}{n}$$

$$s^2 = \frac{\sum (x - \bar{x})^2}{n} \qquad \text{(definitional formula)}$$

or

$$s^2 = \frac{\sum x^2 - \frac{\left(\sum x\right)^2}{n}}{n} \qquad \text{(computational formula)}$$

For population parameters, we use lowercase Greek letters, often with subscripts to differentiate them.

Greek letter mu. μ is a **population's mean.**
Greek letter sigma. σ is a **population's standard deviation.**
σ^2 is a population's variance.
N is a population's size.

Our formulas then become:

$$\mu = \frac{\sum x}{N}$$

$$\sigma^2 = \frac{\sum (x-\mu)^2}{N} \qquad \text{(definitional formula)}$$

or

$$\sigma^2 = \frac{\sum x^2 - \frac{\left(\sum x\right)^2}{N}}{N} \qquad \text{(computational formula)}$$

Now, in a problem calling for a test of significance, we may no longer know both of the population means. In fact, we may know neither of them. Yet, on the basis of sample means (the $\bar{x}$'s) we want to make a generalization about the respective population means (the μ's).

Our null hypothesis is that there is no difference between the two population means.

$$H_0: \mu_1 = \mu_2$$

If the test of significance enables us to reject H_0, we will conclude that

$$H_1: \mu_1 \neq \mu_2$$

Let us examine an example where this kind of null and alternative hypothesis would be formulated. Suppose we wanted to compare Ohio's Democratic members of the House of Representatives to their Republican colleagues, in terms of liberalism. Here liberalism is a score assigned to each representative's voting record by Americans for Democratic Action (ADA), a liberal organization. For Ohio's 21 representatives in the 101st Congress, we have ADA scores for 20 people (all 11 Democrats and 9 of the 10 Republicans).[1] We will treat the 11 Democrats and the 9 Republicans as the populations of all Ohio Democratic and all Ohio Republican representatives, respectively. We calculate the two population means and compare them:

$$\mu_{\text{Democrats}} = 83.18$$

$$\mu_{\text{Republicans}} = 17.22$$

Obviously, the two μ's are unequal to each other, and in fact the μ for the Democrats is much higher (more liberal) than the μ for the Republicans. Since we have compared two population means, no sampling is involved, there is no sampling error, and no test of significance is needed.[2]

COMPARING A SAMPLE MEAN TO A POPULATION MEAN OR OTHER VALUE

Suppose, however, we knew the μ for the Democrats but did *not* know it for the Republicans. Suppose instead that we had access to a random sample of size $n = 3$ of Ohio's Republican representatives and that the sample's mean was 23.33.

$$\bar{x}_{\text{Reps}} = 23.33$$

$$\mu_{\text{Dems}} = 83.18 \qquad \mu_{\text{Reps}} = \text{unknown}$$

We formulate a null hypothesis:

$$H_0\text{: } \mu_{\text{Democrats}} = \mu_{\text{Republicans}}$$

Now, if the null hypothesis is *true,* $\mu_{\text{Republicans}} = \mu_{\text{Democrats}} = 83.18$. The fact that our sample $\bar{x}$ for the Republicans, 23.33, is different from 83.18 would be attributed to sampling error. In other words, due to random chance we drew a sample from a population whose mean was 83.18 and got a sample mean of 23.33.

If, on the other hand, we could reject H_0, we could conclude instead that our sample probably did *not* come from a population whose mean was 83.18, but rather from a population whose mean differed from 83.18.

$$H_1\text{: } \mu_{\text{Democrats}} \neq \mu_{\text{Republicans}}$$

We would then refine our conclusion even more by noticing that since 23.33 is less than 83.18, $\mu_{\text{Republicans}}$ is probably less than 83.18 and Republicans are probably less liberal than the Democrats.

The tests of significance that we would perform to see if H_0 could be rejected are called **one-sample tests** *since we are comparing data from one group's sample to another null-hypothesized value,* in this case data from another group's population. Specifically, depending on the information given to us, we would do either a **one-sample *z* test** or a **one-sample *t* test.** We will examine the former test later in this chapter and discuss both tests in the next chapter.

COMPARING A SAMPLE MEAN TO ANOTHER SAMPLE MEAN

Suppose *both* $\mu_{\text{Republicans}}$ and $\mu_{\text{Democrats}}$ were unknown. We would then have to draw random samples from *both* groups and compare the two sample means. Assume the same figures for the Republican sample as used above. In addition, we draw a sample from the Democrat population, say, a sample of $n = 4$ Democrats with a calculated sample mean of 82.50.

$$\bar{x}_{\text{Dems}} = 82.50 \qquad \bar{x}_{\text{Reps}} = 23.33$$

$$\mu_{\text{Dems}} = \text{unknown} \qquad \mu_{\text{Reps}} = \text{unknown}$$

Our H_0 and H_1 will remain as before:

$$H_0\text{: } \mu_{\text{Democrats}} = \mu_{\text{Republicans}}$$

$$H_1\text{: } \mu_{\text{Democrats}} \neq \mu_{\text{Republicans}}$$

Since our comparison is now between two *sample means, we could perform a test known as the* **two-sample *t* test** (discussed in Chapter 9). We would follow a line of reasoning similar to that used for the one-sample tests discussed earlier. An alternative to the two-sample *t* test is the **one-way analysis of variance** or (**one-way ANOVA,** for short). Generally, though, ANOVA is more commonly used when there are more than two groups to be compared.

COMPARING MORE THAN TWO SAMPLE MEANS

In our Ohio example there are only two parties to compare. What if there were more than two parties? For instance, suppose a similar study had been contemplated for the Canadian House of Commons. Our null hypothesis might look like this:

$$H_0\text{: } \mu_{\text{bloc Québécois}} = \mu_{\text{Liberals}} = \mu_{\text{Reform Party}}$$

As mentioned, *one-way analysis of variance* would be used to compare these means (see Chapter 10).

The data situations, tests of significance, and chapters of this text pertaining to them are summarized below:

Data Situation	*Summary Test of Significance*	*Chapter*
One Sample Mean versus a Population Mean	One-Sample z Test or One-Sample t Test	7, 8 8
One Sample Mean versus Another Sample Mean	Two-Sample t Test	9
Comparing Several Sample Means	One-Way Analysis of Variance	10
Comparing Two Variables in a Cross-Tabulation	Chi-Square Test for Contingency	12

THE TEST STATISTIC

In the case of the z and t tests, the size of the z or t generated is, in part, a function of the distance between the means being compared. In this one-sample case, this is the difference between the sample mean and the population mean. (In other one-sample cases, the sample mean might be compared to some designated value other than a population mean.) In the two-sample case, it is the distance between the two sample means.

Let us we return to the Ohio delegation problem where

$$\mu_{\text{Dems}} = 83.18$$

$$\mu_{\text{Reps}} = \text{unknown}$$

Recall that we drew a sample of $n = 3$ Republicans and got a sample mean of $\bar{x}_{\text{Reps}} = 23.33$. If we also knew the standard deviation of the *population*, σ_{Dems}, we would be in a position to do a one-sample z test. (We will see in the next chapter that if we do not know σ_{Dems} we have to estimate it from our sample data and we would do a one-sample t test instead of a z test.)

Please keep in mind that what follows is an overview of testing for significance as we will be doing in the following four chapters. Thus in chapters to come, the origin and meaning of each formula will be explained.

The formula for the one-sample z test is

$$z = \frac{\bar{x} - \mu}{\sigma / \sqrt{n}}$$

For our specific problem the formula becomes

$$z = \frac{\bar{x}_{\text{Reps}} - \mu_{\text{Dems}}}{\sigma_{\text{Dems}} / \sqrt{n}}$$

Once we know σ_{Dems}, we can calculate z (even though we do not yet know what to do with that information). We find that $\sigma_{\text{Dems}} = 10.5$ and accordingly,

$$z = \frac{\bar{x}_{\text{Reps}} - \mu_{\text{Dems}}}{\sigma_{\text{Dems}} / \sqrt{n}}$$

$$= \frac{23.33 - 83.18}{10.5 / \sqrt{3}}$$

$$= \frac{-59.85}{10.5 / 1.732}$$

$$= \frac{-59.85(1.732)}{10.5}$$

$$= \frac{-103.66}{10.5}$$

$$= -9.872$$

The negative sign on z is due to the fact that $\bar{x}_{\text{Reps}}$ is less than μ_{Dems}. If $\bar{x} > \mu$, then z would be positive.

For now, let us leave our calculated z of -9.872 and look at what happens to z as the distance between $\bar{x}$ and μ increases (and thus the numerator of the z formula increases). We will first imagine a case where $\bar{x}$ and μ are the same.

$$\text{If } \bar{x} = \mu, \qquad z = \frac{83.18 - 83.18}{10.5/\sqrt{3}} = \frac{0}{10.5/1.732} = \frac{0(1.732)}{10.5} = \frac{0}{10.5} = 0$$

Now let us look at what happens to z as $\bar{x}$ gets farther away from μ. We will reduce $\bar{x}$ by increments of 10 units at a time and see how it impacts the absolute value of z.

If $\bar{x}$ = 83.18	$\bar{x} - \mu = 0$	$\lvert z \rvert = 0$
73.18	10	1.649
63.18	20	3.299
53.18	30	4.948
43.18	40	6.598
33.18	50	8.247
23.18	60	9.897
13.18	70	11.546
3.18	80	13.196

If the null hypothesis is true and $\mu_{Dems} = \mu_{Reps}$, we would expect the mean of a sample of Republicans to be very close to the population mean for Republicans. Ideally, they would be the same. If $\mu_{Dems} = \mu_{Reps}$ and $\mu_{Reps} = \bar{x}_{Reps}$, then logically $\bar{x}_{Reps} - \mu_{Dems} = 0$ and z will be zero. Ideally, if the null hypothesis is true, $z = 0$.

However, even if the null hypothesis is true, it still is *quite* likely that the Republican sample mean will be slightly different from the Republican population mean due to sampling error. Thus $\bar{x}_{Reps} - \mu_{Dems}$ could often be slightly different from zero and z could be slightly different from zero as well. Note that as the gap between $\bar{x}_{Reps}$ and μ_{Dems} grows, the likelihood that the null hypothesis is true shrinks. It is *always possible* that the null hypothesis is true, no matter how far $\bar{x}_{Reps}$ is from μ_{Dems} and thus how large a z we get. *But as the gap between* $\bar{x}_{Reps}$ *and* μ_{Dems} *increases and thus z increases, the likelihood that the null hypothesis is true decreases.*

In our Ohio example, nearly 60 points separate our $\bar{x}_{Reps}$ of 23.33 and our μ_{Dems} of 83.18. Our z of −9.8722 is quite large in absolute value, as compared to z scores generally encountered. It is true that given a true null hypothesis, we could get an $\bar{x}$ that different from μ and a z that large due to sampling error. But it is so improbable that in this case an explanation other than sampling error would be much more plausible in accounting for our large z. The alternative explanation is that the population from which the Republican sample was drawn has a mean different from the mean for the Democrat population. In short, it is our *alternative* hypothesis:

$$H_1\text{: } \mu_{\text{Democrats}} \neq \mu_{\text{Republicans}}$$

At what point do we decide that the z obtained is large enough to reject the null hypothesis? While many factors may enter into this decision in applied research, in the confines of the classroom we utilize a common convention called the **.05 level of significance.** *If 5 (or fewer) out of 100 random samples (i.e., 1 out of 20) drawn from a population where the null hypothesis is true would yield a z value equal to or greater than the one we obtained, we will reject the null hypothesis and conclude instead our alternative hypothesis,* H_1*. In doing so, we run a risk not to exceed .05 proportion (5%) that really* H_0 *is true and we are mistakenly rejecting it. The .05 proportion is really the probability that we are falsely rejecting a true null hypothesis.*

PROBABILITIES

Let us briefly examine the nature of **probabilities,** which can be defined as *proportions that reflect the likelihood of a particular outcome occurring.* An easy-to-understand analogy is the batting average used in baseball. Imagine a simplified game in which walks, bean balls, and other anomalies are eliminated so that a batter stepping up to the plate faces one of two possible outcomes. The batter either gets a base hit (be it a single, double, triple, or home run) or does not get a base hit (mighty Casey strikes out). For any given "at bat," we do not know whether the player will get a hit or not, but we can state the probability of a hit occurring by examining the batter's past record. Suppose the batter has come to bat 100 times and has had 30 base hits. The probability that he or she will get a hit when coming to bat is the number of accumulated base hits divided by the total accumulated at bats. In this case:

$$\begin{array}{l}\text{Batting Average}\\ \text{(probability of a base hit)}\end{array} = \frac{\text{base hits}}{\text{at bats}} = \frac{30}{100} = .300$$

Thus there is a .300 probability (a 30% chance) of the batter getting a base hit. Our player is batting 300 (announcers often don't use the decimal). Although this tells us our player's likelihood of getting a hit is a little less than one out of every three at bats, we know for any *given* at bat, a batter will still either get a hit or not—one cannot make .300 of a base hit. Nonetheless, we would prefer to send a batter to the plate whose batting average is .300 than to bring up someone with a .150 batting average!

By using the .05 probability level as a decision point in our test of significance, in effect *we are requiring a "batting average" of .950 before we reject the null hypothesis.* At that point, the risk of error in rejecting the null hypothesis is .05 or 5%. We will continue our discussion of probability in the next chapter.

DECISION MAKING

Recall that in our Ohio problem we had obtained a *z* of −9.872. To know whether we can reject the null hypothesis, we compare the absolute value of the obtained *z*, 9.872, to a series of *critical values* of *z*. These critical values (whose origins we will discuss in the next chapter) are values of *z* for differing levels of significance (probabilities of error in rejecting H_0) beginning with the crucial .05 level.

Probability (Level of Significance)	*Critical Value of z*
.05	1.96
.01	2.58
.001	3.29

Here, the probability is that of falsely rejecting a true null hypothesis. This is also referred to as a **type I error** *or an* **alpha error.**[3]

The procedure is to compare the absolute value of the calculated, or obtained, *z* to the critical values above. Since the .05 level is our decision point, the following two possible overall outcomes exist.

A. $|z_{obtained}| < 1.96$ [4]
In this case, our *z* is less than $z_{critical}$ at the .05 level. We *cannot* reject H_0. We say that the difference (between $\bar{x}_{Reps}$ and μ_{Dems}) is *not* statistically significant and imply that the difference between $\bar{x}$ and μ is the result of sampling error.

B. $|z_{obtained}| \geq 1.96$
In this case, *z* equals or exceeds $z_{critical}$ at the .05 level. We reject

> H_0 (which means we accept H_1). We say that the difference between $\bar{x}$ and μ *is* statistically significant.

In our Ohio problem, $z = -9.872$, $|z| = 9.872$, and $9.872 > 1.96$. Thus we reject the H_0 that there is no difference in liberalism between Democrats and Republicans in Ohio's congressional delegation. The difference between 23.33 and 83.18 is probably not due to sampling error; instead, it probably reflects a real difference between population means.

Now, if $|z_{obtained}| < 1.96$, we have completed our task. If, on the other hand, $|z_{obtained}| \geq 1.96$, we need to take a further step. Remembering that we could be making a mistake in rejecting H_0, we need to report to the reader the likelihood, or odds, that we are making an error. Recall that if $z_{obtained}$ had exactly equaled 1.96, the probability of error would be exactly .05. Thus 5 out of 100 similar-sized random samples drawn from a population where H_0 is true would generate z's of 1.96 or more. If our sample had been one of those five samples, we would make a mistake in rejecting H_0.

Suppose we had obtained a z of exactly 2.58—1 out of 100 samples will yield a z that large due to sampling error. If we then reject H_0, we are falsely rejecting a true null hypothesis. The probability of this happening is .01. We might even have obtained a z of 3.29— the result of sampling error in 1 out of 1000 samples, with a probability of .001.

> *Note:* The larger the z obtained, the smaller the probability of making such an error.

When using the z test, we use these three levels as benchmarks for reporting the probability of a type I error. (Other tests may use additional probability levels below .05—more on that later.) If we had done our z test with a packaged computer program, it would have told us the exact probability of alpha. Having done this by hand, we instead report the probability by the range into which it falls, as follows:

A. If $|z_{obtained}| < 1.96$ and we cannot reject H_0, we report *no* probability.

B. If $|z_{obtained}| \geq 1.96$ but $|z_{obtained}| < 2.58$, we reject H_0 and report $p < .05$; that is, the probability is less than .05 and (by convention) greater than .01. So $p < .05$ tells us that the probability is between .05 and .01.

C. If $|z_{obtained}| \geq 2.58$ but $|z_{obtained}| < 3.29$, we reject H_0 and report $p < .01$. Here, p is less than .01 but greater than .001.

D. If $|z_{obtained}| \geq 3.29$, we reject H_0 and report $p < .001$. We generally stop the process of reporting probabilities of z at the .001 level.

In our Ohio example:

$$|z_{\text{obtained}}| = 9.872 > 1.96, \quad \text{reject } H_0$$

$$= 9.872 > 2.58$$

$$= 9.872 > 3.29, \quad \text{so } p < .001$$

REVIEW

Remember that for the one-sample z test, we are *given* μ and σ for one population. For the random sample drawn from the other population we know the sample's size n and its mean $\bar{x}$. Once we have this information, we use the formula $z = (\bar{x} - \mu)/(\sigma/\sqrt{n})$ to find z, sometimes referred to as z_{obtained}.

We compare $|z_{\text{obtained}}|$ to z_{critical} at the .05 level (i.e., 1.96). If $|z|$ is less than 1.96, we cannot reject H_0. If $|z| \geq 1.96$, we reject H_0. If so, we compare $|z|$ to the critical values of z to determine the probability of a type I error.

EXAMPLES

Suppose we have a scale measuring environmental activism, ranging from 0 to 100 (most active), designed to tap individual attitudes and behavior concerning the environment. Suppose we know that for the population of all those who have graduated from high school but not college, the mean score is 50 with a standard deviation of 10. We then draw a random sample of 100 college graduates, administer the same survey, and calculate the sample means that are presented below. The null and alternative hypotheses are as follows:

$$H_0\text{: } \mu_{\text{high school}} = \mu_{\text{college}}$$

$$H_1\text{: } \mu_{\text{high school}} \neq \mu_{\text{college}}$$

Suppose our sample mean was 51.7. We have all the information needed for a one-sample z test.

$$z = \frac{\bar{x}_{\text{coll}} - \mu_{\text{hs}}}{\sigma_{\text{hs}}/\sqrt{n}} = \frac{51.70 - 50}{10/\sqrt{100}} = \frac{1.70}{10/10} = \frac{1.70}{1} = 1.70$$

Since $1.70 < 1.96$, we cannot reject H_0. We cannot conclude that the mean environmental activism scores of the two populations differ.

What if our sample mean was 52?

$$z = \frac{52 - 50}{10/\sqrt{100}} = \frac{2}{10/10} = \frac{2}{1} = 2.00$$

Since $2.00 > 1.96$, we can reject H_0, but since $2.00 < 2.58$, we can only report $p < .05$.

What if $\bar{x}_{\text{coll}} = 53$?

$$z = \frac{53 - 50}{10/\sqrt{100}} = \frac{3}{10/10} = \frac{3}{1} = 3.00$$

Since $3.00 > 1.96$, we can reject H_0. Then, $3.00 > 2.58$ but $3.00 < 3.29$, so $p < .01$.

What if $\bar{x}_{\text{coll}} = 54$?

$$z = \frac{54 - 50}{10/\sqrt{100}} = \frac{4}{10/10} = \frac{4}{1} = 4.00$$

Since $4.00 > 1.96$ we can reject H_0. Then, $4.00 > 2.58$ and $4.00 > 3.29$, so $p < .001$.

DIRECTIONAL VERSUS NONDIRECTIONAL ALTERNATIVE HYPOTHESES (ONE-TAILED VERSUS TWO-TAILED TESTS)

In the problems we have done so far, our alternative hypotheses have looked like this:

$$H_1: \mu_1 \neq \mu_2$$

Specifically,

$$H_1: \mu_{\text{Dems}} \neq \mu_{\text{Reps}} \quad \text{and} \quad H_1: \mu_{\text{hs}} \neq \mu_{\text{college}}$$

Note that when we use the inequality symbol $\neq$ we allow for two possible conditions.

$$\mu_1 > \mu_2 \qquad \text{or} \qquad \mu_1 < \mu_2$$

We do not specify which of the two conditions is likely, but instead collect our data and calculate z.

In the Ohio example, we were able to reject H_0 with $p < .001$, thus concluding $\mu_{Dems} \neq \mu_{Reps}$. At this point, after all the facts were in, it would be reasonable to say that since $\bar{x}_{Reps} = 23.33$, which is *less* than the μ_{Dems} of 83.18, empirically Republican representatives have lower ADA scores than Democrats in Ohio. Ultimately, then, we are really concluding $\mu_{Dems} > \mu_{Reps}$, which is even more specific than our initial H_1.

When our H_1 has the format $\mu_1 \neq \mu_2$, we refer to it as a **nondirectional alternative hypothesis**—*it does not specify which direction,* $\mu_1 > \mu_2$ or $\mu_1 < \mu_2$, will ultimately be correct. For reasons to be explained in the next chapter, this is often referred to as a **two-tailed alternative hypothesis** or a **two-tailed test of significance.** We will see that the number of tails noted in the expression may not always be correct, making this terminology ambiguous. Nevertheless, these expressions are so much a part of statistics tradition that they are constantly used.

The nondirectional (two-tailed) H_1 is the more traditional and more conservative format for the alternative hypothesis, but it is often replaced by a **directional** or **one-tailed** H_1 (or a **one-tailed test**). To formulate a directional alternative hypothesis, one must be able to discard, *prior to collecting the data,* one of the two possible directions implicit in the H_1, either $\mu_1 > \mu_2$ or $\mu_1 < \mu_2$. When one of the directions is discarded, the remaining direction becomes the alternative hypothesis.

In our Ohio example, the nondirectional H_1 of $\mu_{Dems} \neq \mu_{Reps}$ subsumes two possibilities:

either

$$\mu_{Dems} > \mu_{Reps}$$

or

$$\mu_{Dems} < \mu_{Reps}.$$

What if, in advance of examining the data, we reviewed the logic of these two possible directions. This assumes that we have *prior knowledge* about Democrats and Republicans. Given such knowledge, is it more logical to assume Democrats are more liberal than Republicans or less liberal than Republicans? With the exception of Southern Democrats, all evidence suggests that Democrats are more liberal than Republicans. In fact, recent Republican campaign strategies have been aimed at reinforcing just such an impression. That being the case, can we eliminate in advance the possibility of $\mu_{Dems} < \mu_{Reps}$? If so, we could formulate a directional H_1 as follows:

$$H_1: \mu_{Dems} > \mu_{Reps}$$

(Obviously, if we had evidence that Republicans are the more liberal of the two, our H_1 would be $\mu_{Dems} < \mu_{Reps}$.)

As we will see in the next chapter, picking a directional H_1 gives the advantage of making it easier to reject the null hypothesis. The directional critical value is always less than the nondirectional one. We can see this in the following sets of critical values.

	Critical Value of z	
Probability (Level of Significance)	*One-Tailed (Directional* H_1*)*	*Two-Tailed (Nondirectional* H_1*)*
.05	1.65	1.96
.01	2.33	2.58
.001	3.09	3.29

Suppose that in the high school versus college example we had prior evidence showing that college graduates were more environmentally active than high school-only graduates, so that $\mu_{hs} > \mu_{college}$ was an illogical assumption to make. Accordingly, we form a directional H_1 as follows:

$$H_1: \mu_{hs} > \mu_{college}$$

Note that the computation of z is exactly the same as when H_1 was nondirectional. In the case where $\bar{x}_{college} = 51.70$,

$$z_{obt} = \frac{51.70 - 50}{10/\sqrt{100}} = \frac{1.70}{10/10} = \frac{1.70}{1} = 1.70$$

Before, since $1.70 < 1.96$, we could not reject H_0. Now, however, by making a *directionality assumption* in H_1, we may make use of the lower one-tailed critical values.

$$z_{obt} = 1.70 > 1.65 \qquad \text{Reject } H_0.$$

$$1.70 < 2.33 \qquad p < .05.$$

We are now able to reject H_0, whereas without the directionality assumption, we could not. For that reason, directional alternative hypotheses are widely used.

Despite their wide usage, there are major risks associated with one-tailed alternative hypotheses. For example, is there really prior evidence on which to make a directionality assumption? A researcher may give little, if any, rationale for the direction chosen in the assumption. Moreover, with the use of modern multivariate techniques, there may be dozens of variables being manipulated at once. The more data, the less likely that each pair of means or pair of variables has been systematically examined to justify the directionality of each possible alternative hypothesis. Thus be wary of conclusions from one-tailed tests!

SETTING THE LEVEL OF SIGNIFICANCE

The .05 level usually determines the significance decision because of custom and tradition. At times, researchers may decide to make it harder to reject H_0 by basing the decision on a higher level of significance, such as .01. For example, to be even more certain that a drug being tested was superior to another and that the test results were not due to the effects of sampling, a researcher might use the .01 level. In most social science applications, the need for a higher level of significance is rarely encountered.

We may also go the other way and make it easier to reject the null hypothesis in some circumstances. For instance, $z_{critical}$ at the .10 level (directional) is 1.282. Suppose we obtain a $z = 1.5$. By our earlier standards we need at least 1.65 to reject H_0. But, we can only say that the difference is significant at the .10 level if it was stated earlier (with appropriate justification) that the basis for significance decisions in the study was the .10 level. Otherwise, we could be accused of trying to sneak in as significant something that would normally be considered not statistically significant.

Nevertheless, there are circumstances when a lower level of significance will be acceptable. To understand this, recall that earlier we noted that the size of z is based in part on the difference between $\bar{x}$ and μ. However, the size of z is also determined by the size of the sample. The larger the n, the larger the z. For example, if $n = 100$

$$z = \frac{51.70 - 50}{10/\sqrt{100}} = \frac{1.70\sqrt{100}}{10} = \frac{1.70 \times 10}{10} = 1.70$$

but if $n = 1000$

$$z = \frac{51.70 - 50}{10/\sqrt{1000}} = \frac{1.70\sqrt{1000}}{10} = \frac{1.70(31.622)}{10} = \frac{53.757}{10} = 5.375$$

Using the two-tailed critical values, the z where $n = 100$ is not significant, whereas the z where $n = 1000$ is significant, $p < .001$. Given a big enough n, even trivial differences become statistically significant.

By contrast, suppose our sample n was lower than 100, say, 25.

$$z = \frac{51.70 - 50}{10/\sqrt{25}} = \frac{1.70\sqrt{25}}{10} = \frac{1.70(5)}{10} = 0.85$$

Here, z drops from 1.70 to 0.85 even though $\bar{x}$ and μ were the same.

Keep in mind that survey research costs money and a major factor in the expense is the size of the sample to be interviewed. For example, suppose some local government wants you to do a study of some public service delivery but is only willing to pay you $2000. Because of the dollar limitation, you discover that your sample size cannot exceed 50. Based on earlier studies you feel that you need at least 100 people to get results significant at the .05 level. Since you cannot get funding for $n = 100$, you tell the contracting officer that you will do the survey if the city will accept a lower level of significance, say, .10. The contracting officer—even in the unlikely event that he or she knows what you are talking about—may be willing to accept your suggestion, just to stay within the budget. Finally, you may wish to do a pilot study on a small sample as part of what will eventually be applied to a larger sample. Here you are interested in eliminating ambiguities from your survey document. You accept the .10 level of significance, knowing that in the final study n will be large enough for the .05 level to be used.

In more and more published research you are likely to encounter a trend where the probabilities are simply stated with no statement as to whether the results are statistically significant. Then it is up to you the reader to examine each probability and make your own conclusion about significance.

We have now learned that two factors play a role in determing the magnitude of the z obtained: the size of the difference between $\bar{x}$ and μ and the size of the sample, n. We will examine this subject again in Chapter 9.

DEGREES OF FREEDOM

In the case of the one-sample z test, the critical values of z remain constant. For example, at the .05 level, nondirectional H_1, the critical value of z is always 1.96. With other tests of significance, however, the critical values will vary from one problem to another depending on something called **degrees of freedom,** *d.f.*, or just *df* for short. In these cases, to determine what critical values to use we must first find the degrees of freedom. Beyond z, each variation of t, F, or chi-square will also carry a formula for finding *df.* We will not attempt to define degrees of freedom at this time other than to say that in these tests, *df* is related to the sizes of the samples being used or, in the case of chi-square, the sizes of the tables.

CONCLUSION

STEPS IN SIGNIFICANCE TESTING

In this chapter we have introduced the logic of testing for statistical significance as well as the procedures followed for *any* test of significance. Using the one-sample z test as a reference, let us review these steps in their proper logical order.

1. Before any data are examined, formulate a null hypothesis and an alternative hypothesis. If you are going to use a directional or one-tailed H_1, make sure that there is convincing evidence to justify making the directionality assumption needed for your alternative hypothesis. Also justify any decision to use a level of significance other than .05 for basing your conclusion about whether the obtained statistic is statistically significant.

2. Once the data are collected, make sure that you have or can obtain the information needed to perform the test of significance you have

chosen. For the one-sample *z* test, you will need the mean and standard deviation for one population and the size and mean of the sample you are comparing to that population. In short, you need μ, σ, n, and $\overline{x}$.

3. Calculate the test statistic.

4. Locate the appropriate critical values, which you will compare to the statistic you obtained. Make sure that the values are appropriate for the H_1 you formulated—directional or nondirectional. In cases other than the *z* test, you will first have to determine the appropriate degrees of freedom in order to find the appropriate critical values.

5. Compare the statistic that you obtained (calculated) to the appropriate critical value at the .05 level (unless you have chosen another level for your decision).

6. If your obtained value is less than the appropriate critical value, you cannot reject H_0. You cannot say that the difference is statistically significant.

7. If your obtained value equals or exceeds the appropriate critical value, you can reject H_0 and say that the difference is statistically significant.

8. If a significant difference is found, you still may be making a type I, or alpha, error; that is, falsely rejecting a true null hypothesis. Report the probability that this is the case by comparing your obtained statistic to critical values at progressively higher levels of significance. The level that *p* is less than is the level of the last critical value that your obtained statistic exceeded.

9. Go back to your null and alternative hypotheses and, based on the results of your significance test, restate your conclusion.

EXERCISES

Exercise 7.1

Write a null hypothesis and a nondirectional alternative hypothesis for each of the following.

Example: Scottish voters are more supportive of the Labour Party than English voters. Assume that the variable is a measure of pro-Labour attitudes. Thus

$$H_0: \mu_{Scots} = \mu_{English}$$

$$H_1: \mu_{Scots} \neq \mu_{English}$$

1. Women differ from men on their attitudes toward abortion.
2. Urban residents differ from rural residents on the issue of gun control.
3. The French support state subsidies to their farmers more than do Americans.
4. Israeli Arabs differ from Israeli Jews on the issue of trading land for peace.
5. Liberals in the United States support affirmative action programs more than do conservatives.
6. Australians and New Zealanders differ in their support for a U.S. nuclear presence in their region.
7. French Canadians are less culturally assimilated to the larger English-speaking culture than are the Cajuns of Louisiana.
8. Soviet military officers differed from Soviet civilians in their level of support for perestroika.
9. Californians tend to support proenvironmental legislation more than do New Yorkers.
10. Police officers support the death penalty more than do convicted murderers.

Exercise 7.2

Recall the tests of significance discussed in this chapter:

One-Sample z *Test* or t *Test:* Compares the mean of a sample to the mean of a population.

Two-Sample t *Test:* Compares the mean of one sample to the mean of another sample.

One-Way Analysis of Variance: Compares the means of several samples, generally more than two.

Chi-Square: Tests the significance within a cross-tablulation.

Examine each of the following problems and indicate which of the above tests is most appropriate for that problem.

1. H_1: Members of the House District of Columbia Committee (assuming they are selected at random) are significantly younger than the overall membership of the House of Representatives.

   For the House District Committee the mean age is 35.
   For the House of Representatives the mean age is 45.
   For the House of Representatives the standard deviation is 14.
   The House District Committee has 11 members.

2. H_0: In the population from which a random sample was drawn, respondents' dogmatism scores are unrelated to Socioeconomic Status (SES) classification.
   Assume: Dogmatism is an interval scale.

	SES		
	High	*Medium*	*Low*
Dogmatism	1	3	10
scores	2	5	9
	2	8	8
n equals 18	3	8	10
	4	4	10
	1	6	9

3. H_1: In the population (all countries), there is an association between the number of political parties and that nation's level of political development.

	Level of Development	
	Traditional (Underdeveloped)	*Modern (Developed)*
Mean # Political Parties per Country	3.0	2.8
Sample Size	15.0	25.0
Sample Variance	2.8	6.0

4. A random sample of 25 convicted felons is studied to see if there is a relationship between a convict's income level and the type of sentence received. Treat both variables as nominal dichotomies.

	Sentence		
Income Level	*Fine*	*Jail Term*	*Total*
High	5	10	15
Low	0	10	10
Total	5	20	25

5. H_0: In the population, participation level in professional organizations is unrelated to occupation type.
Assume: Organizational Participation is an interval scale (0 minimum to 10 maximum).

	Occupation Category	
	Professionals	*Nonprofessionals*
Respondents' Organizational Participation Scores:	10	3
	9	5
	9	4
	6	2
	8	0
	9	1
		1
		1
		3
Sample Means:	8.5	2.2
Sample Sizes:	6	9

6. H_0: In the population from which a random sample was drawn, type of occupation is unrelated to job satisfaction.

	Type of Occupation				
Job Satisfaction	*Professional*	*White Collar*	*Blue Collar*	*Farmer*	*Total*
High	35	20	5	5	60
Low	5	10	25	15	60
Total	40	30	30	20	120

7. H_0: In the population, participation in professional organizations is unrelated to type of occupation.
Assume: Organizational Participation is an interval scale (0 minimum to 10 maximum activity).

	Occupation Category		
	Lawyers	*Doctors*	*Other Professionals*
Respondents' Organizational Participation Scores:	10	4	7
	9	5	5
	8	3	8
	10	9	9
	10	1	4
	9		6
Sample Means:	9.3	4.4	6.5
Sample Sizes:	6	5	6

8. Assume that for the entire population of the Irish Republic, the mean age is 30 years. A random sample of 15 members of the Dail (the Lower House of Parliament) yields a mean age of 45 and a standard deviation of 15.
   H_0: There is no difference in mean age between the population of the Irish Republic and the members of the Dail.

Exercise 7.3

Each of the following examples looks like a problem calling for a one-sample *z* test or *t* test. In each case, however, a flaw in the logic of the research design makes a test of significance moot. For each example, identify that flaw.

1. At Breezewood Junior High School there are a total of 50 eighth-graders in two government classes, taught by Mr. Jones and Mrs. Smith, respectively. At the end of the quarter, both classes take a common civics exam.
   H_1: In this instance, in fact, Mrs. Smith's class scores higher than did Mr. Jones' class.

	Class	
	Mr. Jones	*Mrs. Smith*
Mean Exam Score:	86	87
Class Size:	25	25
Class Standard Deviation:	12	14

2. Suppose for the U.S. population as a whole it had been determined that the mean assertiveness score was 50 on a scale ranging from 0 (least) to 100 (most). A researcher wishing to generalize about the Dayton metropolitan area's residents' assertiveness characteristics selects as the experimental group a random sample of 25 jet pilots stationed at a nearby Air Force base. The researcher then administers the assertiveness test to them and obtains a sample mean of 70 and a sample standard deviation of 12.
   H_0: Mu for the Dayton metropolitan area equals mu for the United States.
3. At a certain university, the mean undergraduate grade point average is 2.9, with a standard deviation of 1.2, for the entire undergraduate student body. For the 50 political science majors in the honors program, a random sample of $n = 20$ has a mean of 3.4.

H_1: The mean of the population of political science majors is higher than the mean for all undergraduates.

4. At the same university as in Example 3, the mean grade point average for all psychology majors is 3.3. Is the difference statistically significant when these majors are compared to all undergraduates at the university?

Exercise 7.4

Calculate the one-sample *z* test of significance for each of the following. Just calculate *z*.

1.	$\bar{x} = 10$	$\mu = 12$	$\sigma = 5$	$n = 25$
2.	$\bar{x} = 150$	$\mu = 100$	$\sigma = 25$	$n = 100$
3.	$\bar{x} = 3.1$	$\mu = 2.8$	$\sigma = 1.2$	$n = 36$
4.	$\bar{x} = 32$	$\mu = 30$	$\sigma = 10$	$n = 49$
5.	$\bar{x} = 2.6$	$\mu = 3.0$	$\sigma = 1.4$	$n = 64$

Exercise 7.5

Using the nondirectional (two-tailed) critical values of *z*, examine each obtained *z* below. Reach a conclusion about statistical significance and, if significant at least at the .05 level, state the probability of alpha, using a "$p <$ " statement.

1. $z = 3.30$
2. $z = 2.00$
3. $z = -2.65$
4. $z = -1.90$
5. $z = -2.58$
6. $z = 1.75$
7. $z = -3.10$
8. $z = -1.50$
9. $z = 2.33$
10. $z = -2.50$

Exercise 7.6

Repeat Exercise 7.5 using the directional (one-tailed) critical values of *z*.

Exercise 7.7

Five of the examples in Exercise 7.1 could be directional alternative hypotheses. Identify them and write the appropriate directional alternative hypotheses.

Exercise 7.8

Formulate the null hypotheses and the most appropriate alternative hypotheses (either directional or nondirectional, as you think appropriate) for each of the following. If H_1 is directional, justify the direction.

1. Managers will differ from their employees in their support for business interests.
2. Southern senators will differ from senators in general in terms of their support for conservative policies.
3. A group of people who have watched a video with a decidedly pacifistic message will have a different attitude toward the use of military force than people in general.

Exercise 7.9

Following are the data pertaining to the three items in Exercise 7.8. For each, follow steps 2-9 as outlined in the *Steps in Significance Testing* on pages 197-198. (You have already done step 1 in Exercise 7.8.)

1. For all managers, on a scale ranging from 0 to 100, the mean business support score is 36.1 with a standard deviation of 11.1. For a random sample of 9 employees, the mean business support score is 81.8.
2. For all U.S. senators, the mean conservatism score is 56.2 with a standard deviation of 32.4. For a random sample of seven Southern senators, the mean score is 61.9.
3. For a pacifism scale ranging from 0 (low pacifism) to 10 (high pacifism), the mean pacifism score is hypothesized to be 3.0 with a standard deviation of 0.5. For a random sample of 25 people who watch the video, the sample mean is 2.8.

NOTES

1. From H. Stanley, and R. Niemi, *Vital Statistics on American Politics,* 3rd ed. (Washington, DC: CQ Press, 1992).

Following are the data from which the means were calculated.

District	*Democrats*	*ADA Scores*	*District*	*Republicans*	*ADA Scores*
1	Luken	70	2	Gradison	35
3	Hall	75	4	Oxley	20
9	Kaptur	75	6	McEwen	5
11	Eckart	90	7	DeWine	5
13	Pease	90	8	Lukens	0
14	Sawyer	95	10	Miller	15
17	Traficant	95	12	Kasich	15
18	Applegate	70	15	Wylie	30
19	Feigham	95	16	Regula	30
20	Oakar	90			
21	Stokes	70			
		$\sum x = 915$			$\sum x = 155$

$$\mu_{\text{Dems}} = \frac{915}{11} = 83.18 \qquad \mu_{\text{Reps}} = \frac{155}{9} = 17.22$$

There was no score for Gillmor (R-5th District). Also, the sample $n = 3$ of Republicans mentioned in the text consisted of Gradison, DeWine, and Wylie.

2. Some people argue that such tests may be applied even in comparing two populations. In such a case there is no sampling, but the populations could differ only due to chance randomizations in nature. In this text, however, we will ignore this debate and exclude tests comparing two or more populations.

3. We report the alpha or type I error's probability whenever we reject the null hypothesis. Note that if we do *not* reject the null hypothesis, we could be making another type of error: *failure to reject a false null hypothesis. This is known as a* **type II** or **beta error.** Its probability, however, is *not* reported, even though we should be aware that it exists.

4. The subscripts *obtained* or *obt.* used with Z_{obtained} and other tests in subsequent chapters is often omitted in statistics tests. We use it here, on a selective basis, to assist you in learning this material and differentiating the obtained values from the critical values of the test.

KEY CONCEPTS

normal distribution
standard score
one-sample z test
sampling distribution of sample means
Central Limit Theorem
standard error/standard error of the mean
normality assumption
Law of Large Numbers
one-sample t test
degrees of freedom
$\hat{\sigma}$ / "sigma-hat"
z test for proportions
interval estimation
confidence intervals for means and proportions
conditional probability
addition and multiplication rules of probability
permutations
combinations

8 Probability Distributions and One-Sample *z* and *t* Tests

INTRODUCTION

In the previous chapter, we discussed tests of significance and used the one-sample *z* formula to illustrate the entire procedure.

$$z = \frac{\bar{x} - \mu}{\sigma/\sqrt{n}}$$

In this chapter we turn our attention to the origin of that formula and explain what is taking place when we use it. It is possible to use any statistical formula without such an understanding and simply plug in the numbers as we did in Chapter 7. But if you can visualize what is going on, your understanding will be enhanced since essentially the same process takes place no matter what test of significance is being performed.

The *z* test of significance is based on a frequency distribution known as a normal distribution and is applied to a specific normal curve called the sampling distribution of sample means. When we perform this test, we are actually taking the given sample statistics and population parameters and locating them on the sampling distribution of sample means. In fact, all tests of significance do the same thing, even though their sampling distributions differ from one another.

At the end of this chapter we will discuss the one-sample *t* test, and in subsequent chapters the other commonly used tests of significance will be presented. We will begin by discussing an even simpler *z* formula than the one in the previous chapter and introducing the concept of a normal distribution.

NORMAL DISTRIBUTIONS

Normal distributions *are a family of frequency distributions that when graphed often resemble bells.* Generally they are represented in the form found in Figure 8.1. *Such a curve has three major characteristics: (a) it is unimodal, (b) it is symmetric, and (c) it is asymptotic to the x-axis.* This last characteristic, which becomes very important later on, means that the tails of the curve get closer and closer to the *x*-axis but never reach it. Consequently, no matter how far you get from the mean on the *x*-axis, there will always be a tail continuing beyond that point. The tail never ends, at least not in the mathematical model.

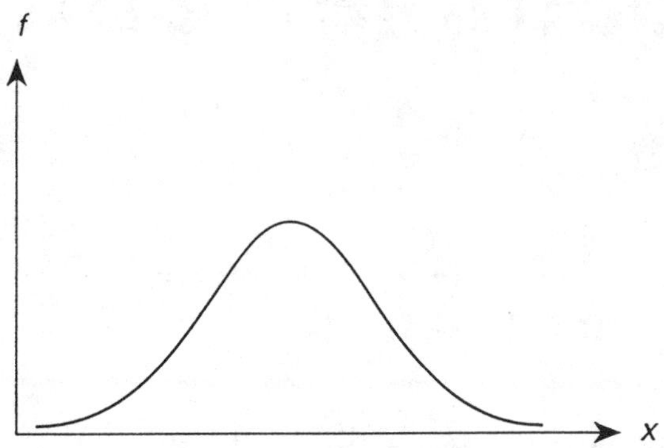

Figure 8.1
The Mathematical Model of the Normal Distribution

The reason for using the term *normal distribution* is that certain characteristics such as human height, weight, or intelligence graph in frequency distributions approximating this bell-shaped pattern. The term *normal* is a bit misleading, however. Not everything in nature is normally distributed, so nonnormal distributions are not really abnormal.

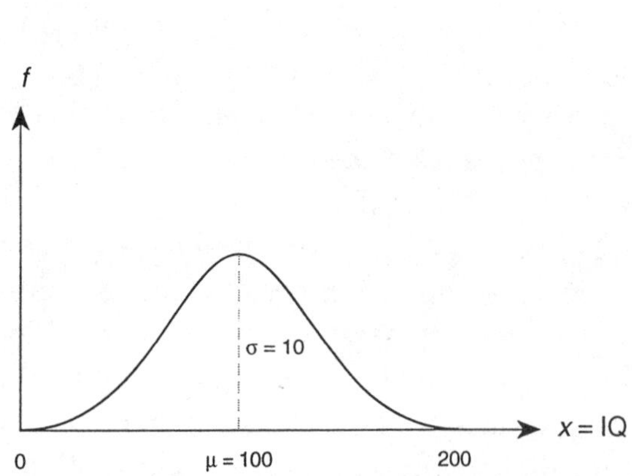

Figure 8.2
IQ as a Normal Distribution

Let us use measures of intelligence — intelligence quotients or IQs — to illustrate the normal distribution (see Figure 8.2). IQs are designed to range from 0 to 200 with a mean of 100. Standard deviations vary by the age of the subjects but are usually around 13 or 14; for computational ease we will use 10. Note that, unlike the mathematical model in Figure 8.1, in Figure 8.2 the tails do end—at IQs of 0 and 200, respectively. This indicates that there are natural upper and lower limits in the actual measurement tool being utilized. For example, say it is a test with 200 questions and someone's IQ is the number of correct answers. Two geniuses each score 200, yet, if one of the two is really twice as smart as the other and the IQ test had 400 questions instead of 200, the one genius would score 400 whereas the other would only get 200.

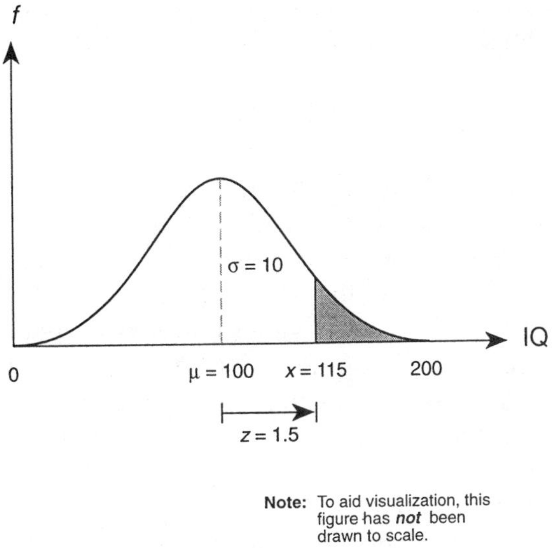

Figure 8.3

Suppose Sandra takes this IQ test and her score, which we designate as x, is 115. We would like to know what proportion of people would likely have higher IQs than Sandra's and what proportion would have lower IQs. (Sorry, this procedure cannot tell us how many will have exactly Sandra's IQ.) It turns out that the area under the curve corresponds to the proportion of people with a particular characteristic. The total area under the curve (1.00 proportion) accounts for all (100%) people. Since a normal curve is symmetric, .50 proportion (50%) of the area under the curve falls below the mean and .50 proportion falls above the mean. Thus half of all people should have IQs below 100 and half should have IQs above it. This proportion of the area also pertains to the probability of randomly selecting a person with a particular characteristic. Since .50 proportion of the area of the curve is below the mean, there is also a .50 probability of randomly selecting a person whose IQ is below 100. Likewise, there is a .50 probability of randomly selecting someone whose IQ is greater than 100.

In Figure 8.3 we have added Sandra's IQ, $x = 115$. The proportion of people with an IQ greater than 115 is the shaded area under the curve in the right tail, from $x = 115$ to $x = 200$. The proportion of people with IQs below 115 is represented by the remaining unshaded area under the curve, from the left of $x = 115$ to $x = 0$. Note that the unshaded area has two components, the .50 proportion of IQs less than 100 plus the area under the curve from 100 to 115.

To find these areas, we use a table of areas under the normal curve that applies to all normal distributions. To use the table, we begin by calculating what is called a **standard score,** which is universally designated by the letter z. (Its relationship to our z test of significance will be explained later.) *To calculate z, we recast the distance from the mean (100) to the value of x we are studying (Sandra's IQ of 115), expressed as standard deviation units.* The distance from the mean of Sandra's IQ is $x - \mu = 115 - 100 = 15$; her IQ is 15 points greater than the mean. To convert that distance into standard deviation units, we divide by the size of the standard deviation, $\sigma = 10$, which was given to us. Since

15/10 = 1.5, we know that Sandra's IQ is 1.5 standard deviation units from the mean. Expressing the whole process in a single equation we get

$$z = \frac{x - \mu}{\sigma} = \frac{115 - 100}{10} = \frac{15}{10} = 1.5$$

We now go to Table 8.1 and see that each page has three blocks of figures and in turn each block has three columns: A, B, C. Column A lists a value of z. Column B shows the area under the curve from the mean out to that specified value of z (note the graphs above each column). Column C shows the area in the tail beyond z. Note that the area in Column B plus the area in Column C always add to .5000. Also note that as z gets larger, the area in Column B gets larger and the area in Column C gets smaller.

The bigger the z, the smaller the tail.

We find our z of 1.5 in the center of the left hand block on the second page of Table 8.1. Note that at z = 1.5 the number in Column B is .4332 and the number in Column C is .0668. The area in the tail, corresponding to the proportion of people with IQs greater than 115, is the number in Column C, .0668—only 6.68% have an IQ higher than Sandra's. To find the proportion with IQs less than 115, we take the area in Column B and add to it .50, the proportion with IQs below the mean.

IQs between 100 and 115:	.4332
IQs between 0 and 100:	.5000
Total =	.9332

Thus .9332 proportion have IQs below 115. Sandra's pretty bright!

If Sandra is bright, George is not; his IQ is only 80. Let us find the proportions of area above and below 80 (see Figure 8.4). First we find z.

$$z = \frac{x - \mu}{\sigma} = \frac{80 - 100}{10} = \frac{-20}{10} = -2.00$$

Here, z is negative since x is less than μ. In Table 8.1 we will look for the absolute value of our z (2.00), but use the graphs at the bottom of the page to see that now the shaded areas are to the left of the mean. On the lowest third of the center column of the second page of the table you will find z = 2.00. The Column B area is .4772 and the Column C area is .0228. Accordingly, only .0228 proportion of people have IQs below George's. To find the other proportion, add the Column B figure to the

Table 8.1 Proportions of Area Under Standard Normal Curve

A z	B	C	A z	B	C	A z	B	C
0.00	.0000	.5000	0.42	.1628	.3372	0.84	.2995	.2005
0.01	.0040	.4960	0.43	.1664	.3336	0.85	.3023	.1977
0.02	.0080	.4920	0.44	.1700	.3300	0.86	.3051	.1949
0.03	.0120	.4880	0.45	.1736	.3264	0.87	.3078	.1922
0.04	.0160	.4840	0.46	.1772	.3228	0.88	.3106	.1894
0.05	.0199	.4801	0.47	.1808	.3192	0.89	.3133	.1867
0.06	.0239	.4761	0.48	.1844	.3156	0.90	.3159	.1841
0.07	.0279	.4721	0.49	.1879	.3121	0.91	.3186	.1814
0.08	.0319	.4681	0.50	.1915	.3085	0.92	.3212	.1788
0.09	.0359	.4641	0.51	.1950	.3050	0.93	.3238	.1762
0.10	.0398	.4602	0.52	.1985	.3015	0.94	.3264	.1736
0.11	.0438	.4562	0.53	.2019	.2981	0.95	.3289	.1711
0.12	.0478	.4522	0.54	.2054	.2946	0.96	.3315	.1685
0.13	.0517	.4483	0.55	.2088	.2912	0.97	.3340	.1660
0.14	.0557	.4443	0.56	.2123	.2877	0.98	.3365	.1635
0.15	.0596	.4404	0.57	.2157	.2843	0.99	.3389	.1611
0.16	.0636	.4364	0.58	.2190	.2810	1.00	.3413	.1587
0.17	.0675	.4325	0.59	.2224	.2776	1.01	.3438	.1562
0.18	.0714	.4286	0.60	.2257	.2743	1.02	.3461	.1539
0.19	.0753	.4247	0.61	.2291	.2709	1.03	.3485	.1515
0.20	.0793	.4207	0.62	.2324	.2676	1.04	.3508	.1492
0.21	.0832	.4168	0.63	.2357	.2643	1.05	.3531	.1469
0.22	.0871	.4129	0.64	.2389	.2611	1.06	.3554	.1446
0.23	.0910	.4090	0.65	.2422	.2578	1.07	.3577	.1423
0.24	.0948	.4052	0.66	.2454	.2546	1.08	.3599	.1401
0.25	.0987	.4013	0.67	.2486	.2514	1.09	.3621	.1379
0.26	.1026	.3974	0.68	.2517	.2483	1.10	.3643	.1357
0.27	.1064	.3936	0.69	.2549	.2451	1.11	.3665	.1335
0.28	.1103	.3897	0.70	.2580	.2420	1.12	.3686	.1314
0.29	.1141	.3859	0.71	.2611	.2389	1.13	.3708	.1292
0.30	.1179	.3821	0.72	.2642	.2358	1.14	.3729	.1271
0.31	.1217	.3783	0.73	.2673	.2327	1.15	.3749	.1251
0.32	.1255	.3745	0.74	.2704	.2296	1.16	.3770	.1230
0.33	.1293	.3707	0.75	.2734	.2266	1.17	.3790	.1210
0.34	.1331	.3669	0.76	.2764	.2236	1.18	.3810	.1190
0.35	.1368	.3632	0.77	.2794	.2206	1.19	.3830	.1170
0.36	.1406	.3594	0.78	.2823	.2177	1.20	.3849	.1151
0.37	.1443	.3557	0.79	.2852	.2148	1.21	.3869	.1131
0.38	.1480	.3520	0.80	.2881	.2119	1.22	.3888	.1112
0.39	.1517	.3483	0.81	.2910	.2090	1.23	.3907	.1093
0.40	.1554	.3446	0.82	.2939	.2061	1.24	.3925	.1075
0.41	.1591	.3409	0.83	.2967	.2033	1.25	.3944	.1056
A -z	B	C	A -z	B	C	A -z	B	C

continued

Table 8.1 Continued

A	B	C	A	B	C	A	B	C
z			z			z		
1.26	.3962	.1038	1.68	.4535	.0465	2.10	.4821	.0179
1.27	.3980	.1020	1.69	.4545	.0455	2.11	.4826	.0174
1.28	.3997	.1003	1.70	.4554	.0446	2.12	.4830	.0170
1.29	.4015	.0985	1.71	.4564	.0436	2.13	.4834	.0166
1.30	.4032	.0968	1.72	.4573	.0427	2.14	.4838	.0162
1.31	.4049	.0951	1.73	.4582	.0418	2.15	.4842	.0158
1.32	.4066	.0934	1.74	.4591	.0409	2.16	.4846	.0154
1.33	.4082	.0918	1.75	.4599	.0401	2.17	.4850	.0150
1.34	.4099	.0901	1.76	.4608	.0392	2.18	.4854	.0146
1.35	.4115	.0885	1.77	.4616	.0384	2.19	.4857	.0143
1.36	.4131	.0869	1.78	.4625	.0375	2.20	.4861	.0139
1.37	.4147	.0853	1.79	.4633	.0367	2.21	.4864	.0136
1.38	.4162	.0838	1.80	.4641	.0359	2.22	.4868	.0132
1.39	.4177	.0823	1.81	.4649	.0351	2.23	.4871	.0129
1.40	.4192	.0808	1.82	.4656	.0344	2.24	.4875	.0125
1.41	.4207	.0793	1.83	.4664	.0336	2.25	.4878	.0122
1.42	.4222	.0778	1.84	.4671	.0329	2.26	.4881	.0119
1.43	.4236	.0764	1.85	.4678	.0322	2.27	.4884	.0116
1.44	.4251	.0749	1.86	.4686	.0314	2.28	.4887	.0113
1.45	.4265	.0735	1.87	.4693	.0307	2.29	.4890	.0110
1.46	.4279	.0721	1.88	.4699	.0301	2.30	.4893	.0107
1.47	.4292	.0708	1.89	.4706	.0294	2.31	.4896	.0104
1.48	.4306	.0694	1.90	.4713	.0287	2.32	.4898	.0102
1.49	.4319	.0681	1.91	.4719	.0281	2.33	.4901	.0099
1.50	.4332	.0668	1.92	.4726	.0274	2.34	.4904	.0096
1.51	.4345	.0655	1.93	.4732	.0268	2.35	.4906	.0094
1.52	.4357	.0643	1.94	.4738	.0262	2.36	.4909	.0091
1.53	.4370	.0630	1.95	.4744	.0256	2.37	.4911	.0089
1.54	.4382	.0618	1.96	.4750	.0250	2.38	.4913	.0087
1.55	.4394	.0606	1.97	.4756	.0244	2.39	.4916	.0084
1.56	.4406	.0594	1.98	.4761	.0239	2.40	.4918	.0082
1.57	.4418	.0582	1.99	.4767	.0233	2.41	.4920	.0080
1.58	.4429	.0571	2.00	.4772	.0228	2.42	.4922	.0078
1.59	.4441	.0559	2.01	.4778	.0222	2.43	.4925	.0075
1.60	.4452	.0548	2.02	.4783	.0217	2.44	.4927	.0073
1.61	.4463	.0537	2.03	.4788	.0212	2.45	.4929	.0071
1.62	.4474	.0526	2.04	.4793	.0207	2.46	.4931	.0069
1.63	.4484	.0516	2.05	.4798	.0202	2.47	.4932	.0068
1.64	.4495	.0505	2.06	.4803	.0197	2.48	.4934	.0066
1.65	.4505	.0495	2.07	.4808	.0192	2.49	.4936	.0064
1.66	.4515	.0485	2.08	.4812	.0188	2.50	.4938	.0062
1.67	.4525	.0475	2.09	.4817	.0183	2.51	.4940	.0060
A	B	C	A	B	C	A	B	C
-z			–z			-z		

continued

Table 8.1 Continued

A z	B	C	A z	B	C	A z	B	C
2.52	.4941	.0059	2.80	.4974	.0026	3.08	.4990	.0010
2.53	.4943	.0057	2.81	.4975	.0025	3.09	.4990	.0010
2.54	.4945	.0055	2.82	.4976	.0024	3.10	.4990	.0010
2.55	.4946	.0054	2.83	.4977	.0023	3.11	.4991	.0009
2.56	.4948	.0052	2.84	.4977	.0023	3.12	.4991	.0009
2.57	.4949	.0051	2.85	.4978	.0022	3.13	.4991	.0009
2.58	.4951	.0049	2.86	.4979	.0021	3.14	.4992	.0008
2.59	.4952	.0048	2.87	.4979	.0021	3.15	.4992	.0008
2.60	.4953	.0047	2.88	.4980	.0020	3.16	.4992	.0008
2.61	.4955	.0045	2.89	.4981	.0019	3.17	.4992	.0008
2.62	.4956	.0044	2.90	.4981	.0019	3.18	.4993	.0007
2.63	.4957	.0043	2.91	.4982	.0018	3.19	.4993	.0007
2.64	.4959	.0041	2.92	.4982	.0018	3.20	.4993	.0007
2.65	.4960	.0040	2.93	.4983	.0017	3.21	.4993	.0007
2.66	.4961	.0039	2.94	.4984	.0016	3.22	.4994	.0006
2.67	.4962	.0038	2.95	.4984	.0016	3.23	.4994	.0006
2.68	.4963	.0037	2.96	.4985	.0015	3.24	.4994	.0006
2.69	.4964	.0036	2.97	.4985	.0015	3.25	.4994	.0006
2.70	.4965	.0035	2.98	.4986	.0014	3.30	.4995	.0005
2.71	.4966	.0034	2.99	.4986	.0014	3.35	.4996	.0004
2.72	.4967	.0033	3.00	.4987	.0013	3.40	.4997	.0003
2.73	.4968	.0032	3.01	.4987	.0013	3.45	.4997	.0003
2.74	.4969	.0031	3.02	.4987	.0013	3.50	.4998	.0002
2.75	.4970	.0030	3.03	.4988	.0012	3.60	.4998	.0002
2.76	.4971	.0029	3.04	.4988	.0012	3.70	.4999	.0001
2.77	.4972	.0028	3.05	.4989	.0011	3.80	.4999	.0001
2.78	.4973	.0027	3.06	.4989	.0011	3.90	.49995	.00005
2.79	.4974	.0026	3.07	.4989	.0011	4.00	.49997	.00003
A −z	B	C	A −z	B	C	A −z	B	C

.50 whose IQs exceed the mean: .4772 + .5000 = .9772 proportion. Poor George! Nearly 98% of all IQs exceed his.

Now we can solve a mystery: the source of the critical values of *z* used in the previous chapter. Although a much more detailed table of areas under the normal curve is needed to find all the critical values of *z* presented in Chapter 7, we can find approximate values using Table 8.1. We simply specify a particular tail area, find it in Column C, and read the corresponding *z* score from Column A.

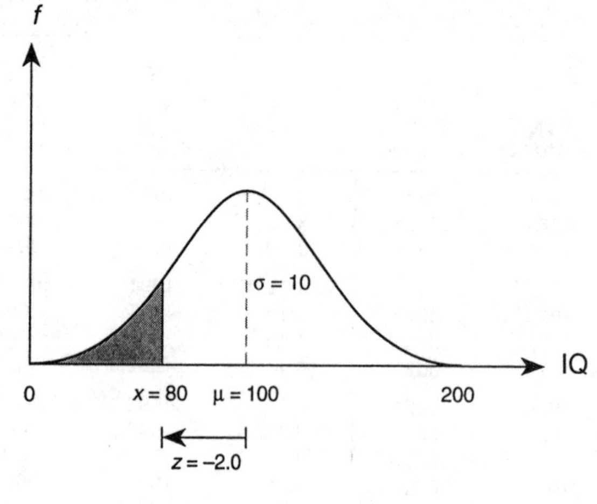

Note: To aid visualization, this figure has ***not*** been drawn to scale.

Figure 8.4

For example, to find the z value for the one-tailed .05 level, we see that in Column C the two closest approximations are .0505 (z = 1.64) and .0495 (z = 1.65). Actually, the mean of the two z values, 1.645, is the true critical value but when we round to two decimal places we get 1.65. Likewise, the closest tail area to .01 is actually .0099 and its z is 2.33. For the .001 level, we find .0010 occurring three times, where z is 3.08, 3.09, and 3.10. If this table had used more decimal places, we would see that 3.09 would be our best-fitting value of z.

For a two-tailed test, we need two tails whose areas added together equal the probability level desired. We take one-half of the probability level as the area to locate in Column C. For instance, at p = .05 we need two equal tails whose areas add to .05, so .05/2 = .025. Finding .0250 in Column C, we see that z = 1.96. (Unfortunately, Table 8.1 is not complete enough for us to find the other two-tailed values of z.)

Figures 8.5 and 8.6 summarize the relationships between z values and probabilities for one and two tails, respectively.

THE ONE-SAMPLE *z* TEST FOR STATISTICAL SIGNIFICANCE

The formula we used in the previous chapter to test for statistical significance, $z = (\bar{x} - \mu)/(\sigma / \sqrt{n})$, is really a reworking of the formula we have been using in this chapter, except that it is applied to a specific type of frequency distribution known as the **sampling distribution of sample means.** *This sampling distribution is the frequency distribution that would be obtained from calculating the means of all theoretically possible samples of a designated size that could be drawn from a given population.* To illustrate this definition, let us imagine a population of 5 people with scores ranging from 1 to 5. Using a sample size of n = 3, What are the different combinations of scores that we would obtain in selecting all possible samples of 3 people out of the original 5? If we consider the order of selection of the people (e.g., 5-4-3 is one sample,

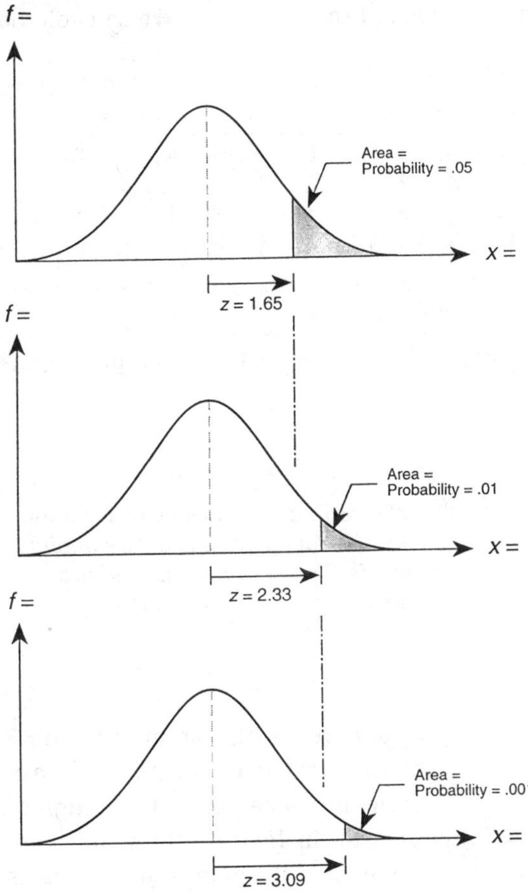

Figure 8.5
Critical Values of z—One Tail

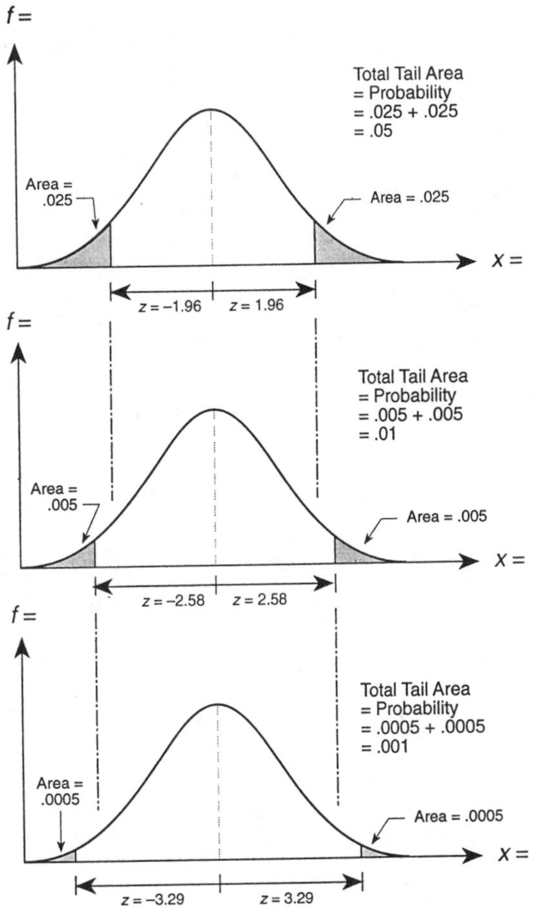

Figure 8.6
Critical Values of z—Two Tails

4-5-3 is another, and 3-4-5 yet another), there are actually 60 possible samples that could be drawn from this population. If we disregard the order of selection, there are only 10 possible combinations of scores.

For the Population; x =	*Selected Samples of Three People Which Could be Drawn From the Population; x =*									
5	5	5	5	5	5	5				
4	4	4	4				4	4	4	
3	3			3	3		3	3		3
2		2		2		2	2		2	2
1			1		1	1		1	1	1

We will work with these 10 samples of 3 people with the 3 scores shown in the columns. For each of these samples, we can calculate the mean.

	X =	X =	X =	X =	X =	X =	X =	X =	X =	X =
	5	5	5	5	5	5	4	4	4	3
	4	4	4	3	3	2	3	3	2	2
	3	2	1	2	1	1	2	1	1	1
$\sum x =$	12	11	10	10	9	8	9	8	7	6
$\bar{x} = \sum x/3 =$	4.0	3.7	3.3	3.3	3.0	2.7	3.0	2.7	2.3	2.0

The frequency distribution of all these possible sample means is as follows:

$\bar{x} =$	$f =$
4.0	1
3.7	1
3.3	2
3.0	2
2.7	2
2.3	1
2.0	1

We graph this frequency distribution, which will approximate *the sampling distribution of sample means,* as a histogram, as shown in Figure 8.7.

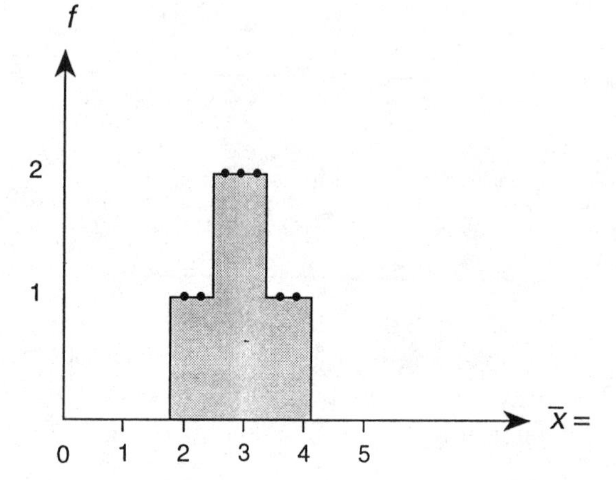

Figure 8.7
The Sampling Distribution of Sample Means Obtained From Ten Samples of Size $n = 3$ From a Specified Population

Note that the histogram in Figure 8.7 has a pattern that begins to resemble a normal curve in the sense that it is unimodal and symmetric. In fact, if either the population from which the samples are drawn is itself normally distributed along the variable *x* and/or the samples drawn from that population are sufficiently large, the sampling distribution of sample means will also be a normal distribution. This characteristic will prove to be very useful to us.

The formal statement of these characteristics comes from what is known as the Central Limit Theorem and a related theorem known as the Law of Large Numbers.

THE CENTRAL LIMIT THEOREM

According to the **Central Limit Theorem,** *if repeated random samples of size* n *are drawn from a population that is normally distributed along some variable* x, *having a mean* μ *and a standard deviation* σ, *then the sampling distribution of all theoretically possible sample means will be a normal distribution having a mean* μ *and a standard deviation* $\sigma/\sqrt{n}$.[1]

If for a particular population some variable (x) is normally distributed and we draw a series of samples of a predetermined size (n) from that population, the Central Limit Theorem tells us that:

1. The sampling distribution of sample means will be a normal distribution.
2. The mean of the sampling distribution of sample means, the mean of all the sample means (designated $\bar{\bar{x}}$), will be equal to μ, the mean of the population from which the samples were originally drawn.
3. The standard deviation of the sampling distribution of sample means will be equal to $\sigma/\sqrt{n}$, the standard deviation of the population from which the samples were drawn divided by the square root of the size of the samples that we were drawing. *This standard deviation of our sampling distribution,* $\sigma/\sqrt{n}$, *is also called the* **standard error of the mean** *or more often just the* **standard error,** *and is sometimes designated with the symbol* $\sigma_{\bar{x}}$.

To illustrate: Suppose your state or province mandates a series of competency examinations in reading, math, and so on, to be taken by all school children in selected grades. Assume that on the math competency exam given to all 9th graders, the mean, μ, is 70 and the standard deviation, σ, is 20. We wish to compare these results to those we would have found if we had studied random samples of 9th graders rather than the whole population. How would the sample means be distributed? Assume we select random samples of size $n = 100$.

According to the Central Limit Theorem, the sampling distribution of sample means would be a normal curve with a mean $\bar{\bar{x}} = \mu = 70$ and a standard deviation (the standard error of the mean) $\sigma/\sqrt{n} = 20/\sqrt{100} = 20/10 = 2.0$. This is graphed in Figure 8.8.

The implications of Figure 8.8 are immense since they suggest that the overwhelming majority of theoretically possible sample means are going to fall very close to the original population's mean. There are tails to the curve in Figure 8.8 above and below the mean and they *are* asymptotic to the x-axis, but they are so tiny they are barely perceivable.

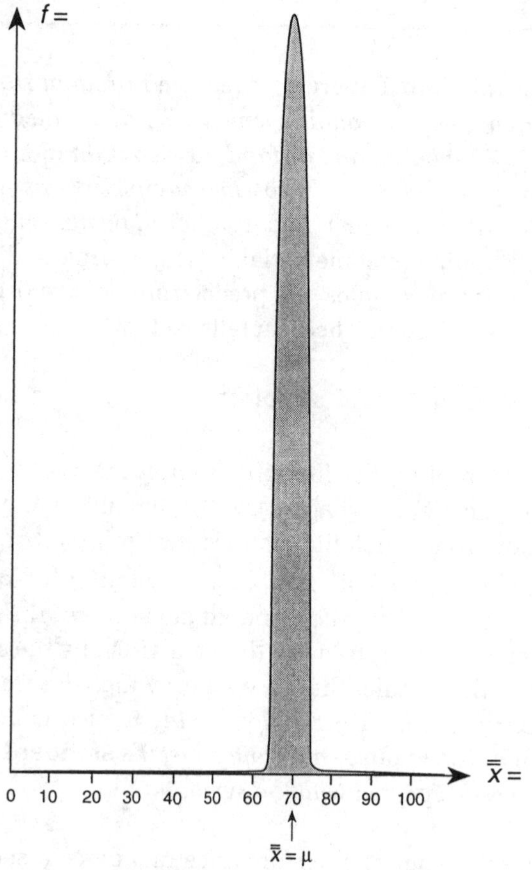

Figure 8.8
The Actual Appearance of a Sampling Distribution of Sample Means for Samples $n = 100$ Drawn From a Population Where $\mu = 70$ and $\sigma = 20$

In fact, we know the following about normal distributions: 68.27% of the area under the normal curve (thus 68.27% of all sample means) falls between $\mu - \sigma_{\bar{x}}$ and $\mu + \sigma_{\bar{x}}$, that is, one standard deviation above and below the mean. Since in this particular example the standard deviation of the sampling distribution (the standard error of the mean) is 2.0, we are observing that 68.27% of all sample means will lie between 68 and 72. We also know that 95.45% of the area under the curve (95.45% of the sample means) falls between $\mu - 2\sigma_{\bar{x}}$ and $\mu + 2\sigma_{\bar{x}}$ and 99.73% of the area falls between $\mu - 3\sigma_{\bar{x}}$ and $\mu + 3\sigma_{\bar{x}}$.

In this particular example:

Area Under the Normal Curve = Percent of all Possible Sample Means	*Range From the Mean in Terms of Standard Errors*	*Range in Terms of Competency Exam Scores*
68.27%	$\mu \pm 1\sigma_{\bar{x}}$	68 – 72
95.45	$\mu \pm 2\sigma_{\bar{x}}$	66 – 74
99.73	$\mu \pm 3\sigma_{\bar{x}}$	64 – 76

With 99.73% of sample means falling between 64 and 76, the remaining 0.27% of all sample means must fall below 64 or above 76. Out of 1000 samples, only 2 or 3 would have means below 64 or above 76 — an extremely improbable, but statistically possible, event.

The Central Limit Theorem becomes most useful to us when we are given, as before, μ and σ for a population and data about one random sample presumably drawn from that population. In that case we are asking how likely it is that from the given population we could draw a random sample whose $\bar{x}$ differs from μ by as much as we observe. If the likelihood is low, we might better conclude that the $\bar{x}$ reflects a population with a mean other than the μ of the population from which we initially assumed that the sample was drawn.

This brings us to the kind of problem presented here and in the previous chapter. Suppose in our competency exam example we have a random sample of 100 9th graders who had been enrolled in a 6-week-long course to prepare for this examination. This sample's mean score is 73. Thus H_0: $\mu_{all} = \mu_{course}$. Assuming advance data on which to make a directionality assumption, we could write

$$H_1: \mu_{all} < \mu_{course}$$

We calculate z using the formula from Chapter 7 and compare $z_{obtained}$ to the critical values of z.

$$z = \frac{\bar{x} - \mu}{\sigma/\sqrt{n}} = \frac{73 - 70}{20/\sqrt{100}} = \frac{3}{20/10} = \frac{3}{2} = 1.50$$

Since $1.50 < 1.65$, we cannot reject H_0. The course appears to have been unsuccessful.

With this formula we are finding our sample's $\bar{x}$ on the $\bar{x}$-axis of the sampling distribution of sample means, finding the distance from that $\bar{x}$ to the mean of the sampling distribution, and converting that distance into standard deviation units (standard scores) based on the standard deviation of the sampling distribution. To see how this works, let us start by converting our simple z formula from symbols to words.

$$z = \frac{x - \mu}{\sigma} =$$

$$= \frac{(1)\ \left|\begin{array}{l}\text{The value of the variable}\\ \text{whose distance from the}\\ \text{mean we wish to find}\end{array}\right| \;-\; (2)\ \left|\begin{array}{l}\text{The mean of our}\\ \text{frequency distribution}\end{array}\right|}{(3)\ \left|\begin{array}{l}\text{The standard deviation of}\\ \text{our frequency distribution}\end{array}\right|}$$

Remember that the frequency distribution is the *sampling distribution* of sample means. Now note:

1. The value of the variable whose distance from the mean (of the sampling distribution) we wish to find is the $\bar{x}$ for those taking the preparatory course.
2. The mean of our frequency distribution (the sampling distribution) $\bar{\bar{x}}$, according to the Central Limit Theorem equals μ for all the 9th graders, and

3. The standard deviation of our frequency distribution, which for a sampling distribution is called the standard error, according to the Central Limit Theorem equals σ for all the 9th graders divided by the square root of our sample size: $\sigma/\sqrt{n}$.

Substituting this information from the Central Limit Theorem for the words in our equation we get

$$z = \frac{(1)[\bar{x}_{\text{course}}] - (2)[\mu_{\text{all}}]}{(3)[\sigma_{\text{all}}/\sqrt{n}]}$$

Simplifying:

$$z = \frac{\bar{x}_{\text{course}} - \mu_{\text{all}}}{\sigma_{\text{all}}/\sqrt{n}}$$

Or, in general terms,

$$z = \frac{\bar{x} - \mu}{\sigma/\sqrt{n}}$$

Thus the formula for the one-sample z test of significance is really a recasting of the basic formula $z = (x - \mu)/\sigma$ to apply to the sampling distribution. The Central Limit Theorem enables us to find a z value on the sampling distribution from data pertaining to the population and the sample. This is illustrated in Figure 8.9, which shows our sampling distribution (not drawn to actual scale) and its components.

We now see why a directional H_1 is dubbed a one-tailed H_1—we only make use of one tail on the sampling distribution. Without a directionality assumption we would move out from the mean of the sampling distribution toward both the left and the right, examine the size of both of the tails by comparing the absolute value of z_{obtained} to z_{critical}, and pay the price of needing a larger z_{obtained} than is needed when using only one tail.

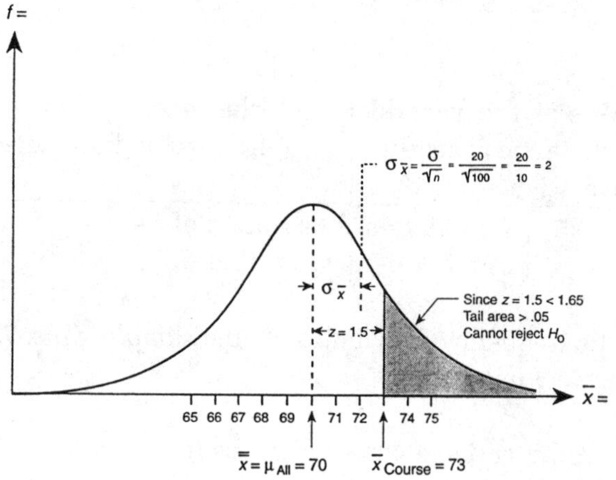

Figure 8.9
The Sampling Distribution of Sample Means for Samples $n = 100$ Based on Competency Exam Data (Hypothetical)

REVIEW

Before we proceed, let us review the fact that in using the Central Limit Theorem, we are working with three separate frequency distributions: the population, the sample, and the sampling distribution of sample means. We are given information about the first two distributions. The Central Limit Theorem then enables us to take data from those two distributions and make use of the properties of the sampling distribution. We know:

1. The frequency distribution for variable x for some population. We assume that this distribution is normal. We know its mean μ and its standard deviation σ.
2. The frequency distribution of a particular random sample that we have drawn. We know its size n and its mean $\bar{x}$. The variable x in our sample is the same variable x in our population.
3. The sampling distribution of sample means. (You never see this distribution; you just make use of it!) There exists a *separate* sampling distribution of sample means for *each* possible sample size (each n). For any given n, this represents the frequency distribution of all possible *sample means* from all possible samples drawn randomly from that population whose mean and standard deviation along variable x are μ and σ, respectively.

For the specific sample that we have drawn, our sample mean $\bar{x}$ will be one point (one value of $\bar{x}$) on that sampling distribution. The Central Limit Theorem enables us to find the distance from the sample's mean to the population's mean, expressed as standard errors or standard deviations of the sampling distribution. Since the sampling distribution is a normal curve, we may determine the probability of our sample's $\bar{x}$ reflecting a population whose mean is μ and, based on that probability, either retain or reject our null hypothesis.

THE NORMALITY ASSUMPTION

Note that the Central Limit Theorem assumes that the population we are studying is normally distributed along variable x. This is called the **normality assumption.** If it is true, the sampling distribution of sample means will be a normal distribution, and we may make use of the z

formula to test for statistical significance. (Note that nothing requires that our sample be normally distributed.) What if we know that the population is *not* normally distributed, or more realistically, what if we have no basis for making a normality assumption about the population in the first place? Even in such cases, if our sample's size is large enough, we may still be able to make use of the Central Limit Theorem due to the Law of Large Numbers.

The **Law of Large Numbers** states that *if the size of the sample, n, is sufficiently large (no less than 30; preferably no less than 50), then the Central Limit Theorem will apply even if the population is* not *normally distributed along variable x.* Thus if n is large enough, the population distribution need not be normal and could, in fact, be anything: skewed, bimodal, trimodal, anything. When n is large enough, we *relax the normality assumption* for our population but the sampling distribution of sample means will still be a normal curve and the Central Limit Theorem will still apply.

How large must n be to relax the normality assumption? The figures given in the above theorem are rather arbitrary; other sources give other cutoffs. In fact, in some texts of statistics for psychology (which often only requires small samples or small experimental and control groups) the minimum sample size is as low as 15, but that is probably too low. Perhaps we ought to put it this way:

If:	*Then*:
$n \geq 100$	It is always safe to relax the normality assumption.
$50 \leq n < 100$	It is almost always safe.
$30 \leq n < 50$	It is probably safe.
$n < 30$	It is probably not safe.

In social science survey research, our sample sizes are generally large enough to make use of the Law of Large Numbers. This is particularly fortunate, since in actual research all too often, the issue of the normality assumption is not adequately addressed.

Let us look at an example. At a small liberal arts college an index of support for civil liberties, ranging from 0 (least supportive) to 10 (most supportive), was pilot tested on the entire student body, yielding a mean of 7.5 and a standard deviation of 1.5. A random sample of 100 students who had been the direct victims or close relatives of victims of serious crimes was also given the test and their mean score was 7.2. May we conclude that for all similar victims the support score for civil liberties differs in general from the population of all students at that college?

Our hypotheses are

$$H_0\text{: } \mu_{\text{all}} = \mu_{\text{victims}}$$

$$H_1\text{: } \mu_{\text{all}} \neq \mu_{\text{victims}} \text{ (no directionality assumed)}$$

Since $n = 100$, we may relax the normality assumption for the population. We have all necessary data for a one-sample z test.

$$z = \frac{\bar{x} - \mu}{\sigma / \sqrt{n}} = \frac{\bar{x}_{\text{victims}} - \mu_{\text{all}}}{\sigma_{\text{all}} / \sqrt{n_{\text{victims}}}} = \frac{7.2 - 7.5}{1.5 / \sqrt{100}}$$

$$= \frac{-.3}{1.5 / 10} = \frac{-.30}{.15} = -2.00$$

We compare the absolute value of z to the two-tailed z_{critical} values:

$$|z| = 2.00 > 1.96\text{; reject } H_0$$

$$2.00 < 2.58,\ p < .05$$

We conclude, therefore, that the civil liberties support score for all serious crime victims at this college is lower than the average for the college as a whole ($p < .05$). (The sampling distribution is shown in Figure 8.10.)

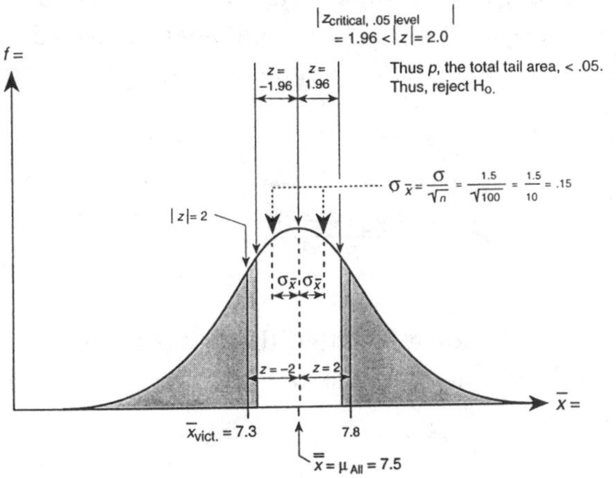

Figure 8.10
The Sampling Distribution of Sample Means for Civil Liberties Support Scores

The One-Sample t Test

We know that to do the one-sample z test we need to know or be able to hypothesize two population parameters, μ and σ. What could we do in the unlikely event that we know μ but not σ? Initially, the sample standard deviation s was assumed to be a good estimate of σ, so s was substituted when σ was unknown. Once the sample mean $\bar{x}$ had been calculated, s was generated using the definitional formula

$$s = \sqrt{\frac{\sum (x - \bar{x})^2}{n}}$$

or one of several possible computational formulas.

However, it was discovered that, particularly when the sample size n was small, calculating z with s produced inaccurate conclusions. A British quality control expert[2] working for a Dublin brewery discovered that by calculating a different estimate of σ from sample data, a better test of significance could be developed. *This new best "unbiased" estimate of* σ*, which we designate* $\hat{\sigma}$ (read as "sigma-hat," because sigma is wearing a hat), *is created when we substitute n – 1 for n in the standard deviation formula.*

$$\hat{\sigma} = \sqrt{\frac{\sum (x - \bar{x})^2}{n - 1}}$$

This new test of significance is called the **t test** to differentiate it from the Z test; note that the formulas are the same except that $\hat{\sigma}$ is substituted for σ.

$$t = \frac{\bar{x} - \mu}{\hat{\sigma} / \sqrt{n}}$$

When n is large, the substitution of $\hat{\sigma}$ for s makes very little difference, but as n gets smaller $\hat{\sigma}$ and s diverge, causing a likewise divergence between t (using $\hat{\sigma}$) and z (using s to estimate σ).

The sampling distributions of t and z also differ. In the case of the z test, the sampling distribution of sample means is a normal curve. Since the value of each sample mean can be expressed as a z score (indicating the distance x is from μ in terms of standard errors), the sampling distribution of sample means is the same as the distribution of all the z scores from all the theoretically possible sample means that make up the sampling distribution. Thus the sampling distribution of z (all the z's from those sample means) is also a normal curve.

If we take the same means in our sampling distribution and calculate t scores instead, the sampling distribution of t (all the t's from those sample means) is a normal distribution only when the sample sizes are above 120. As the sample sizes fall below 120 (give or take), the sampling distribution begins to be flatter than a normal curve (say *platykurtic,* if you want to impress your friends). When the curve is flatter than a normal curve at its peak, the tails are also larger than those of a normal

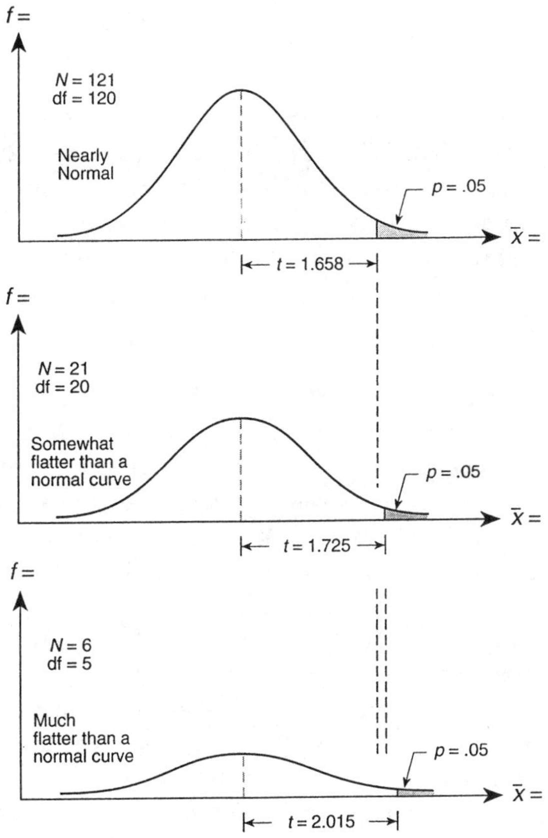

Figure 8.11
Changes in the Sampling Distribution of *t* as Sample Size Decreases

curve. (The effect is similar to pushing a balloon down from its top, thus displacing the air to the sides as we press.) As *n* gets smaller, the peak of the sampling distribution gets flatter and its tails get larger. The important consequence is that as *n* gets smaller, we must go ever-greater distances away from the mean to get a tail area equal to .05 proportion of the area under the curve.

Figure 8.11 shows the changes in the critical values of *t* (.05 level, one-tailed) as *n* decreases. At $n = 121$, the sampling distribution is nearly a normal curve and $t_{critical}$ is 1.658, only slightly larger than the $z_{critical}$ (one-tailed, .05 level), which is 1.65 (actually 1.645 before rounding). In fact, as *n* increases above 121, the critical values of *z* and *t* get ever closer to each other. As *n* gets extremely large, approaching infinity as a limit, the critical values of *z* and *t* become the same. However, as *n* drops *below* 121, the $t_{critical}$ value gets larger. In other words, we have to go farther out to get a tail with .05 of the area under the curve in it. By the time $n = 21$, $t_{critical}$ has gone from 1.658 to 1.725, and at $n = 6$, $t_{critical}$ has risen to 2.015.

In comparing the critical values of *t* to those of *z*, bear in mind that the sampling distribution of *z* is *always* a normal distribution and its critical values remain constant, independent of sample size. By contrast, the critical values of *t* depend on sample size. At best, when *n* is large, the critical values of *t* are almost as small as those of *z*. But as *n* decreases, the critical values of *t* get larger, making it harder to reject the null hypothesis. Thus *if we know* σ *and can therefore do a one-sample* z *test, we always do the* z *test, not the* t *test*. We only do the one-sample *t* test if σ is unknown and we must estimate it with $\hat{\sigma}$. In fact, when *n* gets large (say 30 or more), many statisticians advocate the use of the *z* test, *s* substituting for σ in the formula, rather than the use of the *t* test. But with a smaller *n* where σ is unknown, we must always do the *t* test and retain the normality assumption for the population.

DEGREES OF FREEDOM

Note that in Figure 8.11 under each of the three reported n's—121, 21, and 6—is another number labeled *df*, which is one less than n—120, 20, 5. As we learned in Chapter 7, the *df* stands for **degrees of freedom,** a number we generate to make use of a table of critical t values. In the case of the one-sample t test,

$$df = n - 1$$

We need to find the degrees of freedom in order to find the critical values of t against which we compare our obtained t. As noted, the sampling distribution of t changes from a normal curve as n decreases, and thus the critical values change as well. As we see in Figure 8.11, at 120 degrees of freedom ($n = 121$) we need a t of 1.658 in order to have one tail on the sampling distribution with a .05 area. By the time degrees of freedom drops to five, we need a t of 2.015.

Tables of critical values for all tests of significance beyond the z test require that we first calculate a degrees of freedom figure to make use of the table. Why find *df*? Why not base the tables on n as we did the sampling distributions in Figure 8.11? The simplest answer to the question is that there are several formulas that generate t scores, not just the one presented in this chapter. Likewise, for each of the different t formulas, there is a separate degrees of freedom formula. The formula $df = n - 1$ is only used for the one-sample t test presented here. In the next chapter we will discuss some of the other t formulas, each having its own degrees of freedom formula, but all making use of a common table of critical values of t. Without degrees of freedom, we would need a separate table of critical values for each separate formula.

There is a mathematical meaning to the concept of degrees of freedom, having to do with how many numbers are free to vary in a formula. For instance, if $x_1 + x_2 + x_3 = 10$ and you let any two of the scores vary (say we make $x_1 = 2$ and $x_2 = 5$), then the remaining value of x is fixed. Since $2 + 5 = 7$ and $7 + x_3 = 10$, once x_1 and x_2 are determined, x_3 can only take on one value. In this case, $x_3 = 3$. So three unknowns adding up to a fixed sum has two degrees of freedom. Only two of the unknowns are free to vary. At the level of applied statistics that we cover in this book, it is not really necessary to know the definition of degrees of freedom to make use of the concept. So we will simply move on, referring the curious to more advanced texts. For our purposes, degrees of freedom are simply numbers that we must calculate to make use of critical values tables for t and the other tests of significance to be encountered later.

THE *t* TABLE

The table of critical values of t, found as Table 8.2 and also in the Appendix, is simple to use. At the top are levels of significance for a one-tailed test (a directional H_1) and below it are the corresponding levels for a two-tailed test. Thus $t_{critical}$ one-tailed at the .10 level is the same as $t_{critical}$ two-tailed at the .20 level. The one-tailed probability levels are always one-half of the corresponding two-tailed levels.

Since we always begin by comparing $t_{obtained}$ to $t_{critical}$ at the .05 level, we first isolate the appropriate .05 column for whichever H_1 (one-tailed or two-tailed) we are using. Then we go down the *df* column on the far left until we come to the number that we found in the *df* formula. Noting the values highlighted earlier in Figure 8.11, if *df* is 120, we go all the way down the *df* column until we find 120. We then move across the row until we are under the .05 level for a one-tailed test. At the intersection of the 120 row and the .05 column, we find the critical value of t, 1.658. Likewise, in the same .05 column we find the $t_{critical}$ of 1.725 in the row for 20 degrees of freedom, and 2.015 in the row for 5 degrees of freedom.

Under the $df = 120$ row, we note the symbol for infinity (an eight that has gone down for the count). In this case, "infinity" is any *df* above 120. Here, the sampling distribution has become (or is in the act of becoming) a more perfect normal curve. Note that at this point there is no difference between the critical values of t and those of z.

If you cannot find the *df* that you need in the table, go to the nearest critical value that makes it *harder* to reject H_0. In the case of Table 8.2, move *up* to the next *lower df*. For instance, if the *df* is 35, a number not presented in the table, go up to 30 *df* and use those critical values. Thus if the t obtained in a one-tailed test at 35 *df* were 1.7, you would compare it to the .05 critical value at 30 degrees of freedom, 1.697. Since 1.7 is greater than 1.697, you would reject H_0. What if the obtained t were 1.690? That would be less (barely) than 1.697 and you could not reject H_0 using this table. However, you would be right in assuming that had you known $t_{critical}$ at 35 degrees of freedom there would be a good chance that it would be equal to or less than your t of 1.690. In this case, consult a book of tables for statisticians, which would have a more complete t table than the one used here.[3]

Table 8.2 Distribution of *t*

	Level of significance for one-tailed test					
	.10	*.05*	*.025*	*.01*	*.005*	*.0005*
	Level of significance for two-tailed test					
df	*.20*	*.10*	*.05*	*.02*	*.01*	*.001*
1	3.078	6.314	12.706	31.821	63.657	636.619
2	1.886	2.920	4.303	6.965	9.925	31.598
3	1.638	2.353	3.182	4.541	5.841	12.941
4	1.533	2.132	2.776	3.747	4.604	8.610
5	1.476	2.015	2.571	3.365	4.032	6.859
6	1.440	1.943	2.447	3.143	3.707	5.959
7	1.415	1.895	2.365	2.998	3.499	5.405
8	1.397	1.860	2.306	2.896	3.355	5.041
9	1.383	1.833	2.262	2.821	3.250	4.781
10	1.372	1.812	2.228	2.764	3.169	4.587
11	1.363	1.796	2.201	2.718	3.106	4.437
12	1.356	1.782	2.179	2.681	3.055	4.318
13	1.350	1.771	2.160	2.650	3.012	4.221
14	1.345	1.761	2.145	2.624	2.977	4.140
15	1.341	1.753	2.131	2.602	2.947	4.073
16	1.337	1.746	2.120	2.583	2.921	4.015
17	1.333	1.740	2.110	2.567	2.898	3.965
18	1.330	1.734	2.101	2.552	2.878	3.922
19	1.328	1.729	2.093	2.539	2.861	3.883
20	1.325	1.725	2.086	2.528	2.845	3.850
21	1.323	1.721	2.080	2.518	2.831	3.819
22	1.321	1.717	2.074	2.508	2.819	3.792
23	1.319	1.714	2.069	2.500	2.807	3.767
24	1.318	1.711	2.064	2.492	2.797	3.745
25	1.316	1.708	2.060	2.485	2.787	3.725
26	1.315	1.706	2.056	2.479	2.779	3.707
27	1.314	1.703	2.052	2.473	2.771	3.690
28	1.313	1.701	2.048	2.467	2.763	3.674
29	1.311	1.699	2.045	2.462	2.756	3.659
30	1.310	1.697	2.042	2.457	2.750	3.646
40	1.303	1.684	2.021	2.423	2.704	3.551
60	1.296	1.671	2.000	2.390	2.660	3.460
120	1.289	1.658	1.980	2.358	2.617	3.373
∞	1.282	1.645	1.960	2.326	2.576	3.291

If the obtained t exceeds $t_{critical}$ at the .05 level, you then compare it to the critical values to the right of the .05 column. Following the same procedure used for the z test, you make your probability statement by seeing how many critical values are less than the obtained t. The only difference is that in the t table there are critical values for levels other than .05, .01, and .001. Suppose at 60 *df*, we obtain a *t* value of 3.0 using a nondirectional H_1. Going down the .05 level column, for the two-tailed test, we see at the 60 *df* row a critical value of 2.000. We can reject H_0. We then compare our 3.0 obtained *t* to the critical values to the right of the 2.000 we exceeded. We exceed the 2.390 (.02 level) and the 2.660 (.01 level) but not the 3.460 critical value at the .001 level. Thus we report $p < .01$. Had this been a one-tailed test, we would be reporting $p < .005$.

AN ALTERNATIVE *t* FORMULA

We have been using the following formulas:

$$t = \frac{\bar{x} - \mu}{\hat{\sigma}/\sqrt{n}} \qquad df = n - 1$$

where

$$\hat{\sigma} = \sqrt{\frac{\sum (x - \bar{x})^2}{n - 1}}$$

Suppose you did not have access to $\hat{\sigma}$ but did know the original sample standard deviation of

$$s = \sqrt{\frac{\sum (x - \bar{x})^2}{n}}$$

Rather than recalculating $\hat{\sigma}$, you may make use of the *s* in a modified *t* formula:

$$t = \frac{\bar{x} - \mu}{s/\sqrt{n - 1}} \qquad df = n - 1$$

Again, remember that if you get the standard deviation from either a computer printout or a calculator with a standard deviation function built in, consult the appropriate manual to find out how that standard deviation was calculated to determine whether you have an *s* or a $\hat{\sigma}$. Then pick the appropriate *t* formula to use.

A *z* TEST FOR PROPORTIONS

The formula for the z test for sample means may be modified to test the difference in proportions in a sample compared to the equivalent difference in proportions in a population.

For instance, suppose that in some small community, the proportions of minorities (people of African or Hispanic origin) make up 20% (.20 proportion) of the population. The new school superintendent suspects that minorities are underrepresented among the 100 teachers in her public school system, since there are only 15 minority faculty, 15% or a .15 proportion. For such a problem,

$$z = \frac{P_s - P_p}{\sqrt{P_p Q_p / n}}$$

where

P_s = the proportion of minorities in the sample = .15
P_p = the proportion of minorities in the population = .20
Q_p = the proportion of nonminorities in the population = $1 - P_p = 1 - .20 = .80$
n = the size of the sample or group being studied = 100

Here,

$$H_0 : P_s = P_p$$

$$H_1 : P_s < P_p$$

$$z = \frac{P_s - P_p}{\sqrt{P_p Q_p / n}} = \frac{.15 - .20}{\sqrt{(.20)(.80)/100}} = \frac{-.05}{\sqrt{.16/100}} = \frac{-.05}{\sqrt{.0016}} = \frac{-.05}{.04} = -1.25$$

Using the directional $z_{critical}$ at the .05 level of 1.65, we cannot reject H_0 since 1.25 < 1.65. We cannot conclude that minorities are underrepresented among the teachers.

INTERVAL ESTIMATION

We have already discussed the fact that if we did not know σ, our best estimate of it from sample data would be $\hat{\sigma}$. Likewise, our best estimate of μ would be $\bar{x}$. Suppose we wanted to estimate μ from $\bar{x}$. We know from the sampling distribution of sample means that not all sample means will be exactly equal to μ even though our one $\bar{x}$ is the best estimate of that parameter. *With* **interval estimation,** *we establish an interval of scores called a* **confidence interval** *and we state with a certain level of confidence that the* μ *will fall within the limits of the interval we created.*

For instance, we can see from our sampling distribution that with no directionality assumption, 95% of all sample means lie between μ and $\pm$ 1.96 standard errors. Likewise, 99% lie between μ and $\pm$ 2.58 standard errors. The number of standard errors corresponds to the two-tailed $z_{\text{criticals}}$ at the .05 and .01 levels, respectively. Also, 99.9% of all sample means lie between μ and $\pm$3.29 standard errors and 3.29 is the critical z at the .001 level. Suppose we would be satisfied to find the interval within which 95% of all sample means would fall. *We build an interval around the* $\bar{x}$ and assume that μ will fall within that interval. We call this the *95% confidence interval,* our level of confidence corresponding to the percentage of all means falling within the interval. Thus we are 95% confident that μ will lie in the interval between $\bar{x} - 1.96\sigma_{\bar{x}}$ and $\bar{x} + 1.96\sigma_{\bar{x}}$.

Remembering that we already know that $\sigma_{\bar{x}} = \sigma/\sqrt{n}$, we find our confidence interval by the formula:

$$\bar{x} \pm 1.96(\sigma/\sqrt{n})$$

Suppose $\bar{x} = 55$, $\sigma = 10$, and $n = 64$. The upper limit of our interval would be

$$\bar{x} + 1.96(\sigma/\sqrt{n}) = 55 + 1.96(10/\sqrt{64})$$

$$= 55 + 1.96(10/8)$$

$$= 55 + 2.45$$

$$= 57.45$$

Our lower limit would be

$$\bar{x} - 1.96(\sigma/\sqrt{n}) = 55 - 1.96(10/\sqrt{64})$$

$$= 55 - 1.96(10/8)$$

$$= 55 - 2.45$$

$$= 52.55$$

Thus the 95% confidence interval for estimating μ is 52.55 to 57.45. We know that 95% of all sample means fall within the interval, so we are 95% confident that μ will be between 52.55 and 57.45.

Suppose we wanted a *greater* level of confidence, 99%. The price we would pay for it would be a *wider* confidence interval. For our upper limit

$$\bar{x} + 2.58\ (\sigma/\sqrt{n}) = 55 + 2.58\ (10/\sqrt{64})$$

$$= 55 + 2.58\ (10/8)$$

$$= 55 + 3.23$$

$$= 58.23$$

and for our lower limit

$$\bar{x} - 2.58\ (\sigma/\sqrt{n}) = 55 - 2.58\ (10/\sqrt{64})$$

$$= 55 - 3.23$$

$$= 51.77$$

We are 99% confident that μ falls between 51.77 and 58.23.

If σ is unknown, which is generally the case, we may do exactly the same procedure with the *t* test using either of the following formulas:

$$\mu = \bar{x} \pm t_{\text{critical}}(\hat{\sigma}/\sqrt{n})$$

or

$$\mu = \bar{x} \pm t_{\text{critical}}(s/\sqrt{n-1})$$

The nondirectional $t_{critical}$ at df = $n - 1$ at the .05 level would be used for a 95% confidence level, the $t_{critical}$ at the .01 level would be used for a 99% confidence interval, and so on.

Confidence Intervals for Proportions

Imagine that you are a campaign manager of a presidential candidate in a two-person race. A telephone survey of 900 voters gives your candidate a 53% lead over the opponent. How likely does that percentage lead reflect the electorate? We can construct a confidence interval around the .53 proportion that your candidate received in the sample. The formula we use is

$$P_s \pm 1.96\sqrt{P_p Q_p / n}$$

The 1.96 is the appropriate critical value of z, in this case at the .05 level since we chose a 95% confidence interval.

P_s = your candidate's proportion of support in the sample = .53.

P_p = your candidate's proportion of support in the population, which we estimate with P_s. Thus, $P_s = P_p = .53$.

Q_p = the opponent's proportion of support in the population, which we estimate from the sample by subtracting P_s from 1. Thus, $Q_p = 1 - P_p = 1 - P_s = 1 - .53 = .47$.

n = The number of cases, which must equal or exceed $5/\min(P_s, 1{-}P_s)$, that is, five divided by whichever is smaller, P_s or $1 - P_s$.

Thus the upper limit of our 95% confidence interval is

$$\begin{aligned} P_s + 1.96\sqrt{P_p Q_p / n} &= .53 + 1.96\sqrt{(.53)(.47)/900} \\ &= .53 + 1.96\sqrt{.2491/900} \\ &= .53 + 1.96\sqrt{.00028} \\ &= .53 + 1.96(.0167) \\ &= .53 + .0327 \\ &= .53 + .03 \\ &= .56 \end{aligned}$$

The lower limit would be

$$P_s - 1.96\sqrt{P_pQ_p/n} = .53 - 1.96\sqrt{(.53)(.47)/900}$$

$$= .53 - .03$$

$$= .50$$

So our confidence interval ranges from .50 to .56. Since in percentages this is 50% to 56%, a range of six percentage points, we report that according to our poll, our candidate has a 53% lead, but our margin of error is plus or minus three percentage points. Our candidate could receive as little as 50% or as much as 56%. If we had chosen a 99% confidence interval, our confidence interval would be larger and so would the margin of error reported.

When n is small or if we want to be particularly sure of our estimate, it is safer to make a more conservative estimation of P_p than to use P_s. Here, we assume that each candidate has half of the vote. Thus $P_p = .50$ and $Q_p = .50$. This will yield a larger confidence interval than any other estimate of P_p would generate. By widening the interval, we minimize the risk in making our estimate. Suppose P_s were .53, but $n = 150$ instead of 900. We estimate P_p and Q_p as .50, respectively. For the upper limit

$$P_s + 1.96\sqrt{P_pQ_p/n} = .53 + 1.96\sqrt{(.50)(.50)/150}$$

$$= .53 + 1.96\sqrt{.25/150}$$

$$= .53 + 1.96\sqrt{.00167}$$

$$= .53 + 1.96(.041)$$

$$= .53 + .08$$

$$= .61$$

And our lower limit would be

$$.53 - .08 = .45$$

Here, our 95% confidence interval ranges from .45 to .61 and we have an eight percentage point margin of error.

MORE ON PROBABILITY

Suppose we have developed a scale to be used in a survey. This scale measures the extent to which the respondent is aware of and knowledgeable about HIV and AIDS. Assume that the scale ranges from a low of 0 to a high of 100 and is a normal distribution with a mean of 50 and standard deviation of 15. We may thus apply the z formula to this distribution to determine the proportion of cases falling within a specified range of scores. Let us use this scale to extend our discussion of probability.

We begin by outlining some new notations and defining them.

P(A) = The probability of outcome A occurring.

P(A or B) = The probability of either outcome A or outcome B occurring.

P(A and B) = The probability of both outcomes A and B occurring jointly.

P(A|B) = The probability of outcome A occurring given that outcome B has already occurred (**conditional probability**).

Let us illustrate using our AIDS awareness scale. Suppose outcome *A* is the probability of selecting an individual with an AIDS awareness score of 70 or above. Since $x = 70$, $\mu = 50$, and $\sigma = 15$, we apply the z formula and find $z = 1.33$. Looking at Table 8.1, Column C, we find a probability of .0918. Thus *P(A)* = .0918.

Let outcome *B* be the probability of selecting someone with an AIDS awareness score of 40 or below. Plugging into the z formula, we obtain a z of -0.66 and find a probability of .2546 from Table 8.1. Thus *P(B)* = .2546.

THE ADDITION RULE

Suppose we would like to know the probability of selecting someone whose AIDS awareness score is either 70 or above or 40 or below, *P(A or B)*. Now outcomes *A* and *B* are known as *mutually exclusive outcomes.* If one has an AIDS awareness score above 70 one cannot also have an AIDS awareness score below 40. When outcomes are mutually exclusive, a rule known as the **addition rule** tells us that

$$P(A \text{ or } B) = P(A) + P(B)$$

In this case,

$$P(A \text{ or } B) = P(A) + P(B) = .0918 + .2546 = .3464$$

If we had included a third outcome, outcome *C*, such as awareness between 50 and 55, we would calculate a *z* of 0.33 and, looking this time at column B of Table 8.1, find a probability of .1293.

$$P(C) = .1293$$

Therefore,

$$P(A \text{ or } B \text{ or } C) = P(A) + P(B) + P(C)$$

$$= .0918 + .2546 + .1293$$

$$= .4757$$

When our events are not mutually exclusive but overlap, we must apply a more complex addition rule. Suppose outcome *A* remains a score of 70 and above, and we add another outcome, outcome *D*. If outcome *D* is the probability of selecting a respondent with AIDS awareness between 50 and 75, *z* will be 1.66 and Column B of Table 8.1 will yield a probability of .4515. This time, however, we cannot simply add *P(A)* to *P(D)* to find *P(A or D)* since our outcomes are no longer mutually exclusive. Anyone with a score between 70 and 75 will belong jointly to both outcomes. To account for this, we must expand the addition rule as follows:

$$P(A \text{ or } D) = P(A) + P(D) - \text{P(A and D)}$$

We will see in a moment how *P(A and D)* is determined, but for now assume that we are told that it is .0414. Therefore,

$$P(A \text{ or } D) = P(A) + P(D) - \text{P(A and D)}$$

$$= .0918 + .4515 - .0414$$

$$= .5019$$

Note that in the first example, *P(A or B)*, *A* and *B* had no overlap, so *P(A and B)* = 0. Applying the longer addition rule,

$$P(A \text{ or } B) = P(A) + P(B) - P(\text{A and B})$$

$$= .0918 + .2546 - 0$$

$$= .3464$$

This matches the result found earlier.

THE MULTIPLICATION RULE

As with the addition rule, there are two forms of the **multiplication rule,** *the rule that we use to find P(A and D)*. The simple form of this rule applies when the outcomes or events are independent of one another—when neither event influences the probability of the other event occurring. Symbolically:

$$P(A|D) = P(A) \text{ and } P(D|A) = P(D)$$

In these two events, *A* and *D* are independent; that is, neither event will impact the probability of the other event's occurrence. In our example, determining the probability of selecting someone whose AIDS awareness is 70 or more has no impact on determining the probability of selecting someone with an awareness score between 50 and 75. Two *z* scores are calculated independently of one another.

In the case of *independent events*, the multiplication rule becomes:

$$P(A \text{ and } B) = P(A) \times P(B)$$

In our example;

$$P(A \text{ and } D) = P(A) \times P(D) = (.0918)(.4515) = .0414$$

That was how the value of *P(A and D)* used in the addition rule above was determined. Like the addition rule, the multiplication rule can be extended to more than two independent events.

$$P(A \text{ and } D \text{ and } E) = P(A) \times P(D) \times P(E)$$

What about *nonindependent events*? Let us assume that anyone with a score of 65 or greater has high AIDS awareness. Here $z = 1.00$ and Column C of Table 8.1 shows that the probability of selecting a high awareness person is .1587. If we have a finite group of 13 individuals, we would expect to find $.1587 \times 13$ or 2.06 high awareness scores.

Assume, therefore, that we have 13 people, 2 of whom have high AIDS awareness. What is the probability of making two selections from the group and selecting the two high awareness individuals? The events are nonindependent since the outcome of the first selection has an impact on the second selection. The probability of getting a high awareness scorer on the first draw would be 2/13 or .1538.

If we select a high scorer on the first draw, the probabilities change on the second draw: there are now 12 people, 1 of whom is a high awareness scorer. The probability of selecting a high scorer on the second draw after having selected a high scorer on the first draw, $P(H_2 | H_1)$, is 1/12 or .0833. (Here H stands for high scorer; 1 and 2 for the first and second draws, respectively.)

$$\text{So, } P(H_1 \text{ and } H_2) = P(H_1) \times P(H_2 | H_1)$$

$$= (.1538)\ (.0833)$$

$$= .0128$$

In more general notation,

$$P(A \text{ and } B) = P(A) \times P(B | A)$$

or

$$P(A \text{ and } B) = P(B) \times P(A | B)$$

If two events are *statistically independent* events, then

$$P(A | B) = P(A) \text{ and } P(B | A) = P(B)$$

Thus we return to the simpler formula for the multiplication rule.

$$P(A \text{ and } B) = P(A) \times P(B | A) = P(A) \times P(B)$$

PERMUTATIONS AND COMBINATIONS

Earlier in this chapter we discussed how many different samples of $n = 3$ we could draw from a population where $n = 5$. Recall that there were 60 possible samples when the order of selection was considered. When we drew first a 5 then a 4 then a 3, we considered it a different sample than when we drew first the 4 then the 5 then the 3. *The total possible samples that can be drawn from a population when the order of selection is a factor is called a* **permutation.** *The total possible samples when the order of selection is ignored is called a* **combination.** In the earlier example there were 10 different possible samples when the order of selection was ignored.

The formulas for finding permutations and combinations are presented below, where N is the number of items in the population, K is the size of the sample to be drawn, and $n!$ (read *n factorial*) is a number times each number lower than itself down to 1. (For example, $5! = 5 \times 4 \times 3 \times 2 \times 1 = 120$; $4! = 4 \times 3 \times 2 \times 1 = 24$; $3! = 3 \times 2 \times 1 = 6$; and so on. Note that we never include zero in the multiplication or our answer would always be zero!)

For *permutations,* indicated by the letter P,

$$P_K^N = \frac{N!}{(N-K)!}$$

In our example, $N = 5$ and $K = 3$, so

$$P_K^N = \frac{N!}{(N-K)!} = \frac{5!}{(5-3)!} = \frac{5!}{2!} = \frac{5 \times 4 \times 3 \times 2 \times 1}{2 \times 1} = \frac{120}{2} = 60$$

For *combinations,* indicated by the letter C,

$$C_K^N = \frac{N!}{K!(N-K)!}$$

In our example,

$$C_K^N = \frac{N!}{K!(N-K)!} = \frac{5!}{3!(5-3)!} = \frac{5!}{3!2!} = \frac{5 \times 4 \times 3 \times 2 \times 1}{(3 \times 2 \times 1)(2 \times 1)} = \frac{120}{(6)(2)}$$

$$= \frac{120}{12} = 10$$

CONCLUSION

In Chapters 7 and 8 all the basic elements of tests of statistical significance have been presented in a time-honored sequence, moving from the normal distribution in its basic form to the one-sample z test, and then to the one-sample t test. As stated earlier, every test that follows in this text also follows the same logical assumptions and basic procedures, starting with the formulation of H_0 and H_1, calculating the df (if appropriate), comparing the obtained value to critical values of that statistic, reaching a decision as to whether or not to reject H_0, and if H_0 is rejected, formulating the appropriate probability statement.

However, the tests presented so far have only limited value in that since they are one-sample tests, we are comparing data from that one sample to data from a population. Rarely do we know population parameters such as μ and σ, although it might be possible to estimate them, and rarely do we know if it is valid to assume that these populations, in fact, are normally distributed for the variable in question. More often we are comparing the means of two or more samples and we know no population parameters at all. Often we have problems involving nominal or ordinal levels of measurement when a comparison of means is inappropriate. We cover tests for these purposes in the following chapters.

EXERCISES

Exercise 8.1

An index of cognitive awareness is normally distributed with a mean of $\mu = 8.9$ and a standard deviation of $\sigma = 3.1$.

1. What proportion of people would be expected to have awareness scores of 14 and above?
2. What proportion would have scores between 8.9 and 14?
3. What proportion would have scores below 14?
4. What proportion would have scores of 8.5 and below?
5. What proportion would have scores between the mean and 8.5?
6. What proportion would have scores ranging from 8.5 to 14?
7. What proportion would have scores either less than 8.5 or greater than 14?

8. Remembering that the areas under the normal curve are also the probabilities of randomly selecting someone with a particular characteristic, what is the probability of randomly selecting a person with an awareness score of 12.5 or more?
9. What is the probability of randomly selecting a person with an awareness score between 6 and 8.9?
10. What is the probability of randomly selecting someone whose awareness level is between 6 and 12.5?

Exercise 8.2

Each of the following problems requires either a one-sample z or a one-sample t test. Select the appropriate test and perform it. Assume a nondirectional H_1 unless the wording of the problem suggests otherwise. For each test, indicate whether or not the normality assumption may be relaxed for the population. In doing the t test, make sure you are using the appropriate formula; that is, are you given $\hat{\sigma}$ or s?

1. Suppose you know that for the entire United States, the mean age of the population is 32, with a standard deviation of 14.5 years. Since many retired people move to Florida, you believe that the mean age of all Florida residents is greater than for the United States as a whole. You randomly select a sample of 144 Floridians and obtain a mean sample age of 34.
2. For the Miami metropolitan area, the mean age of a random sample of 25 residents is 36.5, with a standard deviation of $s = 16$ years. Compared to the United States (data given in Part 1), what may we conclude about Miami residents?
3. Suppose the sample size in Part 2 had been $n = 64$. What would your conclusion be?
4. A scale designed to measure support for gun control legislation has been developed. It ranges from 0 to 10, with 10 meaning strongest support for such actions as outlawing "Saturday Night Specials" and semi-automatic weapons. Suppose it has been determined that for the entire population of the state of Maryland, the mean support score is 6.0. A random sample of 100 residents of Maryland's Eastern Shore yields a sample support score of 4.8 with $\hat{\sigma} = 4.0$. What do you conclude?
5. For Baltimore County, a random sample of $n = 81$ has a mean of 7.0 and a standard deviation of $\hat{\sigma} = 4.5$. (For this problem and the

ones that follow, use the population figures given in Part 4.) What do you conclude for each one?

6. For Baltimore City, a random sample of 31 residents produces a mean of 8.0 and a standard deviation of $s = 3.0$.
7. A random sample of 170 members of the National Rifle Association who live in Maryland yields a mean of 1.5 and an $s = 1.25$.
8. To ascertain the attitudes of all residents of the city of Cumberland, Maryland, a random sample of 9 members of that city's police department was interviewed. The sample's mean was 8.5 and its $\hat{\sigma}$ was 3.0.

Exercise 8.3

At a state's maximum security penitentiary, all inmates have taken a battery of psychological tests. Following are the means and standard deviations for several selected indices developed from those tests.

	Index	$\mu =$	$\sigma =$
VIO	Attitudes supporting violence	57.1	15.2
ISO	Feelings of isolation from others	56.2	32.4
RAC	Attitudes of tolerance toward other races	62.5	31.6

A random sample of 50 inmates at low and medium security institutions in the same state yields the following:

Index	$\bar{X} =$
VIO	37.4
ISO	50.7
RAC	71.3

Using one-sample *z* tests (two-tailed), test for significant differences between these two groups for:

1. VIO
2. ISO
3. RAC

What are your conclusions?

Exercise 8.4

Following are the maximum security penitentiary population means for three other indices.

	Index	$\mu =$
REM	Remorse for the victim of the crime	74.9
DET	Determination to commit no further crimes	48.6
ALI	Alienation from societal norms	42.5

For the low and medium security sample, $n = 50$, the statistics are:

Index	$\bar{x} =$	$\hat{\sigma} =$
REM	78.2	16.2
DET	60.2	33.8
ALI	36.6	35.5

Using a nondirectional one-sample *t* test, test for significance and state your conclusions.

1. REM
2. DET
3. ALI

Exercise 8.5

Suppose that for the population, it is known that 51% are women and 49% are men. Suppose random samples of 50 individuals each are drawn from the following occupations, and the proportion of women in each sample is ascertained to be as follows:

Sample	$P_S =$
1. School teachers	.72
2. Nurses	.84
3. College professors	.40
4. Physicians	.35
5. Realtors	.60
6. Law students	.41

For each of the six samples, test for significance (nondirectional) the null hypothesis that the proportion of women in each sample equals the proportion of women in the population.

Exercise 8.6

A mental health assessment instrument designed to measure a person's mental health level on a 30 to 70 scale is known to have a population standard deviation of $\sigma = 12$. A random sample of $n = 25$ yields a mean $\bar{x} = 50$.

1. Generate a 95% confidence interval for estimating μ.
2. Generate a 99% confidence interval.
3. Suppose σ is unknown, but the sample yields a $\hat{\sigma} = 11$. Generate a 95% confidence interval.
4. Suppose σ is unknown, but the sample's standard deviation is $s = 9$. Generate a 99% confidence interval.

Exercise 8.7

A telephone survey of 250 voters shows a local school tax levy passing with 55% of the vote.

1. Construct a 95% confidence interval.
2. Do the same confidence interval, but assume $P_p = Q_p = .50$. What are your conclusions?

Exercise 8.8

In a training session, a group of managers will be asked to fill out an inventory designed to evaluate their ability to solve common management problems. The creators of this inventory have estimated that for the population of all managers, the mean is 10 and the standard deviation is 3. The possible scores on the inventory range from 0 to 20. Assume normality.

1. What is the probability of randomly selecting an individual with an inventory score of 15 or above? (outcome *A*)
2. What is the probability of selecting someone whose score is 7 or below? (outcome *B*)
3. What is the probability of randomly selecting someone with a score either 15 and above or 7 and below?
4. What is the probability of selecting someone with a score between 10 and 11? (outcome *C*)
5. What is the probability of selecting a person whose scores are either 15 and above, 7 or below, or between 10 and 11?
6. If outcome *A* remains a score of 15 or above and outcome *D* is the probability of selecting someone with a score between 10 and 12, what is the probability of randomly selecting a person whose score is either 15 and above or between 10 and 12?
7. Assume that a score of 15 or above is considered an indicator of a very good manager (outcome *A* above). If 43 managers filled out

the inventory, what is the probability of making 2 random selections from this group and obtaining 2 very good managers?

Exercise 8.9

1. How many different samples of size 3 can be drawn from a population of 6? If we disregard the order of selection, how many different samples can be drawn?
2. For a population $N = 8$ and sample size $K = 3$, what is the permutation and the combination?

NOTES

1. Be aware that there are several other ways of wording the Central Limit Theorem and the Law of Large Numbers. In addition, these two are sometimes combined into a single theorem.

2. W. S. Gosset, the expert, published his findings using the pen name *Student.* Thus this test is often called Student's *t.*

3. Unfortunately, it is hard to find more complete t tables that are relatively simple to read. Try H. Arkin and R. Colton (Eds.), *Tables for Statisticians* (College Outline Series), (New York: Barnes & Noble, 1963), p. 121, or H. R. Neave, *Statistical Tables for Mathematicians, Engineers, Economists and the Behavioural and Managerial Sciences* (London and Boston: Allen & Unwin, 1978), p. 41.

KEY CONCEPTS

independent samples/ independently drawn samples
dependent samples
matched pairs *t* test/ dependent samples (or paired difference) *t* test
pooled estimate of common variance
F test for homogeneity of variances
statistical power
type II or beta error
small effects
medium effects
large effects

9 Two-Sample t Tests

INTRODUCTION

Like the one-sample t test, the two-sample t test is a comparison of two means, except that both means are sample means. We no longer know population parameters, though as before, we must assume that the populations are normally distributed along the variable of interest, unless both samples are large enough to relax these normality assumptions. We compare the two sample means to generalize about a difference between the two respective population means. The null and alternative hypotheses are identical to those in the one-sample t test, and H_1 may be either nondirectional or directional. Since we often do not know or have no basis for estimating the population parameters necessary for the one-sample test, the two-sample t test is far more commonly utilized in actual research situations, where sample statistics alone compose the available data.

The two-sample t test is more complicated than the one-sample variety. For one thing, this is really a family of tests, and the researcher must select from this family the formula most appropriate to the data. As we will see, this selection is based on a variety of factors, such as the way in which the samples were selected and whether we may assume that the variances of the populations from which the samples were drawn are equal in magnitude. Second, in most instances the two-sample t formulas are more complex than the one-sample formula and take longer to calculate. Third, there is one instance in which the degrees of freedom formula is so onerous that many textbooks leave it out altogether or, as in this book, provide the formula but also give an easier-to-calculate approximation of it. Despite these difficulties, the calculations may still be made in a reasonable period of time, and because of its widespread usage, it is a particularly important test to understand.

INDEPENDENT SAMPLES VERSUS DEPENDENT SAMPLES

The first kind of two-sample t test we discuss assumes that there are two samples (or groups) being compared and that the samples are **independent;** that is, *the composition of one sample is in no way matched or paired to the composition of the other sample. Thus the two samples reflect two separate populations.* For example, we select a random sample of 50 men and another random sample of 50 women to investigate gender-determined views on social issues. Each sample is selected independently of the other. For each sample, we know its size (n), its mean ($\bar{x}$), and its variance (s^2) or alternatively its $\hat{\sigma}^2$. (For now, we assume we are using s^2 rather than $\hat{\sigma}^2$.)

Sample 1	*Sample 2*
Size n_1	Size n_2
Mean $\bar{x}_1$	Mean $\bar{x}_2$
Variance s_1^2	Variance s_2^2
↑	↑
Population 1	*Population 2*
μ_1 is unknown	μ_2 is unknown
σ_1^2 is unknown	σ_2^2 is unknown

Our H_0 is $\mu_1 = \mu_2$ and our H_1 nondirectional is $\mu_1 \neq \mu_2$ (or if H_1 is directional, either $\mu_1 > \mu_2$ or $\mu_1 < \mu_2$, depending on prior knowledge).

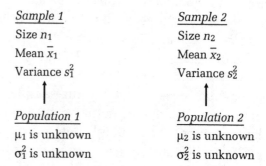

In experimental research, the procedure would be to take a pool of subjects (or as they are often called today, participants) and randomly assign some of the subjects to the experimental group, which will receive some experimental treatment. The remaining participants will compose the control group, which will not receive any treatment. The procedure for selecting the experimental group is identical to drawing a random sample from the available pool of participants. The random assignment of people to the experimental group implies that the control group will also have been randomly assigned. Suppose that out of a pool of 11 people, 6 are chosen to be in an experimental group that will watch a one-minute television commercial for a well-known product. The format of this commercial has been used previously with similar products and has a history of raising the viewers' levels of preference for them. Will it work here? The experimental group watches the ad and then rates the product, assigning a favorability score that ranges from 0 to 10. The five-member control group rates the product without seeing the advertisement.

The random assignment of the two groups should ensure that only viewing versus not viewing the commercial would account for any difference in favorability scores between the groups. A mean is calculated for each group and the two means are compared. If the means differ from one another, is it due to a real difference (as the result of the commercial) or is it due to random error or chance? In other words, if the experimental group's mean is higher than the other, there is a possibility that, by chance, more participants favorable to the advertised product were assigned to the experimental group than the control group. Thus the mean of the experimental group could be higher due to factors other than the commercial they watched.

We expect to find, in any sample that we draw or random assignment that we make, a certain degree of deviation from the population parameters due to sampling error. Our test of significance is designed to tell us whether the differences between the two sample means that we are comparing reflect a difference between their respective population means—a statistically significant difference—or merely reflect the expected sampling variation. In this case, the population means reflect the hypothetical means that would be generated if the experiment were to be repeated infinitely. If the latter case is correct, the observed difference between the two sample means is not statistically significant and we do not have enough evidence to conclude anything other than equality of the two respective population means.

Our experiment is represented as follows:

Group 1 *Experimental Group*	*Group 1* *Control Group*
Size $n_1 = 6$ Mean $\bar{x}_1$ Variance s_1^2	Size $n_2 = 5$ Mean $\bar{x}_2$ Variance s_2^2
↑	↑
Population 1 *All people having seen the ad*	*Population 2* *All people not having seen the ad*
μ_1 is unknown σ_1^2 is unknown	μ_2 is unknown σ_2^2 is unknown

Our H_0 is $\mu_1 = \mu_2$, and since we have prior evidence of the success of other commercials using the same format, our H_1 is directional: $\mu_1 > \mu_2$.

There is another type of research design, used mainly in experimental research, where the samples are **dependent.** Members of one sample are not selected independently, but are instead determined by the makeup of the other sample. We call this a **matched pairs** situation and another type of t test, a **dependent samples *t* test**, is used. To understand what we mean by a matched pair, imagine a situation where we start with a set of pairs of identical twins. One of each pair of twins is assigned to the experimental group and the other member of that pair goes to the control group. *Thus being included in the control group is dependent on one's twin being included in the experimental group.* Then, presumably, each group would share identical inherited traits, making any difference between the groups a function of environmental (as opposed to hereditary) differences; notably, the effects of the experiment. Sometimes the pairs are not twins but are related in other ways (for example, wives and their respective husbands). Or the pairs could be based on other factors such as age and race (for example, if one group includes a 35-year-old Caucasian female, the second group would include another 35-year-old Caucasian female.

A very common matched pairs situation is a before-after or repeated measures experiment. Imagine that each member of a panel was asked to rate each of two candidates contending for the same elective office. Later, the panel watches a 2-hour debate between the contenders. At the end of the debate, each panel member rates the candidates again. The matched pair is each person's "before" score with that person's "after" score. Any difference would presumably be due to the debate watched. We will return to this subject later when the dependent samples t test is discussed.

THE TWO-SAMPLE *t* TEST FOR INDEPENDENTLY DRAWN SAMPLES

In this instance there is no matching, but rather two independently drawn samples or randomly selected groups. The sampling distribution of which we make use is the t distribution, except it is derived from the sampling distribution of the differences between all theoretically possible pairs of sample means. (Consult a more advanced text for more details.) As with the one-sample t test, the difference between our two sample means is divided by the standard deviation of the sampling distribution, the standard error, to find *t*. Now, however, our standard error is the standard deviation of sample mean differences, a different standard error than the $\sigma/\sqrt{n}$ or the $s/\sqrt{n-1}$ used in the one-sample *t* test.

This raises a new complication: The standard error we seek is an estimate based on the variances of our two samples. If the two sample variances are close in magnitude, we may assume that both parent populations from which the samples were drawn have the same variance, that is, $\sigma_1^2 = \sigma_2^2$. If these population variances are the same, we can calculate the standard error of our *t* formula using what we call a **pooled estimate of common variance.** *This is based on a weighted average of our two sample variances being used to estimate the population variance in finding the standard error.* If we can make use of this pooled estimate, we generally will have a greater opportunity to reject the null hypothesis than would be the case when equal population variances cannot be assumed. If, however, we cannot assume equal population variances, that is, $\sigma_1^2 \neq \sigma_2^2$, we must use a different formula for finding the standard error, and thus, a different *t* formula.

To determine which t formula is most appropriate, we can use the ***F test for homogeneity of variances***. (The test is shorter than its title!) Computer programs that run *t* tests will usually run both the equal and unequal population variance *t* formulas and also do an *F* test to help the reader ascertain which obtained *t* value is more accurate.

To avoid confusion, let us stop here and lay out a set of steps for the two-sample *t* test (independent samples) and then illustrate these steps with an example. At the appropriate points, the formulas will be given and explained for the *F* test for homogeneity of variances, the equal population variance *t* test, and the unequal population variance *t* test. First, the steps:

1. Write out H_0 and H_1 for the original problem, the comparison of the two sample means.
2. For each sample, determine its *n*, its $\bar{x}$, and its variance, s^2.
3. To determine which *t* formula to use (that for equal population variances or that for unequal population variances) do the *F* test for homogeneity of variances.
   a. Write out H_0 and H_1 for the *F* test (these are not the same as the ones for the *t* test).
   b. Calculate *F* and its two degrees of freedom.
   c. Compare the obtained *F* to $F_{critical}$ .05 level, taken from a table of critical values of *F*.

d. If $F_{\text{obtained}} \geq F_{\text{critical}}$, assume unequal population variances. If $F_{\text{obtained}} < F_{\text{critical}}$, assume equal population variances.

4. Perform the appropriate t test as determined by the F test.

Example 1. Let us work through these steps, using the example of the TV commercial. Recall that 6 people (the experimental group) will see the commercial and then evaluate the product's favorability. The other 5 (the control group) will evaluate favorability without seeing the commercial. Remember also that because of previous success with the same format for the ad, we expect favorability to rise once the viewing is complete, so our H_1 is directional. Note also that because of the smallness of our groups ($n_1 = 6$ and $n_2 = 5$), we must assume that favorability is normally distributed in our two populations.

1. Write out H_0 and H_1.

$$H_0: \mu_1 = \mu_2$$

$$H_1: \mu_1 > \mu_2$$

(Group 1 is the experimental group.)

2. Determine, n, $\bar{x}$, and s^2 for each sample. Following are the data:

Sample 1 Saw Commercial $x_1 =$	Sample 2 Did Not See Commercial $x_2 =$
10	8
6	3
8	5
7	6
9	7
7	

$n_1 = 6$ $\quad\sum x_1 = 47$

$n_2 = 5$ $\quad\sum x_2 = 29$

$$\bar{x}_1 = \frac{\sum x_1}{n_1} = \frac{47}{6} = 7.83 \qquad \bar{x}_2 = \frac{\sum x_2}{n_2} = \frac{29}{5} = 5.80$$

To get the sample variances using the computational formula, we must square our two sets of scores. (Later on, we will ease your burden by providing you with the variances in advance.)

$x_1^2 =$	$x_2^2 =$
100	64
36	9
64	25
49	36
81	49
49	
$\sum x_1^2 = 379$	$\sum x_2^2 = 183$

$$s_1^2 = \frac{\sum x_1^2 - \frac{\left(\sum x_1\right)^2}{n_1}}{n_1} \qquad s_2^2 = \frac{\sum x_2^2 - \frac{\left(\sum x_2\right)^2}{n_2}}{n_2}$$

$$s_1^2 = \frac{379 - \frac{(47)^2}{6}}{6} \qquad s_2^2 = \frac{183 - \frac{(29)^2}{5}}{5}$$

$$= \frac{379 - \frac{2206}{6}}{6} \qquad = \frac{183 - \frac{841}{5}}{5}$$

$$= \frac{379 - 368.17}{6} \qquad = \frac{183 - 168.2}{5}$$

$$= 1.81 \qquad = 2.96$$

Summarizing our results:

Sample 1 *Saw Commercial*	*Sample 2* *Did Not See Commercial*
$n_1 = 6$	$n_2 = 5$
$\bar{x}_1 = 7.83$	$\bar{x}_2 = 5.80$
$s_1^2 = 1.81$	$s_2^2 = 2.96$

3. We perform the F test for homogeneity of variances.
   a. In the F test we are doing, the null hypothesis is that the population variances are equal and the alternative hypothesis is that they are not.

$$H_0: \sigma_1^2 = \sigma_2^2$$

$$H_1: \sigma_1^2 \neq \sigma_2^2$$

(In the F test, H_1 is always nondirectional.)

b. To calculate F, divide the larger of the two sample variances by the smaller one.

$$F = \frac{s^2_{\text{larger}}}{s^2_{\text{smaller}}}$$

In this case, the larger s^2 is the one for Group 2, the control group. Thus

$$F_{\text{obtained}} = \frac{s_2^2}{s_1^2} = \frac{2.96}{1.81} = 1.64$$

The F has two degrees of freedom, one associated with the numerator and one associated with the denominator. In each case, we subtract one from the sample size. The numerator degrees of freedom is one less than the size of the sample having the larger variance. The denominator degrees of freedom is one less than the size of the sample having the smaller variance. In this case:

$$\text{df numerator} = 5 - 1 - 4$$

$$\text{df denominator} = 6 - 1 = 5$$

c. Our obtained F is then compared to F_{critical}, .05 level, $df = 4$ and 5. Table 9.1, a portion of a fuller F table to be presented later, gives $F_{\text{criticals}}$ for the .05 level only. In the table, n_1 means the numerator degrees of freedom. (n_1 and n_2 here mean degrees of freedom, not sample sizes!) We move along the row until we find the column for the appropriate degrees of freedom. In this case, $n_1 = 4$. We go down the n_2 column until we find the row for our denominator degrees of freedom, $n_2 = 5$. We move along the $n_2 = 5$ row until it intersects the $n_1 = 4$ column. The number at that intersection, 5.19, is our F_{critical}.[1]

d. To reject the H_0 for the F test, F_{obtained} must equal or exceed this F_{critical}. But $F_{\text{obtained}} = 1.64$ and F_{critical} is 5.19. $F_{\text{obtained}} = 1.64 < F_{\text{critical}}$, .05, ($df = 4$ and 5) $= 5.19$. We cannot reject H_0, so we now perform the t test designed for equal population variances, to be presented in step 4.

Table 9.1 Critical Values of *F* (.05 level only) for the *F* Test for Homogeneity of Variances

$n_2 \backslash n_1$	1	2	3	4	5	6	8	12	24	∞
1	161.40	199.50	215.70	224.60	230.20	234.00	238.90	243.90	249.00	254.30
2	18.51	19.00	19.16	19.25	19.30	19.33	19.37	19.41	19.45	19.50
3	10.13	9.55	9.28	9.12	9.01	8.94	8.84	8.74	8.64	8.53
4	7.71	6.94	6.59	6.39	6.26	6.16	6.04	5.91	5.77	5.63
5	6.61	5.79	5.41	5.19	5.05	4.95	4.82	4.68	4.53	4.36
6	5.99	5.14	4.76	4.53	4.39	4.28	4.15	4.00	3.84	3.67
7	5.59	4.74	4.35	4.12	3.97	3.87	3.73	3.57	3.41	3.23
8	5.32	4.46	4.07	3.84	3.69	3.58	3.44	3.28	3.12	2.93
9	5.12	4.26	3.86	3.63	3.48	3.37	3.23	3.07	2.90	2.71
10	4.96	4.10	3.71	3.48	3.33	3.22	3.07	2.91	2.74	2.54
11	4.84	3.98	3.59	3.36	3.20	3.09	2.95	2.79	2.61	2.40
12	4.75	3.88	3.49	3.26	3.11	3.00	2.85	2.69	2.50	2.30
13	4.67	3.80	3.41	3.18	3.02	2.92	2.77	2.60	2.42	2.21
14	4.60	3.74	3.34	3.11	2.96	2.85	2.70	2.53	2.35	2.13
15	4.54	3.68	3.29	3.06	2.90	2.79	2.64	2.48	2.29	2.07
16	4.49	3.63	3.24	3.01	2.85	2.74	2.59	2.42	2.24	2.01
17	4.45	3.59	3.20	2.96	2.81	2.70	2.55	2.38	2.19	1.96
18	4.41	3.55	3.16	2.93	2.77	2.66	2.51	2.34	2.15	1.92
19	4.38	3.52	3.13	2.90	2.74	2.63	2.48	2.31	2.11	1.88
20	4.35	3.49	3.10	2.87	2.71	2.60	2.45	2.28	2.08	1.84
21	4.32	3.47	3.07	2.84	2.68	2.57	2.42	2.25	2.05	1.81
22	4.30	3.44	3.05	2.82	2.66	2.55	2.40	2.23	2.03	1.78
23	4.28	3.42	3.03	2.80	2.64	2.53	2.38	2.20	2.00	1.76
24	4.26	3.40	3.01	2.78	2.62	2.51	2.36	2.18	1.98	1.73
25	4.24	3.38	2.99	2.76	2.60	2.49	2.34	2.16	1.96	1.71
26	4.22	3.37	2.98	2.74	2.59	2.47	2.32	2.15	1.95	1.69
27	4.21	3.35	2.96	2.73	2.57	2.46	2.30	2.13	1.93	1.67
28	4.20	3.34	2.95	2.71	2.56	2.44	2.29	2.12	1.91	1.65
29	4.18	3.33	2.93	2.70	2.54	2.43	2.28	2.10	1.90	1.64
30	4.17	3.32	2.92	2.69	2.53	2.42	2.27	2.09	1.89	1.62
40	4.08	3.23	2.84	2.61	2.45	2.34	2.18	2.00	1.79	1.51
60	4.00	3.15	2.76	2.52	2.37	2.25	2.10	1.92	1.70	1.39
120	3.92	3.07	2.68	2.45	2.29	2.17	2.02	1.83	1.61	1.25
∞	3.84	2.99	2.60	2.37	2.21	2.09	1.94	1.75	1.52	1.00

SOURCE: Table 9.1 is abridged from Table V of R. A. Fisher and F. Yates, *Statistical Tables for Biological, Agricultural and Medical Research* (6th ed.), published by Longman Group UK Ltd., 1974, by permission of the authors and the publisher.
NOTE: Values of n_1 and n_2 represent the degrees of freedom associated with the larger and smaller estimates of variance, respectively. $p = .05$.

4. Since equal population variances may be assumed, we use the following formula in which the denominator is the pooled variance estimate.

$$t = \frac{\bar{X}_1 - \bar{X}_2}{\sqrt{\left(\frac{n_1 s_1^2 + n_2 s_2^2}{n_1 + n_2 - 2}\right)\left(\frac{1}{n_1} + \frac{1}{n_2}\right)}}$$

$$\text{df} = n_1 + n_2 - 2$$

It's easiest to first do the components of the t formula and then put them together.

$$\bar{X}_1 - \bar{X}_2 = 7.83 - 5.80 = 2.03$$

$$\frac{n_1 s_1^2 + n_2 s_2^2}{n_1 + n_2 - 2} = \frac{6(1.81) + 5(2.96)}{6 + 5 - 2} = \frac{10.86 + 14.80}{11 - 2} = \frac{25.66}{9} = 2.85$$

$$\frac{1}{n_1} + \frac{1}{n_2} = \frac{1}{6} + \frac{1}{5} = 0.17 + 0.20 = 0.37$$

$$\sqrt{\left(\frac{n_1 s_1^2 + n_2 s_2^2}{n_1 + n_2 - 2}\right)\left(\frac{1}{n_1} + \frac{1}{n_2}\right)} = \sqrt{(2.85)(0.37)} = \sqrt{1.0545} = 1.0268 = 1.03$$

Thus

$$t = \frac{\bar{X}_1 - \bar{X}_2}{\sqrt{\left(\frac{n_1 s_1^2 + n_2 s_2^2}{n_1 + n_2 - 2}\right)\left(\frac{1}{n_1} + \frac{1}{n_2}\right)}} = \frac{2.03}{1.03} = 1.9708 = 1.971$$

$$t_{\text{obt}} = 1.971$$

The $df = n_1 + n_2 - 2 = 6 + 5 - 2 = 11 - 2 = 9$. We use the critical values of t for a directional H_1. See Table 9.2.

At .05 level, $t_{\text{critical}}\ (df = 9) = 1.833 < 1.971$ reject H_0

At .025 level, $t_{\text{critical}}\ (df = 9) = 2.262 > 1.971$ $p < .05$

Note that if H_1 had been nondirectional,

At .05 level, $t_{\text{critical}}\ (df = 9) = 2.262 > 1.971$

and we would not have been able to reject H_0.

Table 9.2 Distribution of *t*

	Level of significance for one-tailed test					
	.10	*.05*	*.025*	*.01*	*.005*	*.0005*
	Level of significance for two-tailed test					
df	*.20*	*.10*	*.05*	*.02*	*.01*	*.001*
1	3.078	6.314	12.706	31.821	63.657	636.619
2	1.886	2.920	4.303	6.965	9.925	31.598
3	1.638	2.353	3.182	4.541	5.841	12.941
4	1.533	2.132	2.776	3.747	4.604	8.610
5	1.476	2.015	2.571	3.365	4.032	6.859
6	1.440	1.943	2.447	3.143	3.707	5.959
7	1.415	1.895	2.365	2.998	3.499	5.405
8	1.397	1.860	2.306	2.896	3.355	5.041
9	1.383	1.833	2.262	2.821	3.250	4.781
10	1.372	1.812	2.228	2.764	3.169	4.587
11	1.363	1.796	2.201	2.718	3.106	4.437
12	1.356	1.782	2.179	2.681	3.055	4.318
13	1.350	1.771	2.160	2.650	3.012	4.221
14	1.345	1.761	2.145	2.624	2.977	4.140
15	1.341	1.753	2.131	2.602	2.947	4.073
16	1.337	1.746	2.120	2.583	2.921	4.015
17	1.333	1.740	2.110	2.567	2.898	3.965
18	1.330	1.734	2.101	2.552	2.878	3.922
19	1.328	1.729	2.093	2.539	2.861	3.883
20	1.325	1.725	2.086	2.528	2.845	3.850
21	1.323	1.721	2.080	2.518	2.831	3.819
22	1.321	1.717	2.074	2.508	2.819	3.792
23	1.319	1.714	2.069	2.500	2.807	3.767
24	1.318	1.711	2.064	2.492	2.797	3.745
25	1.316	1.708	2.060	2.485	2.787	3.725
26	1.315	1.706	2.056	2.479	2.779	3.707
27	1.314	1.703	2.052	2.473	2.771	3.690
28	1.313	1.701	2.048	2.467	2.763	3.674
29	1.311	1.699	2.045	2.462	2.756	3.659
30	1.310	1.697	2.042	2.457	2.750	3.646
40	1.303	1.684	2.021	2.423	2.704	3.551
60	1.296	1.671	2.000	2.390	2.660	3.460
120	1.289	1.658	1.980	2.358	2.617	3.373
∞	1.282	1.645	1.960	2.326	2.576	3.291

Since H_0 has been rejected, we conclude our alternative hypothesis of $\mu_1 > \mu_2$. As in previous instances, this particular commercial resulted in increased favorability ratings for the product featured in the commercial. Since H_0 has been rejected, we continue to assume that if the entire consumer population had viewed the advertisement, their mean support score μ_1 would also increase.

Remember that there are risks in using directional alternative hypotheses. Not only do we need prior evidence of the assumed direction, as was the case here, we must always make sure that our findings are consistent with the direction assumed. Suppose $\bar{x}_1$ had been 3.77 instead of 7.83, but we retained the $\mu_1 > \mu_2$ alternative hypothesis. The numerator of the t formula would be $3.77 - 5.80 = -2.03$, and t would be -1.971. Using its absolute value of 1.971, we would reject H_0 in the same way as just done. Yet, if anything, our two sample means suggest an H_1 of $\mu_1 < \mu_2$, and we have findings inconsistent with the original H_1. Even though we reject H_0, we cannot conclude $\mu_1 > \mu_2$. In this case, our only option would have been a two-tailed H_1, but we have already seen that, in such a case, we cannot reject H_0.

Note, finally, that it is conceivable—albeit improbable—that really $\mu_1 > \mu_2$ but, due to sampling error, $\bar{x}_1 < \bar{x}_2$. However, we have no basis for knowing that fact when we do our study. Accordingly, if $\bar{x}_1 < \bar{x}_2$ but H_1 said $\mu_1 > \mu_2$, or the reverse, $\bar{x}_1 > \bar{x}_2$ but H_1 stated $\mu_1 < \mu_2$, do *not* proceed with a one-tailed test.

To summarize to this point, we established H_0 and H_1 for our data, found n, $\bar{x}$, and s^2 for each sample, and did the F test for homogeneity of variances to determine the appropriate t formula to use. In this case, the F test led us to use the t formula where equal population variances are assumed. Using the appropriate t test, we were able to reject H_0 with a probability of $p < .05$ and conclude that in the population, viewing the commercial enhances support for the product featured. Now, let us see what would happen if the F test concluded unequal population variances.

Example 2. Suppose Sample 1 remained the same, but the scores for Sample 2 were as follows:

	$x_2 =$
	10
	1
	3
	6
$n_2 = 5$	9
	$\sum x_2 = 29$

$$\bar{x}_2 = \frac{\sum x_2}{n_2} = \frac{29}{5} = 5.80$$

The sample size and sample mean stay the same as before, but the sample variance is now larger.

$$
\begin{array}{rr}
x_2^2 = & \\
\hline
 & 100 \\
 & 1 \\
 & 9 \\
 & 36 \\
 & 81 \\
\hline
\sum x_2^2 = & 227
\end{array}
$$

Thus,

$$s_2^2 = \frac{\sum x_2^2 - \frac{\left(\sum x_2\right)^2}{n_2}}{n_2} = \frac{227 - \frac{(29)^2}{5}}{5} = \frac{227 - \frac{841}{5}}{5}$$

$$= \frac{227 - 168.2}{5} = \frac{58.8}{5} = 11.76$$

Redoing the F test for homogeneity of variances,

$$F = \frac{s_{1g}^2}{s_{sm}^2} = \frac{11.76}{1.81} = 6.497 \qquad \mathrm{df}_n = 5 - 1 = 4 \qquad \mathrm{df}_d = 6 - 1 = 5$$

$$H_0\text{: } \sigma_1^2 = \sigma_2^2$$

At $df = 4$ and 5,

$$F_{\text{critical}}, .05 = 5.19 < 6.497 \qquad \text{reject } H_0 \quad p < .05$$

We may not assume equal population variances.

Where population variances are assumed to be unequal, we use the following formula.

$$t = \frac{\bar{x}_1 + \bar{x}_2}{\sqrt{\frac{s_1^2}{n_1 - 1} + \frac{s_2^2}{n_2 - 1}}} \qquad \text{df = whichever is smaller, } n_1 \text{ or } n_2$$

As before, we first calculate the components.

$$\bar{x}_1 + \bar{x}_2 = 7.83 - 5.80 = 2.03$$

$$\frac{s_1^2}{n_1 - 1} = \frac{1.81}{6-1} = \frac{1.81}{5} = 0.36$$

$$\frac{s_2^2}{n_2 - 1} = \frac{11.76}{5-1} = \frac{11.76}{4} = 2.94$$

$$\sqrt{\frac{s_1^2}{n_1 - 1} + \frac{s_2^2}{n_2 - 1}} = \sqrt{0.36 + 2.94} = \sqrt{3.3} = 1.82$$

$$t = \frac{\bar{x}_1 + \bar{x}_2}{\sqrt{\frac{s_1^2}{n_1 - 1} + \frac{s_2^2}{n_2 - 1}}} = \frac{2.03}{1.82} = 1.115$$

Using as *df* the smaller value of n_1 or n_2, $df = 5$.

For a directional H_1, t_{critical}, .05 level, $(df = 5) = 2.015 > 1.115$. We cannot reject H_0.

The degrees of freedom for this problem, the lesser of n_1 or n_2, is only a substitute for a complex degrees of freedom formula utilized by packaged statistical computer programs. This larger formula is a more accurate approximation but is often too cumbersome for noncomputer applications.

$$df = \frac{\left(\frac{s_1^2}{n_1 - 1} + \frac{s_2^2}{n_2 - 1}\right)^2}{\left[\frac{\left(\frac{s_1^2}{n_1 - 1}\right)^2}{(n_1 - 1)}\right] + \left[\frac{\left(\frac{s_2^2}{n_2 - 1}\right)^2}{(n_2 - 1)}\right]}$$

If we did use it in the above *t* test, we would also work it in stages.

$$\frac{s_1^2}{n_1 - 1} = \frac{1.81}{6-1} = \frac{1.81}{5} = 0.36 \qquad \left(\frac{s_1^2}{n_1 - 1}\right)^2 = (.36)^2 = 0.13$$

$$\frac{s_2^2}{n_2 - 1} = \frac{11.76}{5-1} = \frac{11.76}{4} = 2.94 \qquad \left(\frac{s_2^2}{n_2 - 1}\right)^2 = (2.94)^2 = 8.64$$

$$\left(\frac{s_1^2}{n_1 - 1} + \frac{s_2^2}{n_2 - 1}\right)^2 = (0.36 + 2.94)^2 = (3.30)^2 = 10.89$$

$$\frac{\left(\frac{s_1^2}{n_1 - 1}\right)^2}{(n_1 - 1)} = \frac{0.13}{6 - 1} = \frac{0.13}{5} = .026 \approx .03$$

$$\frac{\left(\frac{s_2^2}{n_2 - 1}\right)^2}{(n_2 - 1)} = \frac{8.64}{5 - 1} = \frac{8.64}{4} = 2.16$$

Thus

$$df = \frac{\left(\frac{s_1^2}{n_1 - 1} + \frac{s_2^2}{n_2 - 1}\right)^2}{\left[\frac{\left(\frac{s_1^2}{n_1 - 1}\right)^2}{(n_1 - 1)}\right] + \left[\frac{\left(\frac{s_2^2}{n_2 - 1}\right)^2}{(n_2 - 1)}\right]} = \frac{10.89}{.03 + 2.16} = \frac{10.89}{2.19} = 4.97$$

A computer would list *df* as 4.97.

If we were doing this by hand, we would round *down* (not up) and use $df = 4$ in our *t* table. Again, this makes it harder to reject H_0 than it would have been had we been able to round up to 5 degrees of freedom. We get essentially the same results as before: t_{critical}, one-tailed, .05 level, $(df = 5) = 2.132 > 1.115$ and we cannot reject H_0. The ad has no effect on favorability scores (in this example, as modified for the assumption of unequal population variances).

ADJUSTMENTS FOR SIGMA-HAT SQUARED ($\hat{\sigma}^2$)

As was the case with the one-sample *t* test, it is possible that instead of s^2 for each sample, $n - 1$ replaced n in the denominator of the formula and consequently what was calculated was $\hat{\sigma}^2$, not s^2. We then would need to modify our *t* formulas accordingly.

In the first example, where population variances were assumed equal, we would find:

Sample 1	*Sample 2*
$n_1 = 6$	$n_2 = 5$
$\bar{x}_1 = 7.83$	$\bar{x}_2 = 5.80$
$\sigma_1^2 = 2.17$	$\sigma_2^2 = 3.70$

We would make the following modification in the t formula for equal population variances:

$$t = \frac{\bar{x}_1 - \bar{x}_2}{\sqrt{\left[\frac{(n_1-1)\sigma_1^2 + (n_2-1)\sigma_2^2}{n_1+n_2-2}\right]\left[\frac{1}{n_1}+\frac{1}{n_2}\right]}} \qquad \begin{aligned} df &= n_1 + n_2 - 2 \\ &= 11 - 2 \\ &= 9 \end{aligned}$$

As before, $\bar{x}_1 - \bar{x}_2 = 2.03$ and $1/n_1 + 1/n_2 = 0.37$. Recalculating the remaining expression to adjust for $\hat{\sigma}_1^2$.

$$\frac{(n_1-1)\sigma_1^2 + (n_2-1)\sigma_2^2}{n_1+n_2-2} = \frac{(6-1)(2.17)+(5-1)(3.7)}{6+5-2}$$

$$= \frac{5(2.17)+4(3.7)}{11-2} = \frac{10.85+14.80}{9}$$

$$= \frac{25.65}{9} = 2.85$$

just as it was in the original formula using s^2.

$$t = \frac{2.03}{\sqrt{(2.85)(0.37)}} = \frac{2.03}{\sqrt{1.0545}} = \frac{2.03}{1.03} = 1.9708$$

$$t = 1.971$$

In the example where unequal population variances were assumed, we would find:

Sample 1	*Sample 2*
$n_1 = 6$	$n_2 = 5$
$\bar{x}_1 = 7.83$	$\bar{x}_2 = 5.80$
$\sigma_1^2 = 2.17$	$\sigma_2^2 = 14.7$

We modify the *t* formula for unequal population variances as follows:

$$t = \frac{\bar{x}_1 - \bar{x}_2}{\sqrt{\frac{\sigma_1^2}{n_1} + \frac{\sigma_2^2}{n_2}}}$$

df = the lesser of n_1 or n_2.

$\bar{x}_1 - \bar{x}_2$ is still 2.03. Recalculating the denominator to adjust for σ^2,

$$\sqrt{\frac{\sigma_1^2}{n_1} + \frac{\sigma_2^2}{n_2}} = \sqrt{\frac{2.17}{6} + \frac{14.7}{5}} = \sqrt{.36 + 2.94} = \sqrt{3.3} = 1.82$$

So, $t = 2.03/1.82 = 1.115$ as it was when we used s_1^2 and s_2^2.

Finally, the alternative longer degrees of freedom formula for unequal variance *t* tests would become:

$$df = \frac{\left(\frac{\sigma_1^2}{n_1} + \frac{\sigma_2^2}{n_2}\right)^2}{\left[\frac{\left(\frac{\sigma_1^2}{n_1}\right)^2}{(n_1 - 1)}\right] + \left[\frac{\left(\frac{\sigma_2^2}{n_2}\right)^2}{(n_2 - 1)}\right]}$$

As before, $df = 4.97$.

INTERPRETING A COMPUTER GENERATED *t* TEST

You can see with this test that the hand calculations become cumbersome. Thus it is no surprise that researchers rely more and more on computers to do their calculating for them. Often they make use of prepared library programs ("canned" programs) for statistical routines such as SAS, SPSS, and Minitab as well as others designed for PCs. Though you need to consult the appropriate manual to understand printouts from the programs available to you, there is enough similarity from one program to another that it is worthwhile to present one here as a model for understanding all of them. Figure 9.1 presents the SAS TTEST printouts for Example 1 in this chapter.

Starting on the left of Figure 9.1, some general statistical information is presented. The dependent variable has been named SCORE by the researcher. For each group the printout lists its size, mean, standard

SAS

TTEST PROCEDURE

VARIABLE: SCORE

GROUP	N	MEAN	STD DEV	STD ERROR	MINIMUM	MAXIMUM	VARIANCES	T	DF	PROB > |T|
1	6	7.83333333	1.47196014	0.60092521	6.00000000	10.00000000	UNEQUAL	1.9377	7.4	0.0916
2	5	5.80000000	1.92353841	0.86023253	3.00000000	8.00000000	EQUAL	1.9897	9.0	0.0778

FOR H0: VARIANCES ARE EQUAL, F′ = 1.71 WITH 4 AND 5 DF, PROB > F′ = 0.5674

Figure 9.1
SAS Printout for Example 1

deviation (SAS uses the $\hat{\sigma}^2$ formula, not s), standard error ($\hat{\sigma}^2/\sqrt{n}$), minimum value, and maximum value. Then comes the label VARIANCES and under it UNEQUAL. To the right of UNEQUAL we find a T value, to its right the degrees of freedom, and to the right of DF, under PROB > |T|, the probability—the exact p value (not a $< .05$ statement but the actual probability). The t, df, and p values in this line are all based on the formula for unequal population variances. On the line below, under EQUAL, we find the t, df, and p values from the formula that assumes equal population variances. The degrees of freedom for the unequal population variances t test is based on the long formula. Recall that for Example 1 we only calculated the t for equal population variances. Our calculated t of 1.971 differs from SAS's 1.9897 since we used fewer decimal places in calculating t. Our df and SAS's are the same, 9. The probability on the printout is the two-tailed probability .0778, not the one-tailed probability that we used in our hand calculations. To get the one-tailed probability, divide the .0778 by 2. Thus the exact directional probability is .0389—less than .05 and greater than .01.

Below the group information, we see the F test for homogeneity of variances. $F' = 1.71$ is greater than the hand calculated F of 1.64 since SAS is dividing $\hat{\sigma}^2_{\text{large}}$ by $\hat{\sigma}^2_{\text{small}}$ whereas we divided s^2's. After a listing of the degrees of freedom, an exact probability of F is given: PROB > F_1 = 0.5674, much greater than .05.

Since exact probabilities are supplied throughout, we can read our test results directly from the printout.

1. Examine the probability of F.
   a. If PROB > F′ exceeds .05, we cannot reject a null hypothesis (for the F test) of equal population variances. Use the t test results to the right of EQUAL.
   b. If PROB > F′ is less than or equal to .05, we reject the F test's H_0 and assume unequal population variances. Use the t test results to the right of UNEQUAL.
2. Examine the appropriate (EQUAL or UNEQUAL) probability of t.
   a. If PROB > |T| exceeds .05 we cannot reject the initial null hypothesis (the one for the t test). The difference is not statistically significant, $\mu_1 = \mu_2$.
   b. If PROB > |T| is less than or equal to .05, we reject the t test's null hypothesis and conclude H_1. (*Note:* If H_1 is directional, divide the PROB > |T| value by 2 before working step 2.) For this specific problem, PROB > F′ = 0.5674 exceeds .05. We

cannot reject H_0: $\sigma_1^2 = \sigma_2^2$ so we will use the *t* test for equal population variances. Looking to the right of the word EQUAL and under PROB > |T|, we find a value of 0.0778. Since our original H_1 was one-tailed, divide the probability by 2. The result is .0389 (as we previously demonstrated). Since .0389 is less than .05, we reject our original H_0 of $\mu_1 = \mu_2$ and conclude H_1: $\mu_1 > \mu_2$.

THE TWO-SAMPLE *t* TEST FOR DEPENDENT SAMPLES

In the case of a before-after experiment or a situation where members of one group are matched to specific members of another group, the samples are not independent, and a separate *t* test must be performed. While we call this test a two-sample *t* test for dependent samples (or matched pairs), actually the test is *a one-sample t test applied to the differences in each pair of scores.* Accordingly, the test is also called a **paired difference *t* test.**

Suppose in Example 1 that before viewing the commercial our 6 subjects rated the product (before scores), then saw the ad and rated the product again. The before and after scores are presented below.

Before	*After*
5	10
5	6
7	8
8	7
7	9
3	7
$\sum x = 35$	$\sum x = 47$
$\bar{x}_b = 5.83$	$\bar{x}_a = 7.83$

For our sample of 6 people, there are 6 pairs of scores, $n_p = 6$, where n_p is the number of pairs. The first subject listed scored 5 before and 10 after seeing the ad.

Note that in all but one case, favorability rose after viewing the commercial. The mean favorability score rose from 5.83 (before) to 7.83 (after the viewing). We are trying to demonstrate that viewing the commercial causes the mean favorability to change. Based on previous positive results with this same format, we expect the change to be an increase in favorability. Thus we use a one-tailed H_1.

$$H_0: \mu_{\text{before}} = \mu_{\text{after}}$$

$$H_1: \mu_{\text{before}} < \mu_{\text{after}}$$

To what extent are we safe in assuming that the increase for our six subjects reflects an increase among the population of all people who would view the commercial? Since the test we will be performing deals with the differences in observed scores, the null and alternative hypotheses could be written in terms of these differences.

$$H_0: \mu_D = 0$$

$$H_1: \mu_D > 0$$

By μ_D we mean the mean of all the differences in scores in the population. If we subtracted everybody's before score from everybody's after score, we would have a difference score for everyone. If a score goes up from before to after, the difference will be a positive number; if the score goes down, the difference will be negative. Since we expect the scores to rise, our directional H_1 is $\mu_D > 0$. If we had made no directionality assumption, our H_1 would be $\mu_D \neq 0$.

We use each difference of scores in our sample (D) as the basis for this *t* test.

$$t = \frac{\overline{D} - \mu_D}{S_D / \sqrt{n_p - 1}}$$

$$df = n_p - 1$$

where

n_p is the number of matched pairs of scores in the sample.

D is the difference (after score – before score) for each pair of scores in the sample.

$\overline{D}$ is the mean of all our sample's difference scores (retaining their + or – signs).

S_D is the sample standard deviation of the difference scores:

$$S_D = \sqrt{\frac{\sum (D - \overline{D})^2}{n_p}}$$

μ_D is the mean of the difference scores for all possible pairs in the population. As per the null hypothesis, μ_D is assumed to equal zero. Accordingly, we may delete it from the formula.

Thus

$$t = \frac{\overline{D}}{S_D / \sqrt{n_p - 1}}$$

$$df = n_p - 1$$

where

$$S_D = \sqrt{\frac{\sum (D - \overline{D})^2}{n_p}}$$

We first find S_D, following these steps:

1. Subtract each before score from each after score to get each D.
2. Add the D's algebraically to get ΣD and divide ΣD by n_p to get $\overline{D}$.
3. Subtract $\overline{D}$ from each D to get $(D - \overline{D})$.
4. Square each $(D - \overline{D})$ to get $(D - \overline{D})^2$.
5. Add together each $(D - \overline{D})^2$ to get $\Sigma(D - \overline{D})^2$.

After finding $\Sigma(D - \overline{D})^2$, plug that figure into the S_D formula.

6. Divide $\Sigma(D - \overline{D})^2$ by n_p to get the variance of sample differences.
7. Take the square root of that variance to reveal S_D.

These steps are performed below:

		Step 1	*Step 3*	*Step 4*	
Before	*After*	$D =$	$D - \overline{D} =$	$(D - \overline{D})^2 =$	
5	10	5	3	9	
5	6	1	−1	1	
7	8	1	−1	1	
8	7	−1	−3	9	
7	9	2	0	0	
3	7	4	2	4	
		$\Sigma D = 13 - 1 = 12$		$\Sigma (D - \overline{D})^2 = 24$	*Step*

Step 2

$$\overline{D} = \frac{\sum D}{n_p} = \frac{12}{6} = 2$$

Steps 6 & 7

$$S_D = \sqrt{\frac{\sum (D - \overline{D})^2}{n_p}} = \sqrt{\frac{24}{6}} = \sqrt{4} = 2$$

8. Having established that $S_D = 2$, we may now complete the t formula.

$$t = \frac{\overline{D}}{S_D/\sqrt{n_p - 1}} = \frac{2}{2/\sqrt{6-1}} = \frac{2}{2/\sqrt{5}} = \frac{2\sqrt{5}}{2} = \sqrt{5} = 2.236$$

Thus

$$t_{\text{obtained}} = 2.236 \qquad df = n_p - 1 = 6 - 1 = 5$$

Examining the one-tailed $t_{\text{criticals}}$ at 5 degrees of freedom:

t_{critical} .05 level = 2.015 < 2.236 reject H_0

t_{critical} .025 level = 2.571 > 2.236 $p < .05$

We conclude that in the population ($p < .05$) viewing the commercial raises favorability ratings for the advertised product.

Many texts also provide alternative computational formulas for the dependent samples test. Since this procedure is not a widely used test in social science disciplines where experimentation is not widespread, we will not cover those computational formulas here.

Statistical Power

When we initially discussed the one-sample z and t tests, we noted that two factors have an influence on the magnitude of the statistic generated: the difference between the two means and the size of the sample.

This can be seen in the original z formula as algebraically transformed below.

$$z = \frac{\bar{x} - \mu}{\sigma/\sqrt{n}} = \frac{(\bar{x} - \mu)\sqrt{n}}{\sigma}$$

An increase in either the size of $\bar{x} - \mu$ or the size of n will enlarge the numerator and thus the final value of z. In the case of the two-sample t test, the same is true except $\bar{x} - \mu$ is replaced by $\bar{x}_1 - \bar{x}_2$ and the sample sizes may apply to either n_1, n_2, or both.

An additional factor impacting our ability to reject the null hypothesis is the nature of the selected test of significance itself. This brings us to the concept of **statistical power,** *the likelihood that our test will reject the null hypothesis when, in fact,* H_1 *really is true*. How likely is our test to reject a null hypothesis when the null hypothesis is false and "ought to be" rejected?

Up to this point we have been concerned about type I or alpha error, the probability of mistakenly rejecting a true null hypothesis. However, there is another type of error, a **type II** or **beta error,** which exists but is not reported. *Beta is the probability that the null hypothesis is really false—*H_1 *is true—but our obtained statistic—*z, t, *and so on—was too low to enable us to reject the* H_0*, even though it "ought to be" rejected*. The relationship between these two types of errors is illustrated in Table 9.3.

The probability of each occurrence is listed in the appropriate cell of the table. If H_0 is really true and we conclude H_1 from our test, we make a type I error whose probability is alpha. If H_0 is really false but we fail to conclude it with our test, we make a type II error whose probability is beta. What if we reach the correct conclusion? Suppose H_0 is true and our test correctly fails to reject it. Since the sum of all possible probabilities is 1 and alpha is the probability of falsely rejecting H_0, the probability of correctly not rejecting H_0 is $1 - \alpha$. Similarly, if H_1 is true and we conclude that fact, the probability of that occurrence is $1 - \beta$. Notice, therefore, that *statistical power as we have defined it is really one minus beta! Power* $= 1 - \beta$.

Table 9.3

In the Population:	*Based on our significance test, we conclude:* H_0 *is True*	H_1 *is True*
H_0 is True	No Error $p = 1 - \alpha$	Type I Error $p = \alpha$
H_1 is True	Type II Error $p = \beta$	No Error $p = 1 - \beta$

We must point out that alpha and beta are related in the sense that by setting alpha low, as we traditionally do, we increase the probability of beta. By making it hard to reject the null hypothesis, we also increase the likelihood of failing to reject null hypotheses that "ought to be" rejected since they are in fact false.

Thus there is a clash of interests between the traditional conservatism of statistical tests (which require a low alpha and thus a large beta) and the objectives of social and behavioral researchers who are seeking to identify the relationships between variables or the real differences between groups in the population. Such differences must be rather large if we are to reject the null hypothesis with our tests of significance. Consequently many relationships between variables or differences between population means that really do exist are not large enough to yield statistically significant results.

For example, in evaluation research Lipsey (1990) found that only 28% of the time do **small effects** *(a hypothesized* $.2\sigma$ *difference between* μ *for the experimental group and* μ *for the control groups*) yield statistical significance. In 72% of the studies no significant results would be detected. Studies of research in other social sciences indicate that between 18% to 34% of the time such small mean differences yield significant results. An exception was sociology at 55%. (Due possibly to larger available sample sizes?) For **medium effects** *(a hypothesized* $.5\sigma$ *difference between the population means*), the percentage of studies yielding significant results was between 52% and 76% of the time, depending on the discipline (with sociology again being higher at 84%). **Large effects** (*a* $.8\sigma$ *population mean difference*) yielded significant results between 71% and 94% of the time for most social sciences. Clearly, only larger mean differences are found to be statistically significant a majority of the time in social research.[2]

Cohen (1977, 1988) has suggested that a reasonable beta value be set at .20. Thus statistical power $(1 - \beta)$ should be .80 at a minimum. Lipsey

(1990) identified four factors that determine statistical power: the test itself, the alpha level, the sample size, and the effect size as estimated (for the two-sample t) by the difference between the two sample means or from other sources. It turns out that of the four factors, the test to be used is determined largely by the type of data available and alpha is determined by statistical tradition. Thus the *sample size* and *effect size* remain just as we demonstrated earlier by the z formula. Of these two, *it can be demonstrated that the effect size has a far larger impact on statistical power than an increase in sample size.* Tables have been developed showing the relationship between statistical power, sample size, and effect sizes, for differing alpha levels. Consult the two works cited in this section for more details on the use of such tables (see Note 2).

A word of caution: Statistical power and effect size appear to be of most concern in the social sciences that must make use of rather small samples or experimental and control groups. Where samples can be larger, such as in sociology, the problems are diminished. You will have to inquire in your own discipline as to the current level of concern about power and effect size. While the low level of statistically significant research results is a matter of concern to us all, to a traditionalist much of this concern may appear to be a justification for the use of levels of significance *lower* than the ones traditionally applied. If so, it will remain controversial.

CONCLUSION

The two-sample t test is one of the oldest tests of statistical significance and one of the most commonly encountered tests. Since comparison of two groups' means is a common method of data analysis and often the groups being compared are random samples, there are many situations that make use of this test. Moreover, the two-sample t test may be used in both experimental and nonexperimental situations, as the examples presented in this chapter demonstrate. In fact, you will find it used in just about every discipline employing statistical techniques.

Now that we have studied the t tests for the difference between the means of two samples (or randomly assigned groups), we can turn to the problem of comparing a larger number of sample means, using the technique of analysis of variance.

EXERCISES

Exercise 9.1

A tolerance index has been developed that is designed to measure one's tolerance of "unpopular" beliefs such as those of a racist or sexist nature. On the scale, 0 means the lowest level of tolerance and 15 the highest level. A random sample of 10 university students (Group 1) is scored along the index. A second sample (Group 2) of students from the same university is a sample of students who had recently attended a workshop on multicultural diversity. Making no directionality assumption in H_1, test a null hypothesis that there is no difference in tolerance between the two populations.

Group 1 (*Control*) $x_1 =$	*Group 2* (*Workshop*) $x_2 =$
2	4
3	4
3	4
4	5
4	5
5	6
5	6
5	7
6	7
6	7

Exercise 9.2

Refer to Exercise 9.1. Suppose you had prior evidence that people attending such workshops generally demonstrated increased tolerance. What are your conclusions with a directional H_1?

Exercise 9.3

The control group of students from Exercise 9.1 is compared to a random sample of military veterans attending the same institution. Making no directionality assumption, test for significance.

Group 2 (Veterans)
$x_2 =$
4
4
4
5
5
7
7
8
8
8

Exercise 9.4

In the SAS printout below, the control group from Exercise 9.1 is compared to a sample of fine arts majors at the same university. Select the appropriate *t* test and state your conclusions about the two populations. (The standard deviations use formulas with $n - 1$ in the denominators.)

```
                        SAS
                  TTEST PROCEDURE
VARIABLE: SCORE
GROUP   N        MEAN       STD DEV     STD ERROR      MINIMUM
--------------------------------------------------------------
1      10  4.30000000  1.33749351  0.42295258  2.00000000
2      10  7.50000000  2.63523138  0.83333333  5.00000000

   MAXIMUM      VARIANCES          T      DF    PROB > |T|
--------------------------------------------------------------
 6.00000000      UNEQUAL      -3.4242   13.3         0.004
10.00000000     EQUAL        -3.4242   18.0         0.003

FOR HO: VARIANCES ARE EQUAL, F' = 4.88, DF = (9,9)
   PROB > F' = 0.0559
```

Exercise 9.5

The control group from Exercise 9.1 is now compared to a random sample of the faculty from the liberal arts college. Test for significance.

EXERCISES

Group 2
(Liberal Arts Faculty)

x_2 =
5
5
5
5
5
15
15
15
15
15

Exercise 9.6

Below is the SAS printout for Exercise 9.5. What is the difference between this and your findings? Does this change your conclusion about significance?

```
VARIABLE: SCORE
GROUP N         MEAN        STD DEV     VARIANCES        T        DF
--------------------------------------------------------------------
1     10   4.30000000   1.33749351     UNEQUAL  -3.3149     10.2
2     10  10.00000000   5.27046277     EQUAL    -3.3149     18.0

PROB > |T|
----------
    0.0077
    0.0039

FOR HO: VARIANCES ARE EQUAL, F' = 15.53 DF = (9,9)
PROB > F' = 0.0004
```

Exercise 9.7

Examine the partial printouts in Exercises 9.4 and 9.6. In comparing the equal variance and unequal variance *t* and *df* values, what do you conclude?

Exercise 9.8

The control group from Exercise 9.1 was then put in a special workshop designed to raise participant tolerance levels by subjecting them to criticism of their own attitudes. The before and after tolerance scores are presented below. Perform the appropriate t test assuming no directionality.

Before $X =$	*After* $X =$
2	3
3	3
3	5
4	7
4	6
5	8
5	7
5	5
6	9
6	8

Exercise 9.9

Another sample of military veterans is also put through the workshop described in Exercise 9.8. Test for significance. Use a two-tailed H_1.

Veterans	
Before $X =$	*After* $X =$
4	3
4	4
4	6
5	7
5	6
6	4
6	8
7	9
7	7
7	8

Exercise 9.10

The liberal arts faculty group from Exercise 9.5 also attend the workshop described in Exercise 9.8. Test for significance (nondirectional).

Faculty

Before	*After*
$X =$	$X =$
5	10
5	10
5	5
5	0
5	0
15	10
15	10
15	15
15	15
15	5

NOTES

1. If the *df* we need does not appear in the table, we use the adjacent *df* value that makes it harder to reject H_0. For example, if n_1 had been 10 instead of 4, we would have selected either 8 or 12. $F_{critical}$ where $n_2 = 5$ is 4.82 where $n_1 = 8$, and 4.68 where $n_1 = 12$. Since 4.82 is the larger value, we use it as $F_{critical}$.

2. The seminal work in this area is to be found in J. Cohen, *Statistical Power Analysis for the Behavioral Sciences*, 2nd ed. (Hillsdale, NJ: Erlbaum, 1988). However, M. W. Lipsey, *Design Sensitivity: Statistical Power for Experimental Research* (Newbury Park, CA: Sage, 1990) is much easier reading for the beginner. It was Cohen who operationalized large effect sizes as .8, medium as .5, and small as .2. These effect sizes are projected population mean differences expressed (as in the *z* test) in standard deviation units.

KEY CONCEPTS

one-way analysis of variance/ one-way ANOVA
F/the *F ratio*
participants or subjects (in an experiment)
grand mean
sum of squares/*SS*
mean square/*MS*
total sum of squares/ SS_{Total}/ SS_{T}
between-groups sum of squares/ SS_{Between}/SS_{B}
within-groups sum of squares/ SS_{Within}/SS_{W}
error sum of squares
between-groups mean square/ MS_{Between}/MS_{B}
degrees of freedom, between-groups/ df_{Between}/df_{B}
within-groups mean square/ MS_{Within}/MS_{W}
degrees of freedom, within-groups/ df_{Within}/df_{W}
ANOVA source table
robustness (of a test of significance)
post hoc procedures/post hoc test of multiple comparisons
Scheffé's Test
Scheffé's critical value
MODEL SS
MODEL MS
ERROR SS
ERROR MS
two-way analysis of variance/ two-way ANOVA
interaction effects

10 One-Way Analysis of Variance

INTRODUCTION

Analysis of variance (ANOVA) is a statistical cousin to the t test. Like the t test it is *a technique for comparing sample means;* but unlike the t test, ANOVA *can be used to compare more than two means.* Analysis of variance is very versatile. It is particularly friendly to experimental applications, where we may be comparing the means of several treatment groups and a control group. Consequently, psychologists rely heavily on this procedure. ANOVA is also useful in nonexperimental situations in the same way that the t test is. Interestingly though, ANOVA has been less widespread in nonexperimental research than many other statistical procedures, despite its great potential.

With ANOVA, because several sample means are usually being compared, *once a null hypothesis has been rejected we need a follow-on, or* **post hoc,** *procedure.* This is because although ANOVA examines all sample means at once, it is possible that some pairs of means may not be significantly different from one another, even though when all means are taken together in their entirety, the null hypothesis may be rejected. Thus the process is a bit like aerial photography. ANOVA gives us a high altitude picture, and if we can reject the null hypothesis, we swoop down for a closer look. The post hoc test provides the low altitude shot.

At the end of the chapter, we will briefly look at some other variations of the ANOVA technique.

HOW ANALYSIS OF VARIANCE IS USED

Analysis of variance is designed, in its nonexperimental application, for two variables; one is interval level of measurement (usually the dependent variable) and the other is grouped data of any level of measurement. Suppose we study a random sample of 8 people. We determine for each of them whether they are from urban or rural areas as well as their score on a "pro-life" index that ranges from 0 (most pro-choice) to 200 (most pro-life) on the issue of abortion. We get the following results:

	Rural (Group 1)		*Urban (Group 2)*
	160		100
	130		110
	150		130
	140		120
$\Sigma x_1 =$	580	$\Sigma x_2 =$	460

$$x_1 = \frac{580}{4} = 145 \qquad x_2 = \frac{460}{4} = 115$$

Clearly, the means of the two groups differ in our sample of 8 subjects. May we assume that they differ in population as a whole? Assuming no directionality, our hypotheses would be:

$$H_0;\ \mu_1 = \mu_2$$

$$H_1;\ \mu_1 \neq \mu_2$$

We could handle this problem using the two-sample *t* test, or we could perform a **one-way analysis of variance (one-way ANOVA).** Unfortunately, analysis of variance does not allow for a directionality assumption (although ANOVA's cousin, the two-sample *t* test, does). Thus we use the nondirectional H_1.

From our sample data, we will calculate a statistic called ***F***, or the ***F* ratio** (named for Fisher, who originally helped developed it). As we did with other tests, we will compare our obtained F to $F_{critical}$ at the .05 level, and if our F exceeds $F_{critical}$, we will reject the null hypothesis.

ANALYSIS OF VARIANCE IN EXPERIMENTAL SITUATIONS

Our first example (abortion stance by area) came from a nonexperimental context, a survey. In many social sciences, experiments have been relatively rare, but in the behavioral sciences, laboratory sciences, and medicine they are quite common. Where experimental designs are common, ANOVA is the most common statistical technique used.

Let us use an example from training and development, a growing subfield of communications that deals with the training of adults, usually in a job-related setting. Suppose an advertising firm is seeking to improve its operation. A training program is being established for employees of this firm to give the employees skills needed to properly advise and assist clients. The firm wishes to develop the most effective training program possible. The trainers are interested in comparing the relative efficiency of day, night, and weekend programs.

Let us assume that the same instructor will present identical material in each of three sections. One section will meet one hour per day, Monday through Friday, at the same time for one week. The second section will meet at night under similar circumstances. The third section will meet on a Saturday for a single long section, including 5 hours of instruction plus time for breaks and meals. There will be 10 people in each section. At the end of the instruction, each person taking the class will rate his or her overall satisfaction with the course on a scale of 0 (dissatisfied) through 5 (completely satisfied).

The 30 "students" are the **participants** or **subjects,** as they used to be called, in the experiment. They are all employees of the same firm and are selected to take the course for professional purposes. There is no random sample being selected here. However, if the trainer suspects that satisfaction will differ among the three classes — day, night, and Saturday — he or she may test that hypothesis.

To do so, a table of random numbers or some computerized random number generator can be used to randomly assign the 30 subjects to 3 groups of 10 people each. This randomization process tends to control for other social or behavioral characteristics. Each group will, for instance, have about the same proportion of men and women as the original group of 30. Each group will have about the same mean age. Each group will have about the same proportion of disgruntled employees who feel forced by their bosses to take the course. (We keep saying "*about* the same," since there will be some differences, slight we hope, equivalent to sampling error, resulting from chance in the randomization process.) Sampling error aside, if we randomly assign subjects, any difference in mean satisfaction scores among the 3 groups will be the result of the time frame for the course (day, night, Saturday) rather than other factors. Thus, for the variable satisfaction:

$$H_0: \mu_{\text{day}} = \mu_{\text{night}} = \mu_{\text{Saturday}}$$

Note that analysis of variance will take as many groups (or categories) as we have, whereas the two-sample *t* test is limited to two groups (or categories). Thus for the above problem we are only able to use one-way analysis of variance.

With more than two categories, it becomes difficult to write H_1 with symbols. In effect, H_1 says that in the population there exists *at least one inequality* that negates the null hypothesis. Any of the following would negate H_0.

$$\mu_{\text{day}} \neq \mu_{\text{night}}$$

$$\mu_{\text{day}} \neq \mu_{\text{Saturday}}$$

$$\mu_{\text{night}} \neq \mu_{\text{Saturday}}$$

It is *not* necessary that all three population means be unequal, although that could be the case:

$$\mu_{\text{day}} \neq \mu_{\text{night}} \neq \mu_{\text{Saturday}}$$

Suppose we obtain the following results:

	x = *Satisfaction Ratings*	
Day	*Night*	*Saturday*
5	5	5
5	4	5
5	4	5
5	3	4
4	3	4
4	2	4
4	2	3
4	2	3
3	1	3
2	0	2
$\sum x_D = 41$	$\sum x_N = 26$	$\sum x_S = 38$

$$\bar{x}_D = \frac{41}{10} = 4.1 \qquad \bar{x}_N = \frac{26}{10} = 2.6 \qquad \bar{x}_S = \frac{38}{10} = 3.8$$

Clearly, the category means of 4.1, 2.6, and 3.8 are different from each other. Are they different enough to conclude that they were not the result of sampling error, in this case due to the randomization process whereby we assigned the subjects to the three categories? If we can reject the null hypothesis, we may conclude that "in the population," that is, for people in general, student satisfaction levels differ by the time and format of the class offered regardless of the instructor, course content, or anything else.

We calculate F and compare it to $F_{critical}$ at the appropriate degrees of freedom. If the F we obtained exceeds $F_{critical}$, .05 level, we reject H_0. We then compare our F to $F_{critical}$ at other levels to form a probability statement.

F—AN INTUITIVE APPROACH

We may think of F as a measure of how well the categories of the independent variable explain the variation in the scores of the dependent variable. If the categories of the independent variable are *totally useless* in explaining the variation of scores of the dependent variable, then F will equal zero. As the categories of the independent variable begin to explain or account for some of the variation in the dependent variable, F begins to get larger. The better the independent variable explains variation in the dependent variable, the greater is the relationship between the variables and the larger F becomes. When the categories of the independent variable explain nearly all the variation in the dependent variable, F grows extremely large, approaching infinity as a limit.

As an illustration, imagine a simplified version of the course satisfaction problem. For simplicity's sake, we will limit the independent variable to two categories, a day class and a night class. Suppose there are six people in each class. The following satisfaction scores emerge.

Case 1

Day	*Night*
5	5
5	5
5	5
3	3
3	3
3	3
$\Sigma x_D = 24$	$\Sigma x_N = 24$

In this case, F— which you haven't yet learned to calculate—would be zero.

$F = 0$

Grand Mean = 48/12 = 4.00

$$\bar{x}_D = \frac{24}{6} = 4.00 \qquad x_N = \frac{24}{6} = 4.00$$

If we were to ignore the existence of the two categories and just calculate the mean of all 12 satisfaction scores, the mean we would get—called the **grand mean**—would be 48/12 or 4.0, the same as the two category means. Since mean satisfaction is the same (4.0) whether or not we know in which category a subject belongs, the categories do not help us predict a subject's score on the dependent variable. Thus the two variables are unrelated. There is no difference between day and night classes in terms of course satisfaction. *F* will be zero.

Now imagine a slight variation in which one more person in the day class has a satisfaction score of 5 and one more person in the night class has a satisfaction score of 3. While the grand mean is unchanged (4.0), the two category means now differ.

Case 2

	Day		*Night*
	5		5
	5		5
	5		3
	5		3
	3		3
	3		3
$\sum x_D =$	26	$\sum x_N =$	22

$F = 1.248$

$$\bar{x}_D = \frac{26}{6} = 4.33 \qquad \bar{x}_N = \frac{22}{6} = 3.67$$

Grand Mean = 48/12 = 4.00

Once we learn to calculate *F*, we will see that for this problem, $F = 1.248$, up from 0 in Case 1. Whereas in Case 1 each category had three scores of 5 and three scores of 3, in Case 2, the day class is slightly more satisfied than the night class: four scores of 5, two scores of 3 in the day class with a mean of 4.33 as opposed to two scores of 5 and four scores of 3 in the night class with a lower mean of 3.67. The two classes now differ somewhat in terms of satisfaction.

We now add one more score of 5 to the day class, replacing a score of 3, and in the night class we replace a score of 5 with a 3.

Case 3

Day	*Night*
5	5
5	3
5	3
5	3
5	3
3	3
$\sum x_D = 28$	$\sum x_N = 20$

$F = 7.994$

$\bar{x}_D = \frac{28}{6} = 4.67$ $\quad \bar{x}_N = \frac{20}{6} = 3.33$ $\quad$ Grand Mean = 48/12 = 4.00

The differentiation between day and night classes, in terms of satisfaction, has grown even more pronounced. Day students are clearly more satisfied than night students. The category means are more widely spread about the grand mean than in Case 2. Though we could be wrong, there is now a five-to-one chance that if someone tells us he or she is in the day class, we would be accurate in predicting the satisfaction score as being 5. (Of the six day students, five have scores of 5; only one does not.) If someone is in the night class, we predict (also with odds of five to one) that the satisfaction score is 3.

Finally, we change the final score of 3 in the day class to 5 and change the final 5 in the night class to 3.

Case 4

Day	*Night*
5	3
5	3
5	3
5	3
5	3
5	3
$\sum x_D = 30$	$\sum x_N = 18$

F is mathematically undefined, but had been getting larger, approaching infinity as a limit.

$\bar{x}_D = \frac{30}{6} = 5.00$ $\quad \bar{x}_N = \frac{18}{6} = 3.00$ $\quad$ Grand Mean = 48/12 = 4.00

In this case, the categories of class type explain *all* the differences in satisfaction scores. Knowing what class one is in gives us perfect predictive ability in terms of satisfaction. If in the day class, a person's satisfaction score is 5; if in the night class, the satisfaction score is 3.

The two variables, satisfaction and class meeting time, are perfectly related.

Notice that in Case 1, all the variations of the scores from the grand mean were actually *within* each of the two categories. The category means did not vary at all from the grand mean. As we progressed through Cases 2 and 3, more and more of the deviations or variations of scores from the grand mean could be explained by the category means. Finally, in Case 4, there were no deviations of scores within the categories. All scores fell at the means of their respective categories. All deviations of scores from the grand mean could be accounted for by the deviations of their respective category means about the grand mean. To see this more clearly, note that algebraically we can break the distance between any score and the grand mean into two components: (a) the distance from that score to its respective category mean, plus (b) the distance from that category mean to the grand mean.

$$
\begin{aligned}
(\text{Score} - \text{Grand Mean}) = {} & (\text{Score} - \text{Category Mean}) \\
& + (\text{Category Mean} - \text{Grand Mean})
\end{aligned}
$$

In Case 1, every score is either a 5 or a 3, both category means are 4.00, and the grand mean is 4.00. Thus for a score of 5 in the day class:

$$(5 - 4) \quad = \quad (5 - 4) \quad + \quad (4 - 4)$$

(5 − 4): 5 = The Score; 4 = The Grand Mean
(5 − 4): 5 = The Score; 4 = The Category Mean
(4 − 4): first 4 = The Category Mean; second 4 = The Grand Mean

$$
\begin{aligned}
(5 - 4) &= (5 - 4) + (4 - 4) \\
(1) &= (1) + (0) \\
1 &= 1 + 0 \\
1 &= 1
\end{aligned}
$$

This part reduces to zero.

The same would hold for a score of 5 in the night class. For a score of 3 in either class:

$$
\begin{aligned}
(3 - 4) &= (3 - 4) + (4 - 4) \\
(-1) &= (-1) + (0) \\
-1 &= -1 + 0 \\
-1 &= -1
\end{aligned}
$$

This part reduces to zero.

Notice that in Case 4 it is (Score – Category Mean) that always reduces to zero. Since all scores in the day section are 5:

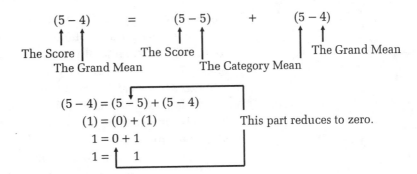

$$(5 - 4) \quad = \quad (5 - 5) \quad + \quad (5 - 4)$$

The Score, The Grand Mean = The Score, The Category Mean + The Category Mean, The Grand Mean

$$(5 - 4) = (5 - 5) + (5 - 4)$$
$$(1) = (0) + (1)$$
$$1 = 0 + 1$$
$$1 = \quad 1$$

This part reduces to zero.

Since all scores in the night section are 3:

$$(3 - 4) \quad = \quad (3 - 3) \quad + \quad (3 - 4)$$

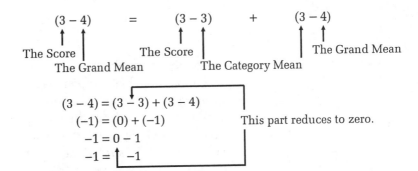

The Score, The Grand Mean = The Score, The Category Mean + The Category Mean, The Grand Mean

$$(3 - 4) = (3 - 3) + (3 - 4)$$
$$(-1) = (0) + (-1)$$
$$-1 = 0 - 1$$
$$-1 = \quad -1$$

This part reduces to zero.

In Case 1, (Category Mean – Grand Mean) is always zero; in Case 4, (Score – Category Mean) is always zero. These are the two extreme cases. In Case 1, the categories explain no variation in the dependent variable and in Case 4, the categories explain all the variation in the dependent variable. Cases 2 and 3 fall in the middle. In Case 2, one third of the distance between a score and the grand mean can be accounted for by the distance between the category mean and the grand mean; two thirds of the distance is within the category, between the score and its category mean. In Case 3, two thirds of the distance between a score and the grand mean can be accounted for by the distance between the category mean and the grand mean; one-third of the distance is within the category, between the score and its category mean.

ANOVA TERMINOLOGY

The logic of analysis of variance is based on partitioning the distance to the grand mean into distances explained by the category means and distances unexplained by the category means, in a manner somewhat analogous to what we have been doing. However, ANOVA *squares* distances to get rid of negative numbers and it works with sums of these squared distances. It also makes use of its own computational formulas. Thus, before learning the technique for calculating *F*, we need to define some terminology.

Since we will ultimately be using *variance estimates* (squared sigma hats), let us return for a moment to the definitional formula for a population variance estimate.

$$\sigma^2 = \frac{\sum (x - \bar{x})^2}{n - 1}$$

The denominator for this estimate, $n - 1$, is its degrees of freedom or *df*. The numerator $\Sigma(x - \bar{x})^2$ can be read: *The sum of the squared deviations of the values of x from the mean.* Each $x - \bar{x}$ is the deviation, the distance of that value of *x* from the mean. These deviations are squared to get rid of negative signs, and the squared deviations are summed. In analysis of variance, "*the sum of the squared deviations of the values of x from the mean*" is shortened to the **sum of squares** and is indicated by the letters *SS*.

When we divide *SS* by *n* for a sample variance or by *df* ($n - 1$) for a variance estimate, we are finding a kind of average amount of squared deviation for each value of *x*: "*The mean squared deviation of a score (a value of x) from the mean of all scores,*" which is shortened to the **mean square** and is indicated by the letters *MS*. A variance or variance estimate is the average, or mean, amount of the sum of squares per unit. Our σ^2 formula is then symbolized as:

$$MS = \frac{SS}{df}$$

What ANOVA does is first to calculate, using the computational formula, the **total sum of squares** (SS_{Total} or SS_{T}), meaning the total of the squared deviations of scores about the grand mean.

$$SS_T = \sum x^2 - \frac{\left(\sum x\right)^2}{n}$$

The total sum of squares (SS_T) is then partitioned (divided) into two components. The first component is the **between-groups sum of squares** ($SS_{Between}$, or SS_B), the portion of the total sum of squares that can be accounted for by the variations of the category means about the grand mean. That is, SS_B is the portion of SS_T that can be accounted for (explained by) the categories.

The second component is the **within-groups sum of squares** (SS_{Within} or SS_W), the portion of the total sum of squares left unexplained by the variations of the category means about the grand mean. This then is the sum of squares within the categories or the squared deviations of scores about their respective category means. It is sometimes called the **error sum of squares.**

In short,

SS_B = the portion of SS_T accounted for by the categories of the independent variable.

SS_W = the portion of SS_T not accounted for by the categories of the independent variable.

These are additive:

$$SS_T = SS_B + SS_W$$

We use SS_B and SS_W to form two separate population variance estimates. The first of these variance estimates, the **between-groups mean square** ($MS_{Between}$ or MS_B) is a variance estimate based on the between-groups sum of squares. MS_B estimates the population variance accounted for by the variation of the category means about the grand mean—the population variance accounted for by the groups or categories of the independent variable. To find MS_B we divide SS_B by the **between groups degrees of freedom** (df_B or $df_{Between}$). Since here we are talking not about the number of respondents but about the number of groups or categories, df_B equals the number of categories (or groups) minus 1.

$$df_B = \text{no. of categories} - 1$$

so,

$$MS_B = \frac{SS_B}{df_B} = \frac{SS_B}{\text{no. of categories} - 1}$$

The second variance estimate is the **within-groups mean square** (MS_{Within} or MS_W), a population variance estimate based on what the categories of the independent variable do *not* explain—the variation of scores within the groups. We take SS_W and divide by the **within groups degrees of freedom** (df_{Within} or df_W). The within-groups degrees of freedom is found by subtracting the between-groups degrees of freedom from the degrees of freedom belonging to $\hat{\sigma}^2$ (our original total variance estimate), which is $n - 1$.
Thus,

$$df_W = df_{\text{Total}} - df_{\text{Between}} = (n - 1) - (\text{no. of categories} - 1)$$

so

$$MS_W = \frac{SS_W}{df_W} = \frac{SS_W}{(n - 1) - (\text{no. of categories} - 1)}$$

Note that like the sums of squares, the degrees of freedom are additive, so that

$$df_{\text{Total}} = df_B + df_W$$

$$\begin{aligned} df_B + df_W &= (\text{no. of categories} - 1) \\ &\quad + [(n - 1) - (\text{no. of categories} - 1)] \\ &= n - 1 = df_{\text{Total}} \end{aligned}$$

Although the *SS*'s and *df*'s are additive, the *MS*'s are not. The mean square total, $\hat{\sigma}^2$, does *not* equal $MS_B + MS_W$.

Finally, we find the *F ratio:*

$$F = \frac{MS_B}{MS_W}$$

The *F* that we obtain is compared to F_{critical} (see Table 10.1).[1] Note that *F* uses two degrees of freedom, df_{Between}, which we locate on the top row, and df_{Within}, which is found on the column on the left-hand side. (Here n_1 and n_2 mean degrees of freedom: $n_1 = df_B$ and $n_2 = df_W$.) There are three pages to this table; one for the .05 level, one for the .01 level,

Table 10.1 Critical Values of *F* for $p = .05$

$n_1 \backslash n_2$	1	2	3	4	5	6	8	12	24	∞
1	161.40	199.50	215.70	224.60	230.20	234.00	238.90	243.90	249.00	254.30
2	18.51	19.00	19.16	19.25	19.30	19.33	19.37	19.41	19.45	19.50
3	10.13	9.55	9.28	9.12	9.01	8.94	8.84	8.74	8.64	8.53
4	7.71	6.94	6.59	6.39	6.26	6.16	6.04	5.91	5.77	5.63
5	6.61	5.79	5.41	5.19	5.05	4.95	4.82	4.68	4.53	4.36
6	5.99	5.14	4.76	4.53	4.39	4.28	4.15	4.00	3.84	3.67
7	5.59	4.74	4.35	4.12	3.97	3.87	3.73	3.57	3.41	3.23
8	5.32	4.46	4.07	3.84	3.69	3.58	3.44	3.28	3.12	2.93
9	5.12	4.26	3.86	3.63	3.48	3.37	3.23	3.07	2.90	2.71
10	4.96	4.10	3.71	3.48	3.33	3.22	3.07	2.91	2.74	2.54
11	4.84	3.98	3.59	3.36	3.20	3.09	2.95	2.79	2.61	2.40
12	4.75	3.88	3.49	3.26	3.11	3.00	2.85	2.69	2.50	2.30
13	4.67	3.80	3.41	3.18	3.02	2.92	2.77	2.60	2.42	2.21
14	4.60	3.74	3.34	3.11	2.96	2.85	2.70	2.53	2.35	2.13
15	4.54	3.68	3.29	3.06	2.90	2.79	2.64	2.48	2.29	2.07
16	4.49	3.63	3.24	3.01	2.85	2.74	2.59	2.42	2.24	2.01
17	4.45	3.59	3.20	2.96	2.81	2.70	2.55	2.38	2.19	1.96
18	4.41	3.55	3.16	2.93	2.77	2.66	2.51	2.34	2.15	1.92
19	4.38	3.52	3.13	2.90	2.74	2.63	2.48	2.31	2.11	1.88
20	4.35	3.49	3.10	2.87	2.71	2.60	2.45	2.28	2.08	1.84
21	4.32	3.47	3.07	2.84	2.68	2.57	2.42	2.25	2.05	1.81
22	4.30	3.44	3.05	2.82	2.66	2.55	2.40	2.23	2.03	1.78
23	4.28	3.42	3.03	2.80	2.64	2.53	2.38	2.20	2.00	1.76
24	4.26	3.40	3.01	2.78	2.62	2.51	2.36	2.18	1.98	1.73
25	4.24	3.38	2.99	2.76	2.60	2.49	2.34	2.16	1.96	1.71
26	4.22	3.37	2.98	2.74	2.59	2.47	2.32	2.15	1.95	1.69
27	4.21	3.35	2.96	2.73	2.57	2.46	2.30	2.13	1.93	1.67
28	4.20	3.34	2.95	2.71	2.56	2.44	2.29	2.12	1.91	1.65
29	4.18	3.33	2.93	2.70	2.54	2.43	2.28	2.10	1.90	1.64
30	4.17	3.32	2.92	2.69	2.53	2.42	2.27	2.09	1.89	1.62
40	4.08	3.23	2.84	2.61	2.45	2.34	2.18	2.00	1.79	1.51
60	4.00	3.15	2.76	2.52	2.37	2.25	2.10	1.92	1.70	1.39
120	3.92	3.07	2.68	2.45	2.29	2.17	2.02	1.83	1.61	1.25
∞	3.84	2.99	2.60	2.37	2.21	2.09	1.94	1.75	1.52	1.00

SOURCE: Table 10.1 is abridged from Table V of R. A. Fisher and F. Yates, *Statistical Tables for Biological, Agricultural and Medical Research* (6th ed.), published by Longman Group UK Ltd., 1974, by permission of the authors and the publisher.
NOTE: Values of n_1 and n_2 represent the degrees of freedom associated with the larger and smaller estimates of variance, respectively.

(Text continued on page 294)

Table 10.1 Continued—Critical Values of F for $p = .01$

$n_1 \backslash n_2$	1	2	3	4	5	6	8	12	24	∞
1	4052	4999	5403	5625	5764	5859	5981	6106	6234	6366
2	98.49	99.01	99.17	99.25	99.30	99.33	99.36	99.42	99.46	99.50
3	34.12	30.81	29.46	28.71	28.24	27.91	27.49	27.05	26.60	26.12
4	21.20	18.00	16.69	15.98	15.52	15.21	14.80	14.37	13.93	13.46
5	16.26	13.27	12.06	11.39	10.97	10.67	10.27	9.89	9.47	9.02
6	13.74	10.92	9.78	9.15	8.75	8.47	8.10	7.72	7.31	6.88
7	12.25	9.55	8.45	7.85	7.46	7.19	6.84	6.47	6.07	5.65
8	11.26	8.65	7.59	7.01	6.63	6.37	6.03	5.67	5.28	4.86
9	10.56	8.02	6.99	6.42	6.06	5.80	5.47	5.11	4.73	4.31
10	10.04	7.56	6.55	5.99	5.64	5.39	5.06	4.71	4.33	3.91
11	9.65	7.20	6.22	5.67	5.32	5.07	4.74	4.40	4.02	3.60
12	9.33	6.93	5.95	5.41	5.06	4.82	4.50	4.16	3.78	3.36
13	9.07	6.70	5.74	5.20	4.86	4.62	4.30	3.96	3.59	3.16
14	8.86	6.51	5.56	5.03	4.69	4.46	4.14	3.80	3.43	3.00
15	8.68	6.36	5.42	4.89	4.56	4.32	4.00	3.67	3.29	2.87
16	8.53	6.23	5.29	4.77	4.44	4.20	3.89	3.55	3.18	2.75
17	8.40	6.11	5.18	4.67	4.34	4.10	3.79	3.45	3.08	2.65
18	8.28	6.01	5.09	4.58	4.25	4.01	3.71	3.37	3.00	2.57
19	8.18	5.93	5.01	4.50	4.17	3.94	3.63	3.30	2.92	2.49
20	8.10	5.85	4.94	4.43	4.10	3.87	3.56	3.23	2.86	2.42
21	8.02	5.78	4.87	4.37	4.04	3.81	3.51	3.17	2.80	2.36
22	7.94	5.72	4.82	4.31	3.99	3.76	3.45	3.12	2.75	2.31
23	7.88	5.66	4.76	4.26	3.94	3.71	3.41	3.07	2.70	2.26
24	7.82	5.61	4.72	4.22	3.90	3.67	3.36	3.03	2.66	2.21
25	7.77	5.57	4.68	4.18	3.86	3.63	3.32	2.99	2.62	2.17
26	7.72	5.53	4.64	4.14	3.82	3.59	3.29	2.96	2.58	2.13
27	7.68	5.49	4.60	4.11	3.78	3.56	3.26	2.93	2.55	2.10
28	7.64	5.45	4.57	4.07	3.75	3.53	3.23	2.90	2.52	2.06
29	7.60	5.42	4.54	4.04	3.73	3.50	3.20	2.87	2.49	2.03
30	7.56	5.39	4.51	4.02	3.70	3.47	3.17	2.84	2.47	2.01
40	7.31	5.18	4.31	3.83	3.51	3.29	2.99	2.66	2.29	1.80
60	7.08	4.98	4.13	3.65	3.34	3.12	2.82	2.50	2.12	1.60
120	6.85	4.79	3.95	3.48	3.17	2.96	2.66	2.34	1.95	1.38
∞	6.64	4.60	3.78	3.32	3.02	2.80	2.51	2.18	1.79	1.00

SOURCE: Table 10.1 is abridged from Table V of R. A. Fisher and F. Yates, *Statistical Tables for Biological, Agricultural and Medical Research* (6th ed.), published by Longman Group UK Ltd., 1974, by permission of the authors and the publisher.
NOTE: Values of n_1 and n_2 represent the degrees of freedom associated with the larger and smaller estimates of variance, respectively.

Table 10.1 Continued—Critical Values of F for $p = .001$

$n_1 \backslash n_2$	1	2	3	4	5	6	8	12	24	∞
1	405284	500000	540379	562500	576405	585937	598144	610667	623497	636619
2	998.5	999.0	999.2	999.2	999.3	999.3	999.4	999.4	999.5	999.5
3	167.5	148.5	141.1	137.1	134.6	132.8	130.6	128.3	125.9	123.5
4	74.14	61.25	56.18	53.44	51.71	50.53	49.00	47.41	45.77	44.05
5	47.04	36.61	33.20	31.09	29.75	28.84	27.64	26.42	25.14	23.78
6	35.51	27.00	23.70	21.90	20.81	20.03	19.03	17.99	16.89	15.75
7	29.22	21.69	18.77	17.19	16.21	15.52	14.63	13.71	12.73	11.69
8	25.42	18.49	15.83	14.39	13.49	12.86	12.04	11.19	10.30	9.34
9	22.86	16.39	13.90	12.56	11.71	11.13	10.37	9.57	8.72	7.81
10	21.04	14.91	12.55	11.28	10.48	9.92	9.20	8.45	7.64	6.76
11	19.69	13.81	11.56	10.35	9.58	9.05	8.35	7.63	6.85	6.00
12	18.64	12.97	10.80	9.63	8.89	8.38	7.71	7.00	6.25	5.42
13	17.81	12.31	10.21	9.07	8.35	7.86	7.21	6.52	5.78	4.97
14	17.14	11.78	9.73	8.62	7.92	7.43	6.80	6.13	5.41	4.60
15	16.59	11.34	9.34	8.25	7.57	7.09	6.47	5.81	5.10	4.31
16	16.12	10.97	9.00	7.94	7.27	6.81	6.19	5.55	4.85	4.06
17	15.72	10.66	8.73	7.68	7.02	6.56	5.96	5.32	4.63	3.85
18	15.38	10.39	8.49	7.46	6.81	6.35	5.76	5.13	4.45	3.67
19	15.08	10.16	8.28	7.26	6.61	6.18	5.59	4.97	4.29	3.52
20	14.82	9.95	8.10	7.10	6.46	6.02	5.44	4.82	4.15	3.38
21	14.59	9.77	7.94	6.95	6.32	5.88	5.31	4.70	4.03	3.26
22	14.38	9.61	7.80	6.81	6.19	5.76	5.19	4.58	3.92	3.15
23	14.19	9.47	7.67	6.69	6.08	5.65	5.09	4.48	3.82	3.05
24	14.03	9.34	7.55	6.59	5.98	5.55	4.99	4.39	3.74	2.97
25	13.88	9.22	7.45	6.49	5.88	5.46	4.91	4.31	3.66	2.89
26	13.74	9.12	7.36	6.41	5.80	5.38	4.83	4.24	3.59	2.82
27	13.61	9.02	7.27	6.33	5.73	5.31	4.76	4.17	3.52	2.75
28	13.50	8.93	7.19	6.25	5.66	5.24	4.69	4.11	3.46	2.70
29	13.39	8.85	7.12	6.19	5.59	5.18	4.64	4.05	3.41	2.64
30	13.29	8.77	7.05	6.12	5.53	5.12	4.58	4.00	3.36	2.59
40	12.61	8.25	6.60	5.70	5.13	4.73	4.21	3.64	3.01	2.23
60	11.97	7.76	6.17	5.31	4.76	4.37	3.87	3.31	2.69	1.90
120	11.38	7.31	5.79	4.95	4.42	4.04	3.55	3.02	2.40	1.56
∞	10.83	6.91	5.42	4.62	4.10	3.74	3.27	2.74	2.13	1.00

SOURCE: Table 10.1 is abridged from Table V of R. A. Fisher and F. Yates, *Statistical Tables for Biological, Agricultural and Medical Research* (6th ed.), published by Longman Group UK Ltd., 1974, by permission of the authors and the publisher.
NOTE: Values of n_1 and n_2 represent the degrees of freedom associated with the larger and smaller estimates of variance, respectively.

and one for the .001 level. If $F_{obtained}$ exceeds $F_{critical}$ at the .05 level, we reject the null hypothesis and go on to the page with $F_{critical}$ at the .01 level. If $F_{obtained}$ exceeds $F_{critical}$ at the .01 level, we go on to compare it to $F_{critical}$ at the .001 level. This is exactly what we did with earlier tests. Probabilities are reported the same way. If ANOVA is done on a computer using a program such as SAS or SPSS, the exact probability will be listed, and thus it will not be necessary to use a critical value of F table.

THE ANOVA PROCEDURE

Before we actually work an F problem through, look at Box 10.1 where the computational steps and all appropriate formulas are given. To calculate F, we must find the following: n for each category, n_{Total}, $\sum x$ for each category, $\sum x_{Total}$, and $\sum x_{Total}^2$, which we get by finding $\sum x^2$ for each category and adding them up. Note that all the examples used so far in this chapter have equal category sizes; this ideal is not necessary and not always possible. Note, too, that though we need not find the category means for the sample to calculate F using these formulas, we do so anyway to better understand the problem we are working. Applying these steps to the first problem presented in this chapter:

x = Pro-Life Index

x_1 = *Rural*	x_2 = *Urban*
160	100
130	110
150	130
140	120

$H_0 : \mu_1 = \mu_2$

$H_1 : \mu_1 \neq \mu_2$

	$x_1 =$		$x_2 =$
	160		100
	130		110
	150		130
$n_1 = 4$	140	$n_2 = 4$	120
	$\sum x_1 = 580$		$\sum x_2 = 460$

$n_{Total} = 4 + 4 = 8$

$$\bar{x}_1 = \frac{\sum x_1}{n_1} = \frac{580}{4} = 145 \qquad \bar{x}_2 = \frac{\sum x_2}{n_2} = \frac{460}{4} = 115$$

$$\sum x_{Total} = \sum x_1 + \sum x_2 = 580 + 460 = 1040$$

$x_1^2 =$	$x_2^2 =$
25,600	10,000
16,900	12,100
22,500	16,900
19,600	14,400
$\sum x_1^2 = 84{,}600$	$\sum x_2^2 = 53{,}400$

$$\sum x_{\text{Total}}^2 = \sum x_1^2 + \sum x_2^2 = 84{,}600 + 53{,}400 = 138{,}000$$

Box 10.1

Procedure for Calculating One-Way Analysis of Variance

1. Calculate the total sum of squares (sum of squared deviations from the grand mean), where:

$$SS_{\text{Total}} = \sum x_{\text{Total}}^2 - \frac{\left(\sum x_{\text{Total}}\right)^2}{n_{\text{Total}}}$$

2. Calculate the between-group sum of squares (sum of squared deviations of category means from the grand mean), where:

$$SS_{\text{Between}} = \frac{\left(\sum x_{\text{cat.1}}\right)^2}{n_{\text{cat.1}}} - \frac{\left(\sum x_{\text{cat.2}}\right)^2}{n_{\text{cat.2}}} + \ldots$$

$$+ \frac{\left(\sum x_{\text{last cat.}}\right)^2}{n_{\text{last cat.}}} - \frac{\left(\sum x_{\text{Total}}\right)^2}{n_{\text{Total}}}$$

Note: + . . . + means continue repeating the $(\Sigma x_{\text{cat}})/n_{\text{cat}}$ for the subsequent categories, if any.

3. Calculate within-groups sum of squares, where:

$$SS_{\text{Within}} = SS_{\text{Total}} - SS_{\text{Between}}$$

continued

Following the steps in Box 10.1:

1. We find the total sum of squares.

$$SS_T = \sum x^2_{Total} - \frac{\left(\sum x_{Total}\right)^2}{n_{Total}} = 138{,}000 - \frac{(1040)^2}{8}$$

$$= 138{,}000 - \frac{1{,}081{,}600}{8} = 138{,}000 - 135{,}200 = 2800$$

BOX 10.1

Continued

4. Calculate between and within mean squares, where:

$$MS_{Between} = \frac{SS_{Between}}{df_{Between}} = \frac{SS_{Between}}{\text{no. of categories} - 1}$$

$$MS_{Within} = \frac{SS_{Within}}{df_{Within}} = \frac{SS_{Within}}{(n_{Total} - 1) - (\text{no. of categories} - 1)}$$

5. Calculate the F ratio, where:

$$F = \frac{MS_{Between}}{MS_{Within}}$$

6. Use the table of F values in your textbook to test the F for significance.

Note: Numerator $df = df_{Between}$ = no. of categories – 1

Denominator $df = df_{Within} = (n_{Total} - 1) - (\text{no. of categories} - 1)$

7. If F is significant and the number of categories is greater than two, perform a post hoc procedure such as Scheffé's test.

8. If F is significant and this is nonexperimental research, measure association with the correlation ratio or r_i. (These will be presented in Chapter 13.)

2. We calculate the between-group sum of squares. Note that the last expression in this formula, $(\Sigma x_{\text{Total}})^2/n_{\text{Total}}$, was already calculated in step 1.

$$SS_{\text{B}} = \frac{\left(\sum x_1\right)^2}{n_1} + \frac{\left(\sum x_2\right)^2}{n_2} - \frac{\left(\sum x_{\text{Total}}\right)^2}{n_{\text{Total}}}$$

$$= \frac{(580)^2}{4} + \frac{(460)^2}{4} - 135{,}200 = \frac{336{,}400}{4} + \frac{211{,}600}{4} - 135{,}200$$

$$= 84{,}100 + 52{,}900 - 135{,}200 = 137{,}000 - 135{,}200 = 1800$$

3. We find the within-group sum of squares.

$$SS_{\text{W}} = SS_{\text{T}} - SS_{\text{B}} = 2800 - 1800 = 1000$$

4. We calculate the mean squares.

$$MS_{\text{B}} = \frac{SS_{\text{B}}}{df_{\text{B}}} = \frac{SS_{\text{B}}}{\text{no. of categories} - 1} = \frac{1800}{2-1} = \frac{1800}{1} = 1800$$

$$MS_{\text{W}} = \frac{SS_{\text{W}}}{df_{\text{W}}} = \frac{1000}{(n_{Total} - 1) - (\text{no. of categories} - 1)}$$

$$= \frac{1000}{(8-1)-(2-1)} = \frac{1000}{7-1} = \frac{1000}{6} = 166.67$$

5. We calculate the F ratio.

$$F = \frac{MS_{\text{B}}}{MS_{\text{W}}} = \frac{1800}{166.67} = 10.799 = 10.80$$

6. We find the degrees of freedom.

$$df_{\text{B}} = \text{no. of categories} - 1 = 2 - 1 = 1$$

$$df_{\text{W}} = (n_{\text{Total}} - 1) - (\text{no. of categories} - 1)$$
$$= (8-1) - (2-1) = 7 - 1 = 6$$

We then find $F_{critical}$ at the .05 level from the first table in Table 10.1, going along the top row to df_B (1) and dropping down that column until it intersects the row for df_W (6):

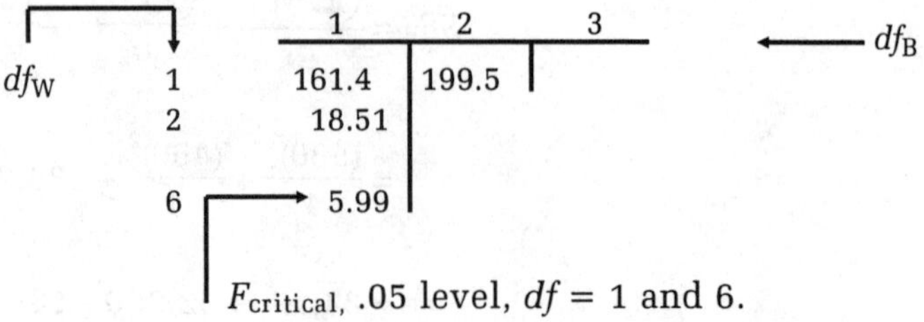

Since $F_{obtained}$ is 10.80 and greater than 5.99, we reject H_0. Using the second table of Table 10.1, we repeat the procedure to find $F_{critical}$ at the .01 level, which is 13.74 and greater than 10.80. Since p is less than .05 but greater than .01, we state our probability of falsely rejecting a true null hypothesis as $p < .05$.

Before moving on, let us return to Table 10.1 for a moment. From time to time, we will calculate a degrees of freedom figure that does not appear on the table, such as $df_B = 7$ or $df_W = 31$. In such a case, we use the adjacent row or column that has the *higher* value of *F*. Thus for $df_B = 7$, we would see which value was higher, *F* at $df = 6$ or *F* at $df = 8$. For example, if $df_B = 7$ and $df_W = 6$, we would have a choice between $F_{critical,}$ *df* 6 and 6 (4.28) or $F_{critical,}$ *df* 8 and 6 (4.15). We use the larger of the two (4.28) as our critical value. If $df_B = 1$ and $df_W = 31$, we have a choice between $F_{critical,}$ *df* 1 and 30 (4.17) or $F_{critical,}$ *df* 1 and 40 (4.08). Again, we select the larger of the two (4.17) as our critical value. If we are very close to rejecting H_0 using this procedure but do not quite make it, our best bet is to use a computer program such as SAS.

We go back once again to our ANOVA problem for which we have now rejected H_0 with a probability of error < .05. We may wish to summarize our findings in what is called an **ANOVA source table.**

Source	*SS*	*df*	*MS*	*F*	*p*
Total	2800				
Between	1800	1	1800	10.80	< .05
Within	1000	6	166.67		

Note that we usually do not report df_{Total} (in this case $n - 1 = 8 - 1 = 7$, or MS_{Total}, which is σ^2, since neither was necessary for finding F.

COMPARING *F* WITH *t*

At this juncture, let us pause to compare F to the two-sample t tests presented previously. Although they are treated in most statistics texts as separate procedures, they are in fact mathematically related. In the case of a two-category independent variable such as the problem just completed, it turns out that if the two-sample t test, assuming equal population variances, had been calculated on the same data, then $F = t^2$. In our problem, calculating the t value would yield 3.287 and squaring that yields 10.80, the same as our F.

If we may do either t or F when comparing two groups, which is preferable? Generally, the two-sample t test is, for two reasons. First, unlike ANOVA, it is possible to make a directionality assumption in a t test's H_1. If we had advance reasons to believe rural residents were more pro-life than urban ones, we could have used the directional critical values of t and reduced our probability of error by one half.

Second, ANOVA as presented here assumes equal population variances. If there is reason to assume unequal population variances, a t test formula must be used. In fact, it is somewhat ironic that the assumption of equal population variances plays such a major role in the t test procedure, since in most actual applications of ANOVA, the researchers merely make the equal variance assumption and do F without testing the assumption. The reason is that statisticians consider F to be **robust,** a term in statistics meaning *accurate even when underlying assumptions (such as equal population variances) are violated.* This is particularly true when all category sizes are the same. Despite this observation about F, if evidence suggests very unequal population variances, this procedure should be avoided.

Finally, note that just as was the case with the t test, ANOVA assumes that the populations from which the categories are drawn are normally distributed along the dependent variable. In our samples, if category sizes are sufficiently large we may relax the normality assumption. In this respect, ANOVA is the same as the t test.

ANALYSIS OF VARIANCE WITH EXPERIMENTAL DATA

Let us now turn our attention to the training of advertising consultants problem. Recall that H_0: $\mu_{day} = \mu_{night} = \mu_{Saturday}$ and that H_1: there exists at least one inequality that negates H_0. We had 10 trainees each in 3 separate classes.

Day $x_1 =$	*Night* $x_2 =$	*Saturday* $x_3 =$
5	5	5
5	4	5
5	4	5
5	3	4
4	3	4
4	2	4
4	2	3
4	2	3
3	1	3
$n_1 = 10$ 2	$n_2 = 10$ 0	$n_3 = 10$ 2
$\sum x_1 = 41$	$\sum x_2 = 26$	$\sum x_3 = 38$
$\bar{x}_1 = 4.1$	$\bar{x}_2 = 2.6$	$\bar{x}_3 = 3.8$

$$n_{Total} = n_1 + n_2 + n_3$$
$$= 10 + 10 + 10$$
$$= 30$$

$$\sum x_{Total} = \sum x_1 + \sum x_2 + \sum x_3 = 41 + 26 + 38 = 105$$

$x_1^2 =$	$x_2^2 =$	$x_3^2 =$
25	25	25
25	16	25
25	16	25
25	9	16
16	9	16
16	4	16
16	4	9
16	4	9
9	1	9
4	0	4
$\sum x_1^2 = 177$	$\sum x_2^2 = 88$	$\sum x_3^2 = 154$

$$\sum x_{Total}^2 = \sum x_1^2 + \sum x_2^2 + \sum x_3^2 = 177 + 88 + 154 = 419$$

Following the steps in Box 10.1:

1.

$$SS_T = \sum x_T^2 - \frac{\left(\sum x_T\right)^2}{n_T} = 419 - \frac{(105)^2}{30} = 419 - \frac{11{,}025}{30}$$

$$= 419 - 367.50 = 51.50$$

2.

$$SS_B = \frac{\left(\sum x_1\right)^2}{n_1} + \frac{\left(\sum x_2\right)^2}{n_2} + \frac{\left(\sum x_3\right)^2}{n_3} - \frac{\left(\sum x_T\right)^2}{n_T}$$

$$= \frac{(41)^2}{10} + \frac{(26)^2}{10} + \frac{(38)^2}{10} - 367.50 = \frac{1681}{10} + \frac{676}{10} + \frac{1444}{10} - 367.50$$

$$= 168.10 + 67.60 + 144.40 - 367.50$$

$$= 380.10 - 367.50 = 12.60$$

3.

$$SS_W = SS_T - SS_B = 51.50 - 12.60 = 38.90$$

4.

$$MS_B = \frac{SS_B}{df_B} = \frac{SS_B}{\text{no. of categories} - 1} = \frac{12.60}{3 - 1} = \frac{12.60}{2} = 6.30$$

$$MS_W = \frac{SS_W}{df_W} = \frac{SS_W}{(n_{\text{Total}} - 1) - (\text{no. of categories} - 1)} = \frac{38.90}{(30 - 1) - (3 - 1)}$$

$$= \frac{38.90}{29 - 2} = \frac{38.90}{27} = 1.44$$

5.

$$F = \frac{MS_B}{MS_W} = \frac{6.30}{1.44} = 4.375$$

Consulting Table 10.1 for F_{critical} at $df = 2$ and 27:

$$F_{\text{critical}, .05} = 3.35 < 4.375 \qquad \text{reject } H_0$$

$$F_{\text{critical}, .01} = 5.49 > 4.375 \qquad p < .05$$

POST HOC TESTING

In rejecting the null hypothesis, we conclude H_1: in the population there exists at least one inequality that negates H_0. Though we can conclude that in the population there is a difference between student satisfaction and the time that the class is offered, we cannot be sure exactly where that difference lies, since only one inequality negates H_0. The possibilities here are:

$$\mu_D \neq \mu_N \neq \mu_S$$

$$\mu_D \neq \mu_N = \mu_S$$

$$\mu_D = \mu_N \neq \mu_S$$

$$\mu_D = \mu_S \neq \mu_N$$

In the context of this specific problem $\mu_D = \mu_N \neq \mu_S$ is probably illogical since $\bar{x}_D$ is closer to $\bar{x}_S$ than to $\bar{x}_N$, although sampling error could conceivably have yielded these means from a population where $\mu_D = \mu_N \neq \mu_S$ is really true. Also, in the context of sampling we might have a situation where we cannot reject the null hypothesis either between μ_N and μ_S or between μ_S and μ_D but we can reject it between μ_N and μ_D.

Why not run a series of two-sample *t* tests between each pair of means and see which null hypotheses could be rejected? The answer is that since the *t* test was predicated on the assumption that a single comparison of means was to be made, if we make more than one comparison, the probability of a type I or alpha error increases with each comparison.

However, there are a number of tests, known as **post hoc tests of multiple comparisons** *that control for such inflated alpha levels and enable us to narrow our conclusion regarding exactly where these population inequalities are.* One such test is **Scheffé's Test** (pronounced as in French: shef-FAY'). There are other more powerful control procedures, but we present this one because of its flexibility and robustness. It can be applied even when the groups being compared have different sizes (some tests assume equal n's) and it is less sensitive to departures from normality and any assumptions of equal population variances than are some other tests.

Scheffé's Test finds the critical difference between any two sample means that is necessary to reject the null hypothesis that their corresponding population means are equal. If $\mu_D \neq \mu_N$, how big must the difference be between $\bar{x}_D$ and $\bar{x}_N$? This difference, **Scheffé's critical value,** may be calculated between each pair of means, and the actual sample mean differences are compared to the critical values. If $|\bar{x}_i - \bar{x}_j|$ for any two categories, i and j, exceeds Scheffé's critical value, we may reject H_0 and conclude $\mu_i \neq \mu_j$.

We begin by presenting the ANOVA source table for the problem just completed.

Source	*SS*	*df*	*MS*	*F*	*p*
Total	51.50				
Between	12.60	2	6.30	4.375	< .05
Within	38.90	27	1.44		

For any two categories, i and j, the following formula a generates Scheffé's critical value.

$$(\bar{x}_i - \bar{x}_j) = \pm\sqrt{(df_B)(F_{\text{critical}})(MS_W)\left(\frac{1}{n_i} + \frac{1}{n_j}\right)}$$

From the source table, we see that $df_B = 2$ and $MS_W = 1.44$. The F_{critical}, .05 level from Table 10.1 is F_{critical}, .05. $df = 2$ and $27 = 3.35$. Finally, since all of our category n's are equal to 10, $n_i = n_j = 10$. Thus *one* critical value will apply to all three mean comparisons. Had our category sizes been unequal we would have had to calculate a separate critical value for each pair of sample means. Plugging into our formula

$$(\bar{x}_i - \bar{x}_j)_{\text{critical}} = \pm\sqrt{(df_B)(F_{\text{critical}})(MS_W)\left(\frac{1}{n_i} + \frac{1}{n_j}\right)}$$

$$= \pm\sqrt{1.9296} = \pm 1.389$$

In other words, the absolute value of any pair of sample mean differences must equal or exceed 1.389 in order to reject H_0. Examining our sample $\bar{x}$'s we see the following.

H_0	$\lvert\bar{x}_i - \bar{x}_j\rvert =$	Scheffé's Critical Value	Conclusion
$\mu_D = \mu_N$	$\lvert 4.1 - 2.6\rvert = 1.5$	> 1.389	Reject H_0
$\mu_D = \mu_S$	$\lvert 4.1 - 3.8\rvert = 0.3$	< 1.389	Cannot reject H_0
$\mu_N = \mu_S$	$\lvert 2.6 - 3.8\rvert = 1.2$	< 1.389	Cannot reject H_0

Thus, although our overall *F* was significant, we have traced that fact to the single explanation of an inequality between the day and night class population means. We cannot conclude that the population mean for the Saturday group differs from either the day or the night classes.

Suppose, however, that $\lvert\bar{x}_N - \bar{x}_S\rvert$ had been larger than 1.389 and we could also have rejected H_0. Our conclusion would be modified: The significant *F* resulted from the difference between the night class's scores, on one hand, and the combined day and Saturday scores on the other. Since the day and Saturday scores are not significantly different, we might conclude that, since the Saturday classes also met during the daytime, it is the day versus night difference that counts, regardless of which day or days of the week that the day class is held.

As noted earlier, there are many post hoc and a priori tests other than Scheffé's. These include Duncan's Multiple-Range Test, the Student-Newman-Keuls' Multiple Range Test, the Least Significant Difference Test, Tukey's Honestly Significant Difference Test, the Bonferroni procedure, and others. There is considerable debate over which test is most appropriate for specific research situations. Consult an advanced research design text to learn more about them.

ANOVA PRINTOUTS

ANOVA printouts resemble the source tables you have seen in this chapter. In general, SPSS's subprograms ONEWAY and ANOVA use the same terminology used in this chapter. In Table 10.2 you will find the SAS ANOVA procedure's source table for the problem we just completed. Note that what we call *Between,* SAS calls **MODEL** and what we call *Within,* SAS calls **ERROR.** SAS's source column first gives information on the MODEL (between) *df, SS,* and *MS.* Below it comes the ERROR (within) *df, SS,* and *MS.* Below the error information, SAS gives the total *df* and *SS.* To the right of the mean square information is the calculated

Table 10.2 SAS ANOVA Procedure Source Table

SAS

ANALYSIS OF VARIANCE PROCEDURE

DEPENDENT VARIABLE: SCORE

SOURCE	DF	SUM OF SQUARES	MEAN SQUARE	F VALUE	PR > F	R-SQUARE	C.V.
MODEL	2	12.60000000	6.30000000	4.37	0.0226	0.244660	34.2945
ERROR	27	38.90000000	1.44074074		ROOT MSE		SCORE MEAN
CORRECTED TOTAL	29	51.50000000			1.20030860		3.50000000

SOURCE	DF	ANOVA SS	F VALUE	PR > F
CLASS	2	12.60000000	4.37	0.0226

F value, 4.37. To the right of *F*, under PR > F, is the exact probability of falsely rejecting a true null hypothesis, $p = 0.0226$.

The *F* value and probability are summarized on the bottom line, which gives the MODEL (between) *df* and *MS*, followed by *F* and its probability. For our purposes here, we may ignore the other information on the table: R-square, C.V., and Root MSE.

The Scheffé's Test results are given in Table 10.3. The three critical values that we calculated earlier are not even listed in this particular version of SAS. Instead, a letter under the title "Scheffé Grouping" identifies each class. The day class (D) and the Saturday class (S) both share the Scheffé grouping letter A. The means of these two groups are not significantly different. Likewise, the Saturday class (S) and the night class (N) share the grouping letter B. Their means are not significantly different. However, the day class (D) and the night class (N) do not share the same letter. Their means *are* significantly different. (Note that the critical probability level, alpha, has been set at .05.)

Our conclusion from the Scheffé's printout mirrors our earlier conclusions. Students prefer the day class to the night class, but we do not have enough information to conclude anything about the Saturday class versus the other two.

Table 10.3 SAS ANOVA Procedure: Scheffé's Test

```
                                   SAS

                   ANALYSIS OF VARIANCE PROCEDURE

SCHEFFE'S TEST FOR VARIABLE: SCORE

NOTE:  THIS TEST CONTROLS THE TYPE I EXPERIMENTWISE ERROR RATE BUT
       GENERALLY HAS A HIGHER TYPE II ERROR RATE THAN REGWF FOR ALL
       PAIRWISE COMPARISONS

         ALPHA = 0.05, DF = 27, MSE = 1.44074
         CRITICAL VALUE OF F = 3.35413
         MINIMUM SIGNIFICANT DIFFERENCE = 1.3903

      MEANS WITH THE SAME LETTER ARE NOT SIGNIFICANTLY DIFFERENT.

         SCHEFFE        GROUPING            MEAN        N       CLASS

                              A            4.1000      10         D
                              A
                  B           A            3.8000      10         S
                  B
                  B                        2.6000      10         N
```

TWO-WAY ANALYSIS OF VARIANCE

We have only scratched the surface of ANOVA, covering topics most germane to social scientists who will generally use nonexperimental techniques. In behavioral science applications, where experiments are more common, a wide variety of elaborate advanced analysis of variance techniques are used by researchers.

In **two-way analysis of variance** *we extend our model to include a second independent variable.* Suppose in our training example, we also wanted to see if subjects' satisfaction with the course could be explained by the nature of their specialization within the firm. Suppose half of the subjects were employees whose missions emphasized advertising strategies and issues, whereas the other half of the subjects were primarily media consultants who were less issue-oriented and more concerned with communications skills and use of mass media. Do the two groups respond similarly to the scheduling of the course?

Suppose the following pattern emerges:

	Time of Class		
	Day	Night	Saturday
	x =	x =	x =
	5	5	5
	5	4	5
for Advertising Strategists x =	5	4	5
	5	3	4
	4	3	4
	$\bar{x}$ = 4.8	$\bar{x}$ = 3.8	$\bar{x}$ = 4.6
	4	2	4
	4	2	3
for Media Specialists x =	4	2	3
	3	1	3
	2	0	2
	$\bar{x}$ = 3.4	$\bar{x}$ = 1.4	$\bar{x}$ = 3.0

Assuming that F is significant when comparing our 6 means, we see the same trend as before: day and Saturday classes are preferred to night classes for both advertising strategists and media specialists. However, we note the consistently higher satisfaction of the strategists. Their satisfaction is higher than their media colleagues regardless of the time of the class. We would need further research to determine why these differences exist; perhaps media people need a different kind of training program.

On the other hand, suppose the following pattern of scores had emerged.

	Time of Class		
	Day	Night	Saturday
	x =	x =	x =
	5	5	5
	4	4	4
for Advertising Strategists x =	4	4	4
	3	3	3
	2	2	2
	$\bar{x}$ = 3.6	$\bar{x}$ = 3.6	$\bar{x}$ = 3.6
	5	3	5
	5	2	5
for Media Specialists x =	5	2	4
	4	1	3
	4	0	3
	$\bar{x}$ = 4.6	$\bar{x}$ = 1.6	$\bar{x}$ = 5.6

Here, the strategists are unaffected by class scheduling. The sample means are the same, 3.6, for each class. We could assume no difference in the populations for the advertising strategists.

For the media specialists, assuming F is significant, not only are there differences in satisfaction from class to class, but the Saturday session is most popular. Also, both day and Saturday mean scores for the media specialists are greater than those for advertising strategists. Only the night class is unpopular among the media people. Such a situation suggests many possible follow-up studies.

Also, *there are situations where the relationship between the dependent variable and one of the independent variables is a function of the levels of the other independent variable.* These are known as **interaction effects,** and two-way ANOVA also measures such effects.

Social scientists have recently been doing greater numbers of experiments than in the past. For example, one now finds structured simulations of decision-making processes under conditions varied by the experimenter for different groups. Accordingly, two-way ANOVA techniques may soon become as important in social science as they are in so many other fields.

CONCLUSION

We have now seen analysis of variance used in both experimental and nonexperimental contexts and discussed Scheffé's Test as well as several other procedures related to ANOVA. In Chapter 14, we will demonstrate another context in which this procedure is applied, namely as part of the regression procedure. At that point, we shall have completed the process of weaving together the two statistical strands—descriptive and inferential—that have run through this text.

EXERCISES

Exercise 10.1

Here is one of the example problems from Chapter 9. Perform ANOVA. Explain why your conclusion differs from the one reached with the two-sample t test.

Saw Commercial	Did Not See Commercial
$x_1 =$	$x_2 =$
10	8
6	3
8	5
7	6
9	7
7	

Exercise 10.2

A scale measuring support for increased gun control legislation (0, no support to 5, most support) is administered to random samples of urban, suburban, and rural voters. Do the three population means differ in terms of support? If so, do Scheffé's Test. What do you conclude?

Urban	Suburban	Rural
$x_1 =$	$x_2 =$	$x_3 =$
5	4	0
4	3	1
3	5	2
4	4	0
5	3	1

Exercise 10.3

The same attitude scale used in the previous exercise is applied to random samples of urban police officers, white-collar workers, and blue-collar workers. Do ANOVA, and if the null hypothesis can be rejected, do Scheffé's Test. What do you conclude?

Police	White Collar	Blue Collar
$x_1 =$	$x_2 =$	$x_3 =$
5	4	1
4	3	3
5	4	2
5	4	0
3	1	1
4	5	
5		

Exercise 10.4

For a random sample of Democrats in the U.S. House of Representatives, liberalism scores were compared by region of the country. Find *F* and if statistically significant, do Scheffé's Test. What do you conclude?

Northwest	*South*	*Midwest*	*Far West*
$x_1 =$	$x_2 =$	$x_3 =$	$x_4 =$
95	80	75	75
90	60	85	95
90	45	55	85
95	65	80	46
80	75	70	85

Exercise 10.5

For a random sample of 25 physicians, scores measuring support for a national health care insurance program were compared by medical specialization. Complete the resulting source table. What are your conclusions?

Source	*SS*	*df*	*MS*	*F*	*p*
Total	32106.00	24			
Between	21462.25	1	____	____	____
Within	10643.75	23	____		

Exercise 10.6

From an experiment measuring the cognitive learning of learning disabled students by various teaching strategies, ANOVA was run. Complete the source table and state your conclusions.

Source	*SS*	*df*	*MS*	*F*	*p*
Total	7339.84	____			
Between	388.09	1	____	____	____
Within	____	23	____		

Exercise 10.7

The same study as in Exercise 10.6 was done with non-learning disabled students. Complete and interpret the source table.

EXERCISES

Source	*SS*	*df*	*MS*	*F*	*p*
Total	______	24			
Between	4093.87	1	____	43.97	____
Within	2141.49	__	____		

Exercise 10.8

ANOVA and Scheffé's Test were run using the *World Handbook* sample presented in Chapter 6 to compare per capita GNP to political rights (high, medium, and low). Review and interpret the following printouts. Note that the particular version of SAS run here presents Scheffé's Test differently from the version presented in this chapter. The three asterisks indicate significantly different category means.

```
DEPENDENT VARIABLE: GNP

SOURCE      DF  SUM OF SQUARES        MEAN SQUARE    F VALUE   PR > F

MODEL        2  152128944.99999900 76064472.49999990 16.12    0.000

ERROR       17   80219150.00000000  4718773.52941177

CORRECTED   19  232348095.00000000
TOTAL

ALPHA = 0.05 CONFIDENCE = 0.95 DF + 17 MSE = 4718774
           CRITICAL VALUE OF F = 3.59153

COMPARISONS SIGNIFICANT AT THE 0.05 LEVEL ARE INDICATED BY '***'


                     SIMULTANEOUS               SIMULTANEOUS
                        LOWER     DIFFERENCE      UPPER
            PRIGHTS   CONFIDENCE   BETWEEN     CONFIDENCE
           COMPARISON    LIMIT       MEANS        LIMIT
           H        -M    -189        3753         7694
           H        -L    3147        5976         8805   ***

           M        -H   -7694       -3753          189
           M        -L   -1658        2223         6105

           L        -H   -8805       -5976        -3147   ***
           L        -M   -6105       -2223         1658
```

EXERCISES

Exercise 10.9

In a certain study, scores earned on a graduate school admissions test were compared between those who had no formal preparation and those taking courses designed to prepare students for the exam. Interpret the printout with regard to statistical significance.

DEPENDENT VARIABLE: SCORE

SOURCE	DF	SUM OF SQUARES	MEAN SQUARE	F VALUE	PR > F
MODEL	1	6.12500000	6.12500000	10.28	0.0032
ERROR	30	17.87500000	0.59583333		
CORRECTED TOTAL	31	24.00000000			

Exercise 10.10

Here is a study of pilot reaction times under two different instrument panel configurations. Interpret the printout.

SOURCE	DF	SUM OF SQUARES	MEAN SQUARE	F VALUE	PR > F
MODEL	1	3.38000000	3.38000000	7.14	0.0120
ERROR	30	14.19500000	0.47316667		
CORRECTED TOTAL	31	17.57500000			

NOTE

1. If you have read the previous chapter, you are already familiar with the use of the *F* table. However, note that here we have tables for the .01 and .001 levels as well as for the .05 level. In the last chapter we used only the .05 level table.

KEY CONCEPTS

measures of association
concordant pairs
discordant pairs
Yule's *Q*
phi coefficient
Goodman and Kruskal's gamma
Goodman and Kruskal's lambda
symmetric versus assymmetric measures of association
proportionate reduction in errors [PRE] measures
lambda symmetric
curvilinearity versus linearity in tables
Pearson's *C* - the contingency coefficient
Cramer's *V*
Kendall's tau-*b*
Kendall's (or Stuart's) tau-*c*
Somer's *d*
Goodman-Kruskal's uncertainty coefficient
correlation coefficient
correlation (association) matrix

11 Measuring Association in Contingency Tables

INTRODUCTION

Generating a table of percentages is a common technique for hypothesis testing, but the results in a table are not always clearly understood and sometimes seem inconsistent. Moreover, except for the extremes of perfect relationship and no relationship, it is not possible to see the exact amount of relationship that exists between the variables. Measures of association, which often accompany tables, seek to resolve this problem by showing the amount and nature of the relationship presented.

Measures of association *are index numbers that generally range in magnitude from 0 (measuring no association) to 1 (meaning perfect association). Often these numbers are signed, with a positive sign (+) denoting a positive relationship between the variables and a negative sign (–) indicating an inverse relationship.* There are dozens of such measures, each designed for specific table sizes and levels of measurement of the variables. Most of these measures are patterned after the Pearsonian correlation coefficient (Pearson's *r*), which we will encounter in Chapter 13.

Most of the computer programs that generate tables (cross-tabs) for social science applications also generate a series of measures of association. Each measure generated may or may not be valid for that cross-tab, depending on features to be discussed shortly.

In this chapter, four of the most generally encountered measures will be presented in detail. They will be sufficient for use in any data situation you are likely to encounter.

MEASURES FOR TWO-BY-TWO TABLES

In a two-by-two table, each variable is a dichotomy, a two-category scale. Although these scales may either be nominal or ordinal levels of measurement, it is not necessary with dichotomies to differentiate nominal from ordinal levels. This phenomenon was mentioned in Chapter 2, where we pointed out that since it is impossible to destroy the sequencing of a dichotomy's categories, even a logically nominal dichotomy can be treated as if it were ordinal level of measurement.

Consider Table 11.1, a two-by-two table where respondent's income (high versus low) is related to respondent's level of participation in fund-raising activities for a charity (high versus low). Note that the table entries are frequencies, not percentages. Every measure of association presented here is calculated from the frequencies, even if the final accompanying table is in percentages. For ease in visualization, let us assume also that the marginal totals are equal in magnitude.

Whenever both variables are ordinal, as is the case in Table 11.1, the categories are listed so that the upper left-hand cell of the table will contain those cases scoring high on both variables and the lower right-hand cell will contain the cases scoring low on each variable. Then, to the extent that the relationship between the two variables is positive, the clustering of cases will be on the main diagonal (upper left

Table 11.1

	Income		
Participation	*High*	*Low*	*Total*
High	50	0	50
Low	0	50	50
Total	50	50	100

Table 11.2

	Attitude on Intervention		
Attitude on Death Penalty	*For*	*Against*	*Total*
For	15	5	20
Against	5	15	20
Total	20	20	40

to lower right). If the relationship is inverse, the upper-right to lower-left cells, the off diagonal, will show the clustering. In the Table 11.1, all 100 respondents or subjects cluster on the main diagonal, indicating a positive relationship. Had that relationship been inverse, the clustering would have been along the off diagonal (high participation with low income; low participation with high income).

When one or both of the variables is measured at only the nominal level, we set up the table so that if the hypothesis is verified, the cases will cluster on the main diagonal. Consider the example used in Chapter 7. Suppose our hypothesis is that one's attitude on the issue of U.S. intervention in Latin America to stop cocaine production and distribution is related to one's attitude on the issue of use of the death penalty on convicted high-level drug dealers ("kingpins") inside the United States. Supporters of intervention also support the death penalty. (See Table 11.2.)

Though not a perfect relationship, 30 of the 40 subjects studied were consistent with the hypothesis, 15 for both intervention and the death penalty and 15 against both intervention and the death penalty. The remaining 10 respondents are inconsistent with the hypothesis, with 5 against intervention but for the death penalty and 5 for intervention but opposed to the death penalty. The measures of association we will calculate for Table 11.1 will have positive signs, denoting that clustering was along the main diagonal. A clustering on the off diagonal would yield a negative measure of association: intervention attitude would be related to capital punishment attitude, but *not* in the predicted direction.

YULE'S *Q*

Yule's *Q* is an easy-to-calculate and easy-to-interpret measure of association for any two-by-two table regardless of level of measurement. To utilize the formula for *Q*, we define the cell entries in our table with alphabetical letters as follows:

a	b
c	d

For Table 11.2, which had the following entries,

15	5
5	15

a = 15, b = 5, c = 5, and d = 15.

For Table 11.1,

50	0
0	50

a = 50, b = 0, c = 0, and d = 50.

The formula for *Q* is:

$$Q = \frac{ad - bc}{ad + bc}$$

Noting that *ad* means *a* times *d* and *bc* means *b* times *c*, *Q* for Table 11.2 would be:

$$Q = \frac{ad - bc}{ad + bc} = \frac{(15)(15) - (5)(5)}{(15)(15) + (5)(5)} = \frac{225 - 25}{225 + 25} = \frac{200}{250} = +0.80$$

The positive sign indicates that the relationship is in the direction predicted by the hypothesis. Since *Q* ranges from 0 (no relationship) to 1.00 (a "perfect" relationship) in magnitude, we may assume that a *Q* of .80 is a fairly strong relationship.

The value of *Q* has a technical interpretation that can be briefly explained at this point. This interpretation is based on pairs of responses. Assuming that in Table 11.2, our hypothesis is that that those favoring intervention will also favor the death penalty and those against intervention will also be against the death penalty, any respondent who is either "for" on both responses or "against" on both responses will be consistent with the hypothesis. *Such consistent pairs of responses are also called* **concordant pairs.** *By contrast, pairs of responses inconsistent with the hypothesis—for intervention and against the death penalty or against intervention but for the death penalty—are called* **discordant pairs. Yule's *Q*** *is the proportionate excess of concordant over discordant pairs of observa-*

Table 11.3

Participation	Income: High	Low	Total
High	25	25	50
Low	25	25	50
Total	50	50	100

tions. Said another way, Q is the net probability of randomly selecting a pair of responses from the table that is consistent with the hypothesis. In this instance, we have a .80 net probability (an 80% chance) of randomly selecting a respondent whose responses are consistent with the hypothesis.

For the data in Table 11.1,

$$Q = \frac{ad - bc}{ad + bc} = \frac{(50)(50) - (0)(0)}{(50)(50) + (0)(0)} = \frac{2500 - 0}{2500 + 0} = \frac{2500}{2500} = +1.00$$

There is a perfect positive relationship between participation and income. Since $bc = 0$, there are no pairs of responses in Table 11.1 that are inconsistent with the hypothesis. Thus we have a 100% chance (a perfect chance) of randomly selecting a respondent whose responses are consistent with the hypothesis.

Using the same variables as in Table 11.1, an example of *no* relationship (see Table 11.3) would be as follows:

$$Q = \frac{ad - bc}{ad + bc} = \frac{(25)(25) - (25)(25)}{(25)(25) + (25)(25)} = \frac{625 - 625}{625 + 625} = \frac{0}{1250} = 0$$

Since $ad = bc$, consistent pairs equal inconsistent pairs, the numerator is 0 and the quotient is 0.

A perfect inverse relationship, all pairs discordant, (see Table 11.4) would look like this:

$$Q = \frac{ad - bc}{ad + bc} = \frac{(0)(0) - (50)(50)}{(0)(0) + (50)(50)} = \frac{0 - 2500}{0 + 2500} = \frac{-2500}{2500} = -1.00$$

Remember that a negative numerator divided by a positive denominator yields a negative quotient.

Table 11.4

Participation	Income High	Income Low	Total
High	0	50	50
Low	50	0	50
Total	50	50	100

Table 11.5

Participation	Income High	Income Low	Total
High	25	0	25
Low	35	45	80
Total	60	45	105

One problem with Q is that the presence of a zero in any cell causes the final quotient to have a value of either +1.00 or −1.00 (i.e., ±1.00, read "plus or minus one" or "positive or negative one"). In Tables 11.1 and 11.4, the relationships were perfect (all cases conformed to the hypothesis) and Q was +1.00 and −1.00, respectively. Sometimes, however, there are exceptions to the hypothesis, but one cell entry of 0 gives Q a magnitude of 1 even though the relationship between the variables is not perfect (see Table 11.5).

$$Q = \frac{ad - bc}{ad + bc} = \frac{(25)(45) - (0)(35)}{(25)(45) + (0)(35)} = \frac{1125 - 0}{1125 + 0} = \frac{1125}{1125} = +1.00$$

Despite a Q of 1.00, the relationship is far from perfect. There were 35 high income respondents with low participation, a contradiction to the hypothesis. Another measure, the phi coefficient (φ), does not share that particular characteristic with Q and is sometimes preferred for that reason, despite a slightly more complicated computation.[1]

THE PHI COEFFICIENT

For the phi coefficient, we use the same lettering format as was used with Q.

a	b	(a + b)
c	d	(c + d)
(a + c)	(b + d)	(a + b + c + d)

The formula for **phi** (pronounced to rhyme with the word "bee" rather than the word "buy") is:

$$\varphi = \frac{ad - bc}{\sqrt{(a+b)(c+d)(a+c)(b+d)}}$$

For Table 11.5, where $Q = +1.00$,

$$\varphi = \frac{ad - bc}{\sqrt{(a+b)(c+d)(a+c)(b+d)}} = \frac{(25)(45) - (0)(35)}{\sqrt{(25)(80)(60)(45)}}$$

$$= \frac{1125}{\sqrt{5{,}400{,}000}} = \frac{1125}{2323.79} = +0.4841$$

Rounding back two decimal places, $\varphi = +0.48$.

Now notice for Table 11.1, where Q was also +1.00 but the relationship was a perfect one, φ reflects that fact.

$$\varphi = \frac{ad - bc}{\sqrt{(a+b)(c+d)(a+c)(b+d)}} = \frac{(50)(50) - (0)(0)}{\sqrt{(50)(50)(50)(50)}}$$

$$= \frac{2500 - 0}{\sqrt{6{,}250{,}000}} = \frac{2500}{2500} = +1.00$$

For the data in Table 11.2, where Q was +.80, phi is:

$$\varphi = \frac{ad - bc}{\sqrt{(a+b)(c+d)(a+c)(b+d)}} = \frac{(15)(15) - (5)(5)}{\sqrt{(20)(20)(20)(20)}}$$

$$= \frac{225 - 25}{\sqrt{160{,}000}} = \frac{200}{400} = +0.50$$

Note that in comparing one table to another, we always compare Q to Q or φ to φ; never do we compare Q in one table to φ in another. Also, while φ will not be misleading the way Q can sometimes be, all measures of association possess certain quirks or defects. This is because in summarizing a table with a single number, some information is naturally lost. All we can do to minimize this problem is make sure that we do not select a measure that is going to mislead our readers.

Phi is actually a variation on a measure known as Pearson's *r*, the correlation coefficient, which we will discuss in Chapter 13.

MEASURES FOR *n*-BY-*n* TABLES

Whenever one of our variables has more than two categories (an *n*-by-*n* table), our selection of a measure of association is complicated by the fact that we must first determine each variable's level of measurement. The appropriate measure of association may then be selected. We learn two measures next, Goodman and Kruskal's gamma (γ) for ordinal-by-ordinal tables and Goodman and Kruskal's lambda (λ) for nominal-by-nominal tables.[2]

GOODMAN AND KRUSKAL'S GAMMA (γ)

Gamma *is a measure designed for ordinal-by-ordinal tables. It may also be used when one of the two variables is a nominal dichotomy* (which we may handle as if it were ordinal). Yule's *Q* is actually a special case of gamma for a two-by-two table, but we treat it here as a separate measure for the sake of simplicity and treat gamma as a measure for tables larger than two-by-two. The formula for gamma is

$$\gamma = \frac{P_s - P_d}{P_s + P_d}$$

P_s and P_d are first determined for the table and then gamma is calculated.

As with *Q*, gamma evaluates the net proportion of concordant pairs of observations—those consistent with the hypothesis—as compared with all pairs of observations. The definitions of concordant and discordant pairs in a larger than two-by-two table is somewhat more complex than in the case of Yule's *Q*. Concordant pairs, (P_s), are ordered the same on each variable, cluster around the main diagonal, and indicate a positive relationship in the table. Discordant pairs, P_d, are ordered higher on one variable than the other, cluster about the off diagonal, and

Table 11.6

Participation	Income High	Income Medium	Income Low
High	5	1	1
Medium	2	4	2
Low	0	1	4

suggest an inverse relationship between the variables. If P_s is greater than P_d, the numerator is a positive number and so is gamma. If $P_s = P_d$, concordant and discordant pairs balance, the numerator becomes 0, and gamma becomes 0. If P_s is less than P_d, the numerator—and gamma—will be negative, and the relationship in the table will be inverse.

10	0	0
0	10	0
0	0	10

$\gamma = +1.00$

5	5	5
5	5	5
5	5	5

$\gamma = 0$

0	2	8
2	8	2
8	2	0

$\gamma = -.94$

Instead of using formulas to find P_s and P_d, we utilize an algorithm, a set of instructions that applies to tables of any size. In Table 11.6, we examine fund-raising participation versus income, with each variable measured as a three-category ordinal scale.

To calculate P_s, we begin by generating a series of sub-tables from Table 11.6. To do this, we begin in the upper left-hand cell. We take note of that cell entry and create a sub-table containing everything that is both below and to the right of the cell entry. We ignore any marginal totals on the table (thus they are left off in Table 11.6). We can also skip any cell whose entry is zero as well as any cell that has nothing below it and to the right. For the upper left-hand cell, we get the following:

5

4	2
1	4

Continuing across the top row of the table:

1

2
4

When we come to the low income, high participation category we see that there is nothing below it and to the right, so we ignore it.

Having exhausted the top row, we move to the far left cell of the next row down and begin again.

2

1	4

Moving one cell to the right:

4

4

Once again, when we move to the far right cell on that row, we find nothing below it and to the right, so we ignore it. This brings us to the far left cell of the bottom row. Noting that for the *entire* bottom row there is nothing below and to the right, we have no more sub-tables to prepare. Thus we have generated the following:

5

4	2
1	4

1

2
4

2

1	4

4

4

Next, we *add* the numbers in each sub-table and *multiply* the sum by the original cell entry above and to the left of the sub-table.

$$\begin{aligned} 5\,(4 + 2 + 1 + 4) &= 5\,(11) = 55 \\ 1\,(2 + 4) &= 1\ \ (6) = \ \ 6 \\ 2\,(1 + 4) &= 2\ \ (5) = 10 \\ 4\,(4) &= 4\ \ (4) = 16 \end{aligned}$$

The sum of the four resulting products is P_s.

$$\begin{array}{r} 55 \\ 6 \\ 10 \\ +\ 16 \\ \hline P_s = 87 \end{array}$$

To find P_d, we go "through the looking glass." We start at the upper-right hand cell of the table and move *left* on the row. This time we look to see what is *below* us and also to the *left*. In short, we follow in reverse the procedure used to find P_s.

4*

0	

2

0	1

1

2
0

1

2	4
0	1

(*We could have left this sub-table out since there is only a zero below and to the left of the cell entry.)

We add the numbers in each subtable, multiply the total by the cell entry above and to the right of the subtable, and add these products to find P_d. These calculations are presented in the order that the subtables were generated moving right to left theough the above subtables.

$$\begin{aligned} 1\,(2 + 4 + 0 + 1) &= 1\,(7) = 7 \\ 1\,(2 + 0) &= 1\,(2) = 2 \\ 2\,(0 + 1) &= 2\,(1) = 2 \\ 4\,(0) &= 4\,(0) = \underline{0} \\ P_d &= 11 \end{aligned}$$

Since $P_s = 87$ and $P_d = 11$,

$$\gamma = \frac{P_s - P_d}{P_s + P_d} = \frac{87 - 11}{87 + 11} = \frac{76}{98} = +0.7755 = +\,0.78$$

Gamma is inappropriate for nominal level data since it assumes that the table is composed of two variables, each with categories appearing in logical sequence, as was the case in Table 11.6. With nominal variables there is no logical sequence to the categories. Accordingly, for nominal level data, we need a measure of association that will remain the same for any given table no matter how the categories are ordered. To illustrate the problem, let us recast in Table 11.7 the data from Table 11.6 but with the income categories no longer in logical sequence.

Let us calculate gamma for this table.

For P_s:

5

2	4
4	1

1

4
1

2

4	1

2

1

$$\begin{aligned} 5\,(2 + 4 + 4 + 1) &= 5\,(11) = 55 \\ 1\,(4 + 1) &= 1\,(5) = 5 \\ 2\,(4 + 1) &= 2\,(5) = 10 \\ 2\,(1) &= 2\,(1) = +\,\underline{2} \\ P_s &= 72 \end{aligned}$$

Table 11.7

	Income		
Participation	*High*	*Low*	*Medium*
High	5	1	1
Medium	2	2	4
Low	0	4	1

For P_d:

2
0

4	
0	4

1
2
0

1	
2	2
0	4

$$1\,(2 + 2 + 0 + 4) = 1\,(8) = 8$$
$$1\,(2 + 0) = 1\,(2) = 2$$
$$4\,(0 + 4) = 4\,(4) = 16$$
$$2\,(0) = 2\,(0) = +\ 0$$
$$P_d = \overline{26}$$

Therefore

$$\gamma = \frac{P_s - P_d}{P_s + P_d} = \frac{72 - 26}{72 + 26} = \frac{46}{98} = +0.469 = +0.47$$

Compare this result to the gamma of 0.78 obtained from Table 11.6. A measure of association designed for nominal-by-nominal data would have yielded the *same* result for either table; it would be insensitive to the ordering of the categories.

GOODMAN AND KRUSKAL'S LAMBDA (λ)

Lambda *is designed for a table where at least one variable is nominal and is not a dichotomy.* If gamma is inappropriate for the given table, we may use lambda. Since lambda is designed for nominal-by-nominal tables, it carries *no* sign (+ or –). Directionality of the relationship must be explained by reference to the variables in the table.

Unlike the other measures discussed in this chapter, lambda is an **asymmetric measure of association,** *meaning that the value of the measure depends on whichever variable was dependent.* In fact, there are actually *three* lambdas for any given table: one for when the variable

Table 11.8

Occupation	*Nationality* Russian	Ukrainian	Belorussian	*Total*
Professional	10	15	20	45
Blue Collar	20	15	5	40
Farmer	10	20	5	35
Total	40	50	30	120

whose categories are the table's rows is the dependent variable, one for when the variable whose categories are the table's columns is the dependent variable, and one an average of the first two. All our previously discussed measures were **symmetric;** *there was only one measure for any given table and it did not matter which variable was dependent or independent.* In Table 11.6, for instance, gamma was +.78 whether we were explaining income based on participation or whether we were explaining participation based on income. With lambda, we have to decide *in advance* what the dependent variable is, income or participation.

LAMBDA—COLUMN VARIABLE DEPENDENT

In Table 11.8 we are studying a group of recent immigrants from the former U.S.S.R. who have come to North America. We have a three-category nominal scale for occupation against a three-category nominal scale for nationality grouping in the former Soviet Union. Gamma for this table is inappropriate; therefore we calculate lambda.

Let us first assume that the dependent variable is nationality and we wish to predict nationality based on knowledge of the respondent's occupation. To find the appropriate lambda we first determine two figures, called E_1 and E_2. Lambda is based on a strategy for predicting respondents' scores on the dependent variable. To find E_1, we play a game in which we know initially only the marginal totals of the dependent variable: We do *not* know the data in the cells of the table itself.

	Nationality Russian	Ukrainian	Belorussian	*Total*
Total	40	50	30	120

The strategy is to place all subjects in the largest category of nationality and see how many assignment errors we will make. The largest nationality category is Ukrainian, with 50 of the 120 respondents so listed. If we were to put all 120 subjects in the Ukrainian category, we would have correctly assigned the 50 respondents who actually were Ukrainian but would have incorrectly assigned 70 people, the 40 Russians and the 30 Belorussians. The total number of assignment errors, 70, is E_1. Note that we make fewer errors assigning all 120 to Ukrainian than we would make assigning them to either of the other two categories of the dependent variable. If we assigned everyone to the Russian category we would incorrectly be assigning 50 Ukrainians and 30 Belorussians to the wrong category, and we would make 80 assignment errors. If we put all 120 in the Belorussians category we would correctly assign 30 but incorrectly assign 90 to the wrong category. Thus the lambda strategy is to put everyone in the *largest* category and then tally up the number of errors of assignment made.

$$E_1 = 40 \text{ Russians} + 30 \text{ Belorussians} = 70 \text{ total errors.}$$

The logic behind lambda is that the greater the relationship between the two variables in the table, the fewer will be our assignment errors when we know the information in the complete table. E_2 is the total number of assignment errors made when we know all of the data in the table. In effect, we assign each respondent to a category of the dependent variable based on knowledge of that respondent's category of the independent variable. We assign by each category of the independent variable, keeping tabs of the errors in assignment made. When we are done with all categories of the independent variable, we add up all the errors made to get E_2.

Let us take the first category of the independent variable.

		Nationality		
Occupation	*Russian*	*Ukrainian*	*Belorussian*	*Total*
Professional	10	15	20	45

The largest nationality category for the professional group is Belorussian. If we assign all 45 professionals to Belorussian, we will correctly assign 20 people, but we will incorrectly assign 10 Russians and 15 Ukrainians. Thus we will make a total of 25 assignment errors for that category of our independent variable. Now let us look at the next occupational category.

	Nationality			
Occupation	*Russian*	*Ukrainian*	*Belorussian*	*Total*
Blue Collar	20	15	5	40

Here, the largest nationality category is Russian, with 20 of the 40 blue-collar workers being Russians. If we put all 40 in the Russian category, we will correctly assign 20 but incorrectly assign the 15 Ukrainian and 5 Belorussian blue-collar workers; Thus 20 errors are made for this category of occupation. For the last occupational category:

	Nationality			
Occupation	*Russian*	*Ukrainian*	*Belorussian*	*Total*
Farmer	10	20	5	35

Here the largest category is Ukrainian. Putting all 35 farmers in the Ukrainian slot, we will correctly assign 20 but incorrectly assign 15 (10 Russians and 5 Belorussians).

Now we tally up:

Occupational Category	*Assignment Errors Made*
Professional	25
Blue Collar	20
Farmer	15
Total	E_2 = 60

We made 70 prediction errors (E_1) knowing only the marginal totals for nationality. Knowing all the information in the table, we reduced the number of assignment errors to 60. Lambda is the **proportionate reduction in error (PRE).** We find the total reduction in errors (70 – 60 = 10) and express it as a proportion of E_1, thus:

$$\lambda = \frac{E_1 - E_2}{E_1} = \frac{70 - 60}{70} = \frac{10}{70} = .1428 = .14$$

Knowing the data in the table, we reduce our assignment errors by a proportion of .14 or 14%.

LAMBDA—ROW VARIABLE DEPENDENT

To find the lambda where occupation is dependent, we follow the same procedures but work column-by-column.

Occupation	*Total*
Professional	45
Blue Collar	40
Farmer	35
Total	120

If we put all 120 respondents in the largest occupational category, we will correctly assign 45 people but incorrectly assign a total of 75 people (40 blue collar and 35 farmers). Thus $E_1 = 75$.

We now go column-by-column through the table to find E_2.

Occupation	*Russian*		*Ukrainian*		*Belorussian*		
Professional	10		15		20		
Blue Collar	20		15		5		
Farmer	10		20		5		
Total =	40		50		30		
E_2 = Assignment Errors =	20	+	30	+	10	=	60

If we put all 40 Russians in their largest category of occupation (blue collar), we will be right for 20 but wrong for another 20 (10 professionals and 10 farmers). Thus we make 20 assignment errors here. For the Ukrainians, the largest category is farmer. Calling all 50 Ukrainians farmers, we will be right for 20 but wrong for 30. For the Belorussians, the largest category is professional. Putting all 30 Belorussians in that category results in 10 assignment errors. Thus

$$E_2 = 20 + 30 + 10 = 60$$

Since E_1 was 75

$$\lambda = \frac{E_1 - E_2}{E_1} = \frac{75 - 60}{75} = \frac{15}{75} = .20$$

Thus we get a 20% reduction in error when predicting occupational type knowing respondents' nationality categories.

Sometimes we may have both lambdas but have no basis for calling one or the other variable the dependent variable. In such instances we can "average" the two lambdas to get a **lambda symmetric.** Adding the two lambdas together and dividing by two, we get

$$\lambda_{\text{symmetric}} = \frac{.14 + .20}{2} = \frac{.34}{2} = .17$$

Note also that in general, lambdas tend to produce lower numbers than the other measures of association covered in this chapter. For the problem in Table 11.6 where gamma was .78, we could calculate lambdas for the table and compare them to the gamma. Although this is an ordinal-by-ordinal table, finding lambda *is* permissible since it assumes a lower level of measurement (nominal-by-nominal). Although it is not appropriate to calculate gamma for a nominal-by-nominal table, since gamma assumes a higher level of measurement. Let us find lambdas for Table 11.6.

		Income			
Participation	*High*	*Medium*	*Low*	*Total=*	*Assignment Errors*
High	5	1	1	7	2
Medium	2	4	2	8	4
Low	0	1	4	5	1
Total =	7	6	7	20	$E_2 = 7$

For predicting income from participation, we assign all 20 respondents to the largest income category. Both high income and low income have marginal totals of 7; it does *not* matter which one we use. Either way, we correctly assign 7 and incorrectly assign 13. Thus $E_1 = 13$. Going row-by-row, for the 7 respondents with high participation, the biggest category of income is high. Putting all 7 in high income, we correctly assign 5 but incorrectly assign 2. For the 8 medium participation respondents, the largest income category is medium. Assigning all 8 to medium income, we make four errors. Finally, putting all 5 with low participation in the biggest income category, low, we make one error. Thus:

$$\lambda = \frac{E_1 - E_2}{E_1} = \frac{13 - 7}{13} = \frac{6}{13} = .46$$

When we treat participation as the dependent variable, we first assign all 20 to the largest marginal category for participation, medium. We correctly assign 8, but incorrectly assign 12; $E_1 = 12$. Going column-by-column, we note the following assignment errors.

Participation	*High*	*Medium*	*Low*	*Total*
High	5	1	1	7
Medium	2	4	2	8
Low	0	1	4	5
Total =	7	6	7	20
E_2 = Assignment Errors =	2 +	2 +	3 = 7	

$$E_2 = 2 + 2 + 3 = 7$$

$$\lambda = \frac{12 - 7}{12} = \frac{5}{12} = .42$$

Therefore,

$$\lambda_{\text{symmetric}} = \frac{.46 + .42}{2} = \frac{.88}{2} = .44$$

Recall that the same table produced a gamma of .78.

Curvilinearity

Sometimes when gammas and lambdas are calculated for the same ordinal-by-ordinal table, we get seemingly contradictory results. Consider Table 11.9. We begin by calculating gamma.

For P_s 15

5	5
15	0

5

15	0

Table 11.9

Radicalism	*Income* High	*Medium*	*Low*	*Total =*
High	15	0	15	30
Medium	5	5	5	15
Low	0	15	0	15
Total	20	20	20	60

$$15(5 + 5 + 15 + 0) = 15(25) = 375$$
$$5(15 + 0) = 5(15) = \underline{\ 75}$$
$$P_s = 450$$

For P_d

5

0	15

15

5	5
0	15

$$15(5 + 5 + 0 + 15) = 15(25) = 375$$
$$5(0 + 15) = 5(15) = \underline{\ 75}$$
$$P_d = 450$$

Therefore,

$$\gamma = \frac{P_s - P_d}{P_s + P_d} = \frac{450 - 450}{450 + 450} = \frac{0}{900} = 0$$

Using gamma, we were unable to find a relationship in the table. But note what happens when we find lambda. For income dependent,

$$E_1 = 40$$

$$E_2 = 15 + 10 + 0 = 25$$

$$\lambda = \frac{E_1 - E_2}{E_1} = \frac{40 - 25}{40} = \frac{15}{40} = .375 = .38$$

For radicalism dependent,

$$E_1 = 30$$

$$E_2 = 5 + 5 + 5 = 15$$

$$\lambda = \frac{E_1 - E_2}{E_1} = \frac{30 - 15}{30} = \frac{15}{30} = .50$$

Therefore,

$$\lambda_{\text{symmetric}} = \frac{.38 + .50}{2} = \frac{.88}{2} = .44$$

Table 11.10

	Income			
Traditionalism	*High*	*Medium*	*Low*	*Total*
High	0	5	0	5
Medium	0	5	0	5
Low	10	0	10	20
Total	10	10	10	30

For lambdas, a .44 is quite large. This suggests that the two variables are related, even though gamma was 0. This apparent contradiction has to do with the *nature* of the relationship between the variables. Gamma is sensitive only to *linear* relationships, where the cases in the table cluster along a straight diagonal line, either the main diagonal or the off diagonal. The clustering in Table 11.9 is not in a straight line but rather in a curve.

	High	*Medium*	*Low*
High	15	0	15
Medium	5	5	5
Low	0	15	0

Note in particular where the largest cell entries in each category fall.

	High	*Medium*	*Low*
High	15	0	15
Medium	5	5	5
Low	0	15	0

We call such a relationship **curvilinear** *to differentiate it from a straight-line, or* **linear,** *relationship.* In the case above, high levels of radicalism are associated with *both* high and low income and low levels of radicalism are associated with medium income levels.

Table 11.10 gives another example of curvilinearity. The curve is reversed, with both high and low income associated with *low* traditionalism and middle income associated with *high* traditionalism. If we calculate gamma, we will find it to be 0, whereas lambda is .50 whichever variable is dependent.

If a relationship is curvilinear, a measure of association designed for nominal-by-nominal tables is more likely to indicate that fact than an ordinal-by-ordinal measure. This is because such a measure is sensitive to all relationships, not just those clustering along a diagonal. Accordingly, if you have an ordinal-by-ordinal table and you calculate a gamma that is small, it is a good idea to inspect the table again. If curvilinearity appears to exist in the table, do a nominal-by-nominal measure such as lambda.

OTHER MEASURES OF ASSOCIATION

In addition to the measures discussed above, there are dozens of others. In the two most common sets of statistical computer programs for the social sciences, SPSS and SAS, cross-tab programs calculate phi, lambda (asymmetric as well as symmetric), and gamma. (Remember that for a 2×2 table, gamma is the same as Yule's *Q*. Thus all the measures discussed here are generated by these two computer programs.) In addition, SPSS and SAS generate other measures. For our purposes, it is only necessary to be aware of the existence of these measures and their approximate equivalence to the ones we have learned.

Pearson's *C* (the contingency coefficient) and **Cramer's *V*** are similar to the phi coefficient but apply to tables larger than 2×2 as well. *C* and *V* for a 2×2 table are the same as phi. (It is possible, using a different formula than the one we presented, to calculate phi for a table larger than 2×2, but *C* and *V* are easier to interpret and, therefore, preferable to phi if the table is larger than 2×2.)

Kendall's tau-*b* and **Kendall's (or Stuart's) tau-*c*** are similar to gamma and, like gamma, are symmetric measures of association. If the number of rows differs from the number of columns, tau-*c* is preferred to tau-*b*. If the number of rows equals the number of columns, tau-*b* is preferred.

Somer's *d* is similar to gamma, but unlike gamma it is asymmetric, as was the case with lambda.

Goodman-Kruskal's uncertainty coefficient is functionally similar to Goodman and Kruskal's lambda.

Finally, all of the above asymmetric measures may be made symmetric by "averaging" the two asymmetric measures for a given table, the way we did with lambda.

INTERPRETING A CORRELATION MATRIX

When a common measure of association or, as we will sometimes call it, a **correlation coefficient,** *is calculated for each pair of a series of variables, the results are often presented in a special table known as a* **correlation matrix.** Consider the example below where there are six variables and gamma is calculated between each pair.

The following correlation matrix (these are gammas) resulted from a study of the costs of vandalism in student dormitories in a public university. The variables are:

Damage: Dollars of damage in a dorm per student

GPA: Mean grade point average of all dorm residents

% Male: Percentage of dorm residents who are male

% in State: Percentage of dorm residents whose families reside in state

Age: Mean age of dormitory residents

Beer: Mean number of cans of beer consumed by the residents in one school year

Data were collected on 20 dormitory units with each variable recorded as either high, medium, or low and the dormitory assigned to the appropriate ranked category.

Damage	*f* =
High	5
Medium	10
Low	5
Total	20

Thus, out of 20 dormitories studied, 5 had high levels of damage, 10 had medium levels, and 5 had low levels. These three-category grouped ordinal variables were then used to form 3×3 tables (one for each pair of variables), and from each table a gamma was calculated. The results are presented in a form resembling a distance table on a road map. To find the gamma between any pair of variables, find one variable on a row and the other on a column and look to see what coefficient appears at the intersection of the row and column.

Notice two aspects of Table 11.11. First, the "main diagonal" contains only 1.00 where each variable's row intersects the same variable's

Table 11.11

	Damage	*GPA*	*% Male*	*% in State*	*Age*	*Beer*
Damage	1.00	–.74	.68	.01	–.66	.85
GPA	–.74	1.00	–.66	.02	.50	–.66
% Male	.68	–.66	1.00	–.03	–.01	.50
% in State	.01	.02	–.03	1.00	.02	.20
Age	–.66	.50	–.01	.02	1.00	.01
Beer	.85	–.66	.50	.20	.01	1.00

column. The gamma (or any measure) between a variable and itself is always 1.00, a perfect association. A second aspect is that everything above and to the right of the main diagonal is a mirror image of what is below and to the left of that diagonal. To find the association between damage and GPA we can either look at the intersection of the damage row and the GPA column (finding a –.74) or look at the intersection of the damage column and the GPA row (also finding a –.74).

	Damage	*GPA*
Damage	1.00	–.74
GPA	–.74	1.00

The two intersections have the same number in them since gamma is symmetric. Had we used an asymmetric measure such as lambda, and designated the variables in the rows as independent variables and the variables in the columns as dependent variables, the number at the intersection of the damage row and GPA column would be the lambda for predicting GPA from damage. The lambda for predicting damage from GPA would be at the intersection of the GPA row and damage column. The two lambdas would not necessarily be the same, as was the case with the two gammas.

When we have as our coefficients gammas or other symmetric measures, we often save space by eliminating the coefficients above and to the right of the main diagonal, since they are redundant. See Table 11.12. If you look for a gamma and find only a blank space, such as the damage row and the GPA column, reverse the row and column. Going down the damage column to the GPA row, we find a gamma of –.74.

Now we can select some variable as a dependent variable and see which of the five remaining variables are associated with it and are, therefore, plausible independent variables.

Table 11.12

	Damage	*GPA*	*% Male*	*% in State*	*Age*	*Beer*
Damage	1.00					
GPA	–.74	1.00				
% Male	.68	–.66	1.00			
% in State	.01	.02	–.03	1.00		
Age	–.66	.50	–.01	.02	1.00	
Beer	.85	–.66	.50	.20	.01	1.00

What are the likely characteristics of the students in a dorm in which high damage was recorded? Damage is the dependent variable. We look at Table 11.12 to see the gammas between damage and the other variables. In this case they are easily found in the damage column. Looking down that column we see: Damage 1.00 (obviously), GPA –.74, % Male .68, % in State .01, Age –.66, and Beer .85. It is often useful to list these beginning with the highest positive correlation, working down toward zero, and then out again to the largest negative coefficient:

Damage versus Beer	.85
Damage versus % Male	.68
Damage versus % in State	.01
Damage versus Age	–.66
Damage versus GPA	–.74

Beer consumption has the highest positive gamma with Damage (.85). The greater the beer consumption, the greater the damage. The .68 with % Males suggests that the greater the percentage of males in the dorm, the greater the damage. The gamma of .01 between % in State and Damage suggests little association, so we exclude that variable. Age and GPA are inversely related to Damage. The greater the damage, the younger the students and the lower their GPA.

To summarize, dorms with high damage rates tend to be those with higher levels of drinking, greater proportions of male residents, younger students, and students with low GPAs.

What would be the "ideal" composition of residents if one wanted to minimize damage? To answer this question, simply reverse the conclusion to the first question. Dorms with low levels of damage would have low rates of beer consumption, a higher percentage of female residents, older students, and students with higher GPAs.

What are the characteristics of a dormitory whose residents have relatively high GPAs? Try to answer this one on your own.

CONCLUSION

As mentioned earlier, since association measures summarize entire cross-tabs, the nuances found in the tables often are not reflected in the association measures generated from them. Measures of association are also dependent on the size of the table, the levels of measurement of the variables, and whether or not the relationship in the table is linear. Nevertheless, though they must be used with care, these measures of association are valuable tools. The information in Table 11.12, for instance, summarizes fifteen meaningful cross-tabulations and can be generated easily by a computer. For this reason, correlation matrices are and will continue to be valuable tools for data analysis.

We return to the topic of association in Chapter 13, which deals with correlation-regression analysis.

EXERCISES

Exercise 11.1

Calculate Q and φ for the table in Exercise 6.7.

Exercise 11.2

Calculate Q and φ for the table in Exercise 6.11.

Exercise 11.3

For the table in Exercise 6.3, calculate gamma, lambda for each variable dependent, and lambda symmetric.

Exercise 11.4

For the table in Exercise 6.4, calculate gamma and all three lambdas. Which measure is most appropriate for this table?

Exercise 11.5

Calculate gamma for the table in Exercise 6.5.

Exercise 11.6

For the table in Exercise 6.6, calculate gamma and the lambda for deaths from political violence as the dependent variable. Do these findings confirm the observations you made when initially doing Exercise 6.6?

Exercise 11.7

For the table in Exercise 6.7, the SAS program generates the following measures of association.

Table E11.7

PHI COEFFICIENT	−0.382
CONTINGENCY COEFFICIENT	0.357
CRAMER'S V	−0.382
STATISTIC	VALUE
GAMMA	−0.680
KENDALL'S TAU-B	−0.382
STUART'S TAU-C	−0.377
SOMER'S D C|R	−0.386
SOMER'S D R|C	−0.378
PEARSON CORRELATION	−0.382
SPEARMAN CORRELATION	−0.382
LAMBDA ASYMMETRIC C|R	0.333
LAMBDA ASYMMETRIC R|C	0.250
LAMBDA SYMMETRIC	0.294
UNCERTAINTY COEFFICIENT C|R	0.109
UNCERTAINTY COEFFICIENT R|C	0.111
UNCERTAINTY COEFFICIENT SYMMETRIC	0.110

Given that the table is ordinal-by-ordinal, what measures presented in this chapter would be most appropriate for interpreting the table? (Disregard the Pearson and Spearman correlations. Remember, too, that Yule's *Q* is really gamma for a 2-by-2 table.) Did you get these right in Exercise 11.1?

Exercise 11.8

In Exercise 6.8, the CRIGHTS by DEATHS relationship was examined, controlling for ADJUST. Here are some of the association measures generated for the tables in that exercise. What does this tell you?

Table E11.8.1 ADJUST = H

PHI COEFFICIENT	0.408
CONTINGENCY COEFFICIENT	0.378
CRAMER'S V	0.408
STATISTIC	VALUE
GAMMA	1.000
KENDALL'S TAU-B	0.408
STUART'S TAU-C	0.320
SOMER'S D C|R	0.333
SOMER'S D R|C	0.500
PEARSON CORRELATION	0.408
SPEARMAN CORRELATION	0.408
LAMBDA ASYMMETRIC C|R	0.000
LAMBDA ASYMMETRIC R|C	0.000
LAMBDA SYMMETRIC	0.000
UNCERTAINTY COEFFICIENT C|R	0.237
UNCERTAINTY COEFFICIENT R|C	0.176
UNCERTAINTY COEFFICIENT SYMMETRIC	0.202

Table E11.8.2 ADJUST = M

PHI COEFFICIENT	0.200
CONTINGENCY COEFFICIENT	0.196
CRAMER'S V	0.200
STATISTIC	VALUE
GAMMA	1.000
KENDALL'S TAU-B	0.200
STUART'S TAU-C	0.111
SOMER'S D C|R	0.200
SOMER'S D R|C	0.200
PEARSON CORRELATION	0.200
SPEARMAN CORRELATION	0.200
LAMBDA ASYMMETRIC C|R	0.000
LAMBDA ASYMMETRIC R|C	0.000
LAMBDA SYMMETRIC	0.000
UNCERTAINTY COEFFICIENT C|R	0.074
UNCERTAINTY COEFFICIENT R|C	0.074
UNCERTAINTY COEFFICIENT SYMMETRIC	0.074

Table E11.8.3 ADJUST = O (low)

PHI COEFFICIENT	−0.775
CONTINGENCY COEFFICIENT	0.612
CRAMER'S V	−0.775

STATISTIC	VALUE
GAMMA	−1.000
KENDALL'S TAU-B	−0.775
STUART'S TAU-C	−0.750
SOMER'S D C|R	−0.750
SOMER'S D R|C	−0.800
PEARSON CORRELATION	−0.775
SPEARMAN CORRELATION	−0.775
LAMBDA ASYMMETRIC C|R	0.667
LAMBDA ASYMMETRIC R|C	0.750
LAMBDA SYMMETRIC	0.714
UNCERTAINTY COEFFICIENT C|R	0.575
UNCERTAINTY COEFFICIENT R|C	0.549
UNCERTAINTY COEFFICIENT SYMMETRIC	0.562

Exercise 11.9

Following are the φ's and *Q*'s for the two tables in Exercise 6.9.

Table 2 GNP = M	Table 3 GNP = 0
$\varphi = .091$	$\varphi = .200$
$Q = .200$	$Q = 1.000$

Do these measures confirm your observations when you read the tables?

Exercise 11.10

For the "tables" in Exercise 6.10, pick a table and try calculating *Q* and φ for it. What happens? (Remember that if a number is divided by zero, the quotient is undefined.)

Exercise 11.11

Here are the φ's and *Q*'s for the tables in Exercises 6.11 and 6.12.

	$\varphi =$	$Q =$
Ex. 6.11 (No Controls)	0.459	0.818
Ex. 6.12 Table 1 (World = F)	0.527	1.000
Ex. 6.12 Table 2 (World = T)	−0.316	−1.000

How do these correspond to your original conclusions when you did the exercises in Chapter 6? Did you get φ and Q right in Exercise 11.2?

Exercise 11.12

Following is a correlation (association) matrix of selected variables from the *World Handbook* (see Chapter 6).

Matrix of Gammas for Selected Variables

	GNP	*Phones*	*Agric.*	*P. Rights*	*Deaths*	*Sanctions*
GNP	1.00					
Phones	.96	1.00				
Agric.	−.92	−.91	1.00			
P. Rights	.64	.66	−.60	1.00		
Deaths	−.79	−.79	.66	−.68	1.00	
Sanctions	−.22	−.17	.08	−.12	.76	1.00

1. Examine the gammas between GNP, Phones, and Agric. Then look at how each of the three correlate with P. Rights, Deaths, and Sanctions. Is there evidence that GNP, Phones, and Agric. are all really measuring the same underlying variable (known as a factor)? What would you name that factor?
2. Look at Deaths and Sanctions. Do they also appear to be measuring a common factor?

(Note: We will do more with this type of matrix in the exercises for Chapter 13.)

NOTES

1. It is also argued that for Q to be valid there must be at least 5 in each cell and the ratio of marginal totals should not exceed 30:70. If these criteria are not met, find phi in lieu of Q.

2. L. A. Goodman and W. H. Kruskal, "Measures of Association for Cross-Classifications," *Journal of The American Statistical Association, 49*, 732-764.

KEY CONCEPTS

- chi-square test for contingency/contingency chi-square/chi-square test for independence
- observed frequencies/f_o
- expected frequencies/f_e
- Yates' correction for continuity
- Fisher's Exact Test
- asymptotic standard error (ASE)
- research significance versus statistical significance

12 The Chi-Square Test

INTRODUCTION

In this chapter we turn our attention back to tests of significance to examine the chi-square test for contingency, a measure appropriate for cross-tabulations. The variables in such tables may be any level of measurement—nominal, ordinal, or interval—which is one reason why this test is so popular.

Although today we generally prefer to create indices of measurement that approximate interval level data and to which the previously covered tests of significance would apply, there are still many instances where such index construction is not feasible. Accordingly, we fall back on chi-square when nominal or ordinal data must be analyzed. Thus it is still useful for us today, but in the early days of quantitative social research, its use was even more widespread.

There are many other tests of significance besides chi-square designed for less than interval level data. Some are even more powerful than chi-square. However, surprisingly few of these other tests appear in the research literature. This is partly because some of those tests are designed for ungrouped rankings, individual ordinal data, which we rarely use. But probably the major reasons for the popularity of the chi-square test are its versatility and relative ease of calculation.

THE CONTEXT FOR THE CHI-SQUARE TEST

In Table 12.1 we reintroduce the death penalty/drug intervention problem presented in Chapter 11. The relationship from our random sample of 40 students at a specific college has already been confirmed ($Q = .80$ and $\varphi = .50$). How safe are we in generalizing that the relationship holds for all students in the college from which our sample was drawn?

In effect, our null hypothesis postulates that within the entire population of students in that college, there is no relationship between support for intervention against foreign cocaine production and support of the death penalty for domestic drug kingpins. That being the case, the Q or φ found for the sample of 40 students must be the result of sampling error, and ideally, if we had studied the entire population, Q and φ should both be zero. We thus could express H_0 in words as follows:

H_0: In the population the two sets of attitudes are unrelated.

We could also express H_0 symbolically, selecting either of our two calculated measures of association; for instance

$$H_0\colon \phi_{\text{population}} = 0$$

Our H_1 could be likewise symbolically expressed.

$$H_1\colon \phi_{\text{population}} \neq 0$$

Note that our H_1 is nondirectional, in that it allows for either of two possibilities: $\phi_{\text{pop}} > 0$ or $\phi_{\text{pop}} < 0$. (Later on we will work with directional alternative hypotheses for this test.)

If we can reject H_0, then the relationship for our sample holds for the entire population at the college, and the Q (or φ) calculated from the

Table 12.1

	Attitude on Drug Intervention		
Attitude on Death Penalty	*For*	*Against*	*Total*
Supports	15	5	20
Opposes	5	15	20
Total	20	20	40

sample becomes a practical estimate of the relationship in the entire population. We attempt to reject H_0 by means of a test of significance known as the **chi-square test for contingency** (or the **contingency chi-square,** the **chi-square test for independence,** or most often simply, the **chi-square test**). We often see this designated using the square of the Greek letter chi as the symbol, χ^2, which unfortunately is difficult to distinguish from a capital X^2. Also note that chi is pronounced "kye."

Before we begin trying to reject our null hypothesis we ask ourselves *what kind of a table would we have expected to get if H_0 were true?* The cell entries in Table 12.1 are what we actually observed in our study. They are called **observed frequencies** (or f_o's where f stands for frequency and the subscript is the letter o, for observed). Now we ask: assuming (a) that H_0 is true and (b) that the marginal totals in the observed frequency table actually reflect those marginals in the population (no. of intervention supporters = no. of opposers, and so on), what are our **expected frequencies** (f_e's)? (You guessed it: f means frequency and e means expected.)

EXPECTED FREQUENCIES

To find the expected frequency for any cell, multiply its row marginal total by its column marginal total and divide by the grand total.

$$f_e = \frac{\text{row marginal} \times \text{column marginal}}{\text{grand total}}$$

The expected frequencies are calculated below and presented in Table 12.2.

Cell	
Supports, For	$f_e = \frac{20 \times 20}{40} = \frac{400}{40} = 10$
Supports, Against	$f_e = \frac{20 \times 20}{40} = \frac{400}{40} = 10$
Opposes, For	$f_e = \frac{20 \times 20}{40} = \frac{400}{40} = 10$
Opposes, Against	$f_e = \frac{20 \times 20}{40} = \frac{400}{40} = 10$

Several things should be noted. First, the f_e's are all equal because the marginal totals were all equal. This is rarely the case. Suppose we had the marginals shown in Table 12.3.

Table 12.2

Death Penalty	Expected Frequencies Intervention For	Against	Total
Supports	10	10	20
Opposes	10	10	20
Total	20	20	40

Table 12.3

Death Penalty	Observed Frequencies Intervention For	Against	Total
Supports	13	2	15
Opposes	5	20	25
Total	18	22	40

Then the f_e's would be:

Cell	
Supports, For	$f_e = \frac{15 \times 18}{40} = \frac{270}{40} = 6.75$
Supports, Against	$f_e = \frac{15 \times 22}{40} = \frac{330}{40} = 8.25$
Opposes, For	$f_e = \frac{25 \times 18}{40} = \frac{450}{40} = 11.25$
Opposes, Against	$f_e = \frac{25 \times 22}{40} = \frac{550}{40} = 13.75$

Though logic would seem to suggest that we round these to whole numbers (because we cannot have .75 or .25 of a person), from a mathematical perspective it is preferable to keep the decimal places. Thus the expected frequencies are as shown in Table 12.4.

Note that the marginal totals and grand total are the same in the f_e table as in the f_o table. If they ever differ slightly, it will be due to rounding error, as a consequence of having decimal places among the expected frequencies. Note also that while the f_e's are not all the same number, as they were in Table 12.2, the rows and columns in an f_e table

Table 12.4

	Expected Frequencies Intervention		
Death Penalty	*For*	*Against*	*Total*
Supports	6.75	8.25	15.00
Opposes	11.25	13.75	25.00
Total	18.00	22.00	40.00

Table 12.5 Expected Frequencies as Percentages of Column Totals

	Intervention	
Death Penalty	*For*	*Against*
Supports	37.5%	37.5%
Opposes	62.5	62.5
Total	100.0%	100.0%
(*n* =)	(18)	(22)

Table 12.6 Expected Frequencies as Percentages of Row Totals

	Intervention			
Death Penalty	*For*	*Against*	*Total*	*(n =)*
Supports	45.0%	55.0	100.0%	(15)
Opposes	45.0	55.0	100.0%	(25)

are proportional to one another. If we percentaged Table 12.4 by column, we would get the results shown in Table 12.5.

Note in Table 12.5 that the death penalty supporters are 37.5% of both those for and those against intervention and death penalty opponents are 62.5% of both groups. We would get similar results if we percentaged by row totals as in Table 12.6.

If we interpret Tables 12.5 and 12.6 as we interpreted percentage tables earlier in this text we would conclude from the f_e's that there is *no* relationship between the variables. That is exactly what the null hypothesis states. In fact, if we go back to the expected frequencies in Table 12.2 and Table 12.4 and calculate Q or φ for each, they would be 0. Since earlier we stated our null hypothesis in terms of φ, let us calculate φ and see.

For Table 12.2,

$$\varphi = \frac{ad - bc}{\sqrt{(a+b)(c+d)(a+c)(b+d)}} = \frac{(10)(10) - (10)(10)}{\sqrt{(20)(20)(20)(20)}}$$

$$= \frac{100 - 100}{\sqrt{160,000}} = \frac{0}{400} = 0$$

For Table 12.4,

$$\varphi = \frac{ad - bc}{\sqrt{(a+b)(c+d)(a+c)(b+d)}} = \frac{(6.75)(13.75) - (8.25)(11.25)}{\sqrt{(15)(25)(18)(22)}}$$

$$= \frac{92.81 - 92.81}{\sqrt{148,500}} = \frac{0}{385.36} = 0$$

You can confirm that Q is also 0 for each of the tables.

A final point about the expected frequencies. In calculating each f_e, we used the formula

$$f_e = \frac{\text{row marginal} \times \text{column marginal}}{\text{grand total}}$$

Applying this to the upper left (*Supports, For*) cell, we calculated

$$f_e = \frac{20 \times 20}{40} = \frac{400}{40} = 10$$

We did this in all four cells to get Table 12.2, but we really only needed to generate one f_e. We could have subtracted from row and column marginals to get the other three expected frequencies.

10		20
		20
20	20	40

The upper row marginal of 20 minus the *Supports, For* f_e of 10 gives us the *Supports, Against* f_e: 20 − 10 = 10. Now we have:

10	10	20
		20
20	20	40

Subtracting the *Supports, For* f_e of 10 from its column marginal of 20 yields the *Opposes, For* f_e: 20 – 10 = 10. The same can be done in the *Against* category, also getting 10. Thus:

f_e

10	10	20
10	10	20
20	20	40

For the problem in Table 12.3, we also need only calculate one f_e. Having found out that the f_e for *Supports, For* is:

$$\frac{15 \times 18}{40} = \frac{270}{40} = 6.75$$

we could subtract from its row marginal of 15: 15 – 6.75 = 8.25.

6.75	8.25	15.00
		25.00
18.00	22.00	40.00

Subtracting the 6.75 from its column marginal total: 18.00 – 6.75 = 11.25.

6.75	8.25	15.00
11.25		25.00
18.00	22.00	40.00

We can get the last f_e (*Opposes, Against*) from either its row marginal 25.00 – 11.25 = 13.75 or its column marginal 22.00 – 8.25 = 13.75. Thus:

6.75	8.25	15.00
11.25	13.75	25.00
18.00	22.00	40.00

We have reconstructed Table 12.4.

Therefore, for a two-by-two table, we need only calculate one f_e, and the others can then be obtained via subtraction from the marginal totals. This characteristic is used later in our actual significance test. The number of f_e's needed before you can get the rest from the marginal totals is *the test's degree(s) of freedom*, symbolized *df*, and must be calculated before we conclude our test of significance. Note the resemblance to the logic in the formula used to explain degrees of freedom in Chapter 8, when the concept was first discussed in depth. Obviously, our 2×2 table examples have one degree of freedom, $df = 1$. For any size table, we may obtain the degrees of freedom from the following formula:

$$df = (\text{number of rows} - 1) \times (\text{number of columns} - 1)$$

for a 2×2 table, $df = (2 - 1)\ (2 - 1) = (1 \times 1) = 1$
for a 2×3 table, $df = (2 - 1)\ (3 - 1) = (1)(2) = 2$
for a 3×3 table, $df = (3 - 1)\ (3 - 1) = (2)(2) = 4$ and so on.

Let us summarize our first problem.

f_o

15	5	20
5	15	20
20	20	40

$\varphi = .50$

f_e

10	10	20
10	10	20
20	20	40

$\varphi = 0$

$df = (2 - 1)(2 - 1) = (1)(1) = 1 \quad H_0\text{: } \varphi_{\text{population}} = 0$

OBSERVED VERSUS EXPECTED FREQUENCIES

If the null hypothesis is true, we would expect that our observed frequencies would be identical to the expected frequencies. However, as the result of sampling error, we might often encounter some slight deviation.

Thus, if H_0 is true, we would expect this:

f_e

10	10	20
10	10	20
20	20	40

But we might actually get this:

f_o

11	9	20
9	11	20
20	20	40

f_o's deviate ± 1 unit from the f_e's.

Sometimes, less often, we might even get this:

f_o

12	8	20
8	12	20
20	20	40

f_o's deviate ± 2 units from the f_e's.

The greater the deviation, the less it is likely to be the result of sampling error. Though extremely rare, we could even get the following:

f_o

20	0	20
0	20	20
20	20	40

f_o's deviate ± 10 units from the f_e's.

In this case, even though H_0 is true, based on sampling error we conclude a perfect relationship between the variables.

What the chi-square test does is to look at the deviation between each f_o and its respective f_e. These deviations are squared to get rid of negative signs, just as we did with the standard deviation. Each squared deviation is divided by its expected frequency, forming a kind of proportion of the f_e (although conceivably each "proportion" might exceed one) and the "proportions" are added together to yield a single chi-square value for each table. The greater the deviation between f_o's and f_e's, the larger the chi-square. Following are the chi-square values for these tables in order of increasing f_o versus f_e deviation.

Table (f_o)			$f_o - f_e$ *Deviation Per Cell*	*Chi-Square Obtained*
10	10	20	0	0
10	10	20		
20	20	40		
11	9	20	± 1	0.4
9	11	20		
20	20	40		
12	8	20	± 2	1.6
8	12	20		
20	20	40		
15	5	20	± 5	10.0
5	15	20	(This is our	
20	20	40	original problem.)	
20	0	20	± 10	40.0
0	20	20		
20	20	40		

We will turn shortly to the technique for calculating the chi-square values. For now, note that when f_o's equal their respective f_e's (no deviation) chi-square is 0. As f_o's begin to deviate from their f_e's, chi-square begins to increase. The larger the f_o versus f_e deviation, the larger chi-square gets. However, as previously mentioned, the larger the f_o's deviate from the f_e's, and thus the larger the chi-square value, the smaller becomes the probability of drawing a random sample and getting that large a chi-square as the result of sampling error. There is always some probability of drawing a sample from a population where H_0 is true and getting a large chi-square. At some point however, we should feel safe enough to reject the null hypothesis and conclude instead that our chi-square obtained was not the result of sampling error, but instead reflected the fact that the variables in the population are indeed related.

As with our other tests of significance, we base our decision on whether or not to reject the null hypothesis by comparing the obtained chi-square to the critical value of chi-square at the .05 level. As we will shortly see, based on a table of critical values of chi-square, for a nondirectional H_1 at the .05 level with one degree of freedom (a 2×2 table), chi-square critical is 3.84. If a chi-square obtained from a 2×2 table equals or exceeds 3.84, we can reject the null hypothesis. Returning to the tables with increasing f_o versus f_e deviations presented earlier and comparing the obtained chi-square to 3.84, we can reach a decision about statistical significance.

Table				*$f_o - f_e$ Deviation Per Cell*	*Chi-Square Obtained*	*Decision*
f_o	10	10	20	0	0	Not Significant
	10	10	20			
	20	20	40			
f_o	11	9	20	± 1	0.4	Not Significant
	9	11	20			
	20	20	40			
f_o	12	8	20	± 2	1.6	Not Significant
	8	12	20			
	20	20	40			
f_o	15	5	20	± 5	10.0	Significant
	5	15	20	(This is our		(Bingo! 10.0
	20	20	40	original problem.)		exceeds 3.84.
						Reject H_0.)

f_o

20	0	20	± 10	40.0	Very Significant
0	20	20			
20	20	40			

USING THE TABLE OF CRITICAL VALUES OF CHI-SQUARE

In making our decision to reject the null hypothesis or not, we make use of a table that provides the mathematically determined critical values of chi-square, at selected probability levels, for a wide range of degrees of freedom. See Table 12.7 and the appendices at the end of the book. In the table, the degrees of freedom are listed down the column on the far left side—not unlike the table for the *t* test—and the tail areas, or significance levels, are listed along the top row. For example, go to the line corresponding to the degrees of freedom for the table and look under the probability, the significance level. Thus, at one degree of freedom, chi-square critical at the .10 level is 2.71. Remember, we do not use that level to decide whether or not to reject H_0 in most classroom situations (but we often use it in nonacademic settings). For us, the crucial critical value of chi-square is the one at the .05 level of significance found in Table 12.7.

If the obtained chi-square that we calculated is less than chi-square critical at the .05 level, as in previous tests we cannot reject the null hypothesis. If the obtained chi-square equals or exceeds chi-square critical at the .05 level, we do reject the null hypothesis. The remainder of this procedure also parallels those of other tests of significance. We compare the chi-square obtained to the other critical values of chi-square appearing on the table (.01 and .001 levels) and state our probability based on the level of significance of the final critical value exceeded by the obtained chi-square.

Earlier, we characterized a chi-square value of 40 obtained from a 2×2 (1 *df*) table as "very significant." Now, note that at one degree of freedom, the critical value of chi-square at the highest level of significance in the table, the .001 level, is only 10.83. This is less than 40. Thus $p < .001$. In that same exercise we are informed that for the original problem (Table 12.1) we would obtain a chi-square of 10. We note the critical values at one degree of freedom on Table 12.7. The obtained chi-square of 10 exceeds 3.84, which is the critical value of chi-square at the .05 level, so we reject the null hypothesis. At the .01 level, the critical value is 6.64, which is also less than 10. However, 10 is less than

Table 12.7 Critical Values of Chi-Square

df	*.10*	*.05*	*.01*	*.001*
1	2.71	3.84	6.64	10.83
2	4.60	5.99	9.21	13.82
3	6.25	7.81	11.34	16.27
4	7.78	9.49	13.28	18.47
5	9.24	11.07	15.09	20.52
6	10.64	12.59	16.81	22.46
7	12.02	14.07	18.48	24.32
8	13.36	15.51	20.09	26.12
9	14.68	16.92	21.67	27.88
10	15.99	18.31	23.21	29.59
11	17.28	19.68	24.72	31.26
12	18.55	21.03	26.22	32.91
13	19.81	22.36	27.69	34.53
14	21.06	23.68	29.14	36.12
15	22.31	25.00	30.58	37.70
16	23.54	26.30	32.00	39.25
17	24.77	27.59	33.41	40.79
18	25.99	28.87	34.80	42.31
19	27.20	30.14	36.19	43.82
20	28.41	31.41	37.57	45.32
21	29.62	32.67	38.93	46.80
22	30.81	33.92	40.29	48.27
23	32.01	35.17	41.64	49.73
24	33.20	36.42	42.98	51.18
25	34.38	37.65	44.31	52.62
26	35.56	38.88	45.64	54.05
27	36.74	40.11	46.96	55.48
28	37.92	41.34	48.28	56.89
29	39.09	42.56	49.59	58.30
30	40.26	43.77	50.89	59.70
40	51.80	55.76	63.69	73.40
50	63.17	67.50	76.15	86.66
60	74.40	79.08	88.38	99.61
70	85.53	90.53	100.42	112.32

SOURCE: Table 12.7 is abridged from R. A. Fisher and F. Yates, *Statistical Tables for Biological, Agricultural and Medical Research* (6th ed.), published by Longman Group UK Ltd., 1974, by permission of the authors and the publisher.

10.83, the critical value at the .001 level. Accordingly, we report in this case only that $p < .01$.

If we use a computer program to calculate chi-square, the exact probability of alpha would be generated for us by the program. Below is the information for another example as reported by *SAS*.

```
STATISTIC          DF          VALUE          PROB
CHI-SQUARE         1           11.250         0.001
```

This tells us that from a one degree of freedom table an obtained chi-square value of 11.250 was generated. The probability of falsely rejecting a true null hypothesis is *exactly* 0.001. (It's actually a bit lower, but it rounds up to .001 in this case.)

With this information, note that all we really have to do is examine the probability generated. If that number is *less* than .05, then chi-square obtained was greater than chi-square critical at the .05 level, and we may reject the null hypothesis. If the probability on the printout was *greater* than .05, then chi-square obtained was less than chi-square critical at the .05 level, and we cannot reject H_0. Remember, the bigger the chi-square, the smaller the probability, and vice versa.

To summarize, for a specified *df* level:

If	*Decision*	*Report*
$\chi^2_{obt.} < \chi^2_{crit.}$, .05 level	χ^2 not significant. Do not reject H_0.	No *p* statement.
$\chi^2_{crit.}$, .01 level $> \chi^2_{obt.} > \chi^2_{crit.}$, .05 level	χ^2 is significant. Reject H_0.	$p < .05$
$\chi^2_{crit.}$, .001 level $> \chi^2_{obt.} > \chi^2_{crit.}$, .01 level	χ^2 is significant. Reject H_0.	$p < .01$
$\chi^2_{obt.} > \chi^2_{crit.}$, .001 level	χ^2 is significant. Reject H_0.	$p < .001$

Before turning our attention to the actual calculation of a chi-square value, let us briefly review the steps taken in running a test of significance in general, and the chi-square test in particular.

1. Before any data are examined, formulate the null hypothesis and the alternative hypothesis.
2. Examine the data and calculate the appropriate test of significance. For chi-square, this will require observed frequencies, expected frequencies, the chi-square value itself, and the degrees of freedom.
3. Using a table of critical values, locate the appropriate critical values for the degrees of freedom.
4. Compare the calculated test of significance, in this case the obtained chi-square value, to the critical value at the .05 level.
5. If the obtained value is less than the critical value at the .05 level, we cannot reject the null hypothesis.

6. If the obtained value exceeds the critical value at the .05 level, reject the null hypothesis. If the obtained value is statistically significant, examine the other critical values to determine what the probability statement will be.
7. If the data are statistically significant, you may use these sample statistics to estimate characteristics of the population (the *population parameters*). Since for Table 12.1 we have been informed that its chi-square value of 10.0 is statistically significant, $p < .01$, and we may reject H_0, we now conclude H_1: In the population, attitudes toward intervention against foreign drug producers and attitudes concerning the death penalty for domestic drug kingpins are indeed related. Thus the $\varphi = .50$ value we determined from our sample is the best estimate of what φ would be for our population of all students in the college that we are studying.

CALCULATING THE CHI-SQUARE VALUE

Let us now see how to determine the obtained chi-square. We have already formulated our hypotheses.

$$H_0\text{: } \varphi_{\text{population}} = 0$$

$$H_1\text{: } \varphi_{\text{population}} \neq 0$$

We have also been given the observed frequencies.

f_o

15	5	20
5	15	20
20	20	40

Using the formula

$$f_e = \frac{\text{row marginal} \times \text{column marginal}}{\text{grand total}}$$

and also using subtraction from marginal totals, we generated the f_e's.

f_e

10	10	20
10	10	20
20	20	40

Also, we have calculated the degrees of freedom.

$$\begin{aligned} df &= (\#\text{rows} - 1)\ (\#\text{columns} - 1) \\ &= (2 - 1)\ (2 - 1) \\ &= (1)(1) \\ &= 1 \end{aligned}$$

Now it is time to see how the chi-square value of 10.0 that we were given was actually calculated. The formula used was

$$\chi^2 = \sum \frac{(f_o - f_e)^2}{f_e}$$

We will use the following steps.

1. Make a column labeled f_o = and enter the observed frequencies starting with the upper left cell, moving across the top row until completed. Then go down to the left side of the next row and continue across, and so on.
2. Do the same thing with the f_e's, making sure each f_e is matched to its respective f_o. Thus:

f_o =	f_e =	|
15	10	|
5	10	|
5	10	|
15	10	|

The vertical line is optional; it is just a guide to where you begin calculating the components of the chi-square value.

3. Make a column labeled $f_o - f_e$ = and subtract each f_e from its respective f_o. In a 2×2 table (but only in a 2×2 table), these differences will be either + or – the same value.
4. Make a column labeled $(f_o - f_e)^2$ = and enter the square of the number in the column generated in step 3.

5. Make a column labeled $\frac{(f_o - f_e)^2}{f_e} =$, taking each $(f_o - f_e)^2$ from the column from step 4 and setting up a division problem where the divisor is that row's f_e.
6. Do each $\frac{(f_o - f_e)^2}{f_e}$ division.
7. Add up all the quotients from step 6, thus producing $\sum \frac{(f_o - f_e)^2}{f_e}$ which is in fact chi-square.

Thus:

$f_o =$	$f_e =$	$f_o - f_e =$	$(f_o - f_e)^2 =$	$\frac{(f_o - f_e)^2}{f_e} =$
15	10	5	25	25/10 = 2.5
5	10	−5	25	25/10 = 2.5
5	10	−5	25	25/10 = 2.5
15	10	5	25	25/10 = 2.5

$$\chi^2 = \sum \frac{(f_o - f_e)^2}{f_e} = 10.0$$

Recall that we then compare the obtained value to the critical values at 1 *df* and conclude that we can reject the null hypothesis, $p < .01$.

There is a special formula for calculating chi-square for a 2×2 table that does not require the calculation of the f_e's. We do not present it here for two reasons. First, it saves very little calculation time. Second, as we will see later, we need to examine the f_e's to make sure that chi-square is a valid test for the data, and the special chi-square formula does not generate the needed expected frequencies.

Note that the formula we are using here is based on actual frequencies and not percentages.

Let us try calculating chi-square for the problem presented in Table 12.3.

f_o	*For*	*Against*	*Total*
Supports	13	2	15
Opposes	5	20	25
	18	22	40

Recall that we found for the upper left cell that

$$f_e = \frac{\text{row marginal} \times \text{column marginal}}{\text{grand total}}$$

$$= \frac{15 \times 18}{40} = \frac{270}{40} = 6.75$$

By either reapplying this formula to the other cells or by subtraction from row or column marginals, we found the following expected frequencies.

f_e

6.75	8.25	15.00
11.25	13.75	25.00
18.00	22.00	40.00

Setting up and calculating chi-square:

$f_o =$	$f_e =$	$f_o - f_e =$	$(f_o - f_e)^2 =$		$\frac{(f_o - f_e)^2}{f_e} =$
13	6.75	6.25	39.06	39.06/6.75 =	5.78
2	8.25	−6.25	39.06	39.06/8.25 =	4.73
5	11.25	−6.25	39.06	39.06/11.25 =	3.47
20	13.75	6.25	39.06	39.06/13.75 =	2.84

$$\chi^2 = \sum \frac{(f_o - f_e)^2}{f_e} = 16.82$$

df = (#rows − 1) × (#columns − 1) = (2 − 1) (2 − 1) = (1)(1) = 1

Referring to the critical values of chi-square table (Table 12.7) at 1 *df*, we find

		$\chi^2_{critical}$		$\chi^2_{obtained}$	
$\chi^2_{critical,\ .05}$	=	3.84	<	16.82	reject H_0
$\chi^2_{critical,\ .01}$	=	6.64	<	16.82	
$\chi^2_{critical,\ .001}$	=	10.83	<	16.82	so $p < .001$

YATES' CORRECTION

We noted earlier that chi-square increases from a table with no relationship in it (chi-square = 0) to one with a perfect relationship (chi-square = 40). For this table whose marginal totals were

		20
		20
20	20	40

we went from this table: $\chi^2 = 0$ to this table: $\chi^2 = 40$

f_o

10	10	20
10	10	20
20	20	40

20	0	20
0	20	20
20	20	40

The $f_o - f_e$ deviations went from 0 per cell to ±10 per cell. Given that set of marginal totals, only 11 possible tables could be generated: $f_o - f_e = 0$, 1, 2, 3, . . . , 10. Thus only 11 chi-square values can be calculated for the given marginal totals. Because the actual values of chi-square that can be generated in a low degrees of freedom table are so few, some statisticians have argued that the chi-squares obtained only approximate the smooth curve of the theoretical sampling distribution of chi-square. This could result in generating an obtained chi-square large enough to reject the null hypothesis whereas, if the table had more cells, as shown below, chi-square obtained would not have been large enough to reject the null hypothesis.

	Attitude on Drug Intervention				
Attitude on Death Penalty	*Strongly Favorable*	*Moderately Favorable*	*Unsure*	*Moderately Opposed*	*Strongly Opposed*
Supports					
Opposes					

For this reason, some—but by no means all—statisticians will adjust the chi-square for a 1 *df* table by applying a factor that lowers the value of chi-square obtained, making it harder to reject the null hypothesis. This adjustment is known as **Yates' correction for continuity.** (Some writers call it *Yates' correction for discontinuity.* It's the same thing, depending on whether you see the glass as half-full or half-empty. We are correcting the discontinuity to achieve the effects of continuity.) The correction takes the absolute value of each $f_o - f_e$ and subtracts one-half a point before going on to square the number.

Therefore, for our first problem where uncorrected chi-square was 10.0, we use the following format to get the corrected chi-square.

$f_o =$	$f_e =$	$f_o - f_e =$	$\lvert f_o - f_e \rvert =$	$\lvert f_o - f_e \rvert - .5 =$	$[\lvert f_o - f_e \rvert - .5]^2 =$
15	10	5	5	4.5	20.25
5	10	−5	5	4.5	20.25
5	10	−5	5	4.5	20.25
15	10	5	5	4.5	20.25

$$\frac{[\lvert f_o - f_e \rvert - .5]^2}{f_e} =$$

20.25/10 = 2.025
20.25/10 = 2.025
20.25/10 = 2.025
20.25/10 = 2.025

$$\chi^2_{\text{corrected}} = \sum \frac{[\lvert f_o - f_e \rvert - .5]^2}{f_e} = 8.100$$

At one degree of freedom,

$$\chi^2_{\text{critical, .05}} = 3.84 < 8.1 \qquad \text{Reject } H_0$$

$$\chi^2_{\text{critical, .01}} = 6.64 < 8.1 \quad p < .01$$

$$\chi^2_{\text{critical, .001}} = 10.83 > 8.1$$

In this case, even though when corrected for continuity chi-square fell from 10.0 to 8.1, our original rejection of H_0 at the .01 level remains valid.

For the second problem, where uncorrected chi-square was 16.82, corrected chi-square falls to 14.22 (do the calculations to confirm this). We may still reject H_0 at the .001 level. Although the use of the corrected chi-square is debated among statisticians, for purposes of consistency we will usually assume that it is appropriate for two-by-two tables.

VALIDITY OF CHI-SQUARE

To be a valid test of significance, chi-square usually requires that most expected frequencies be 5 or larger. This is always true for a two-by-two table. If larger than a two-by-two table, there a few exceptions are allowed as long as (a) no f_e is less than one and (b) no more than 20% of the f_e's are less than 5. For most of the tables we encounter, this means the following:

Table Size	*No. of f_e's*	*20% of f_e's*	*Actual Whole Number of Cells Where the f_e's Can Be Greater Than 1 and Less Than 5*
2×2	4	0.8	0
2×3	6	1.2	1
3×3	9	1.8	1
3×4	12	2.4	2
4×4	16	3.2	3
4×5	20	4.0	4
5×5	25	5.0	5

Remember, these limitations apply to *expected frequencies,* not observed frequencies.

When a computer program calculates a chi-square value, it will usually issue a warning that chi-square may be invalid when too many f_e's are below the level of tolerance. (Unfortunately, these same programs often

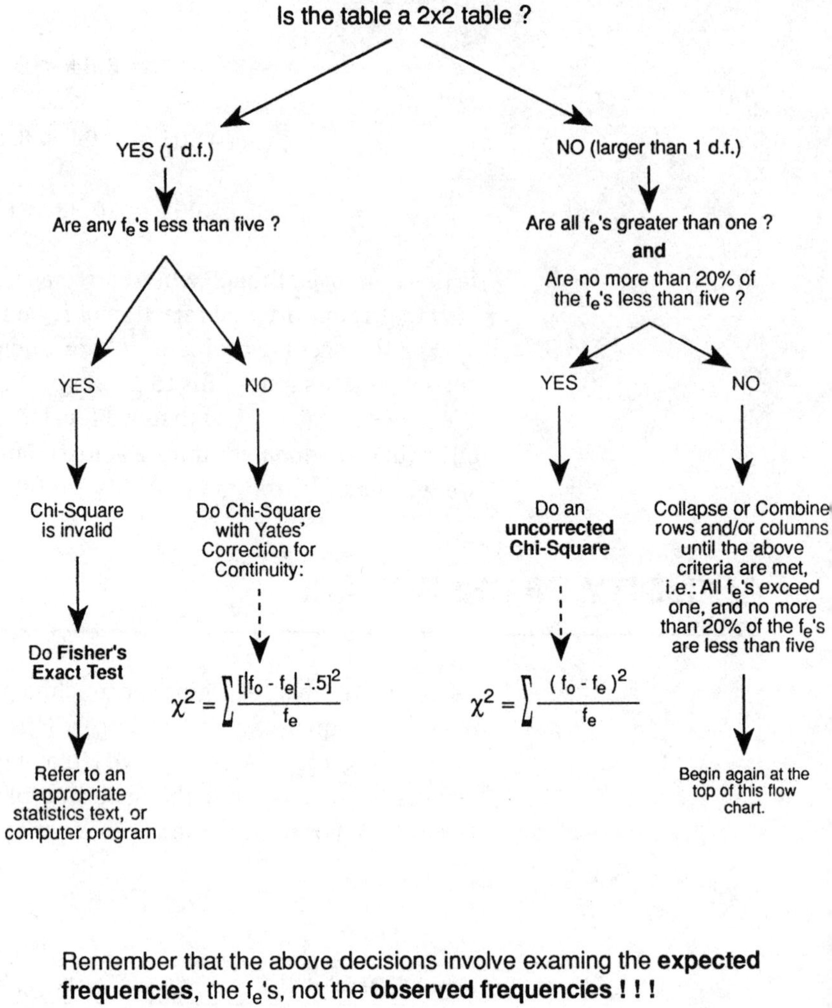

Figure 12.1
The Chi-Square Test for Contingency Flow Chart

do not show the f_e's, unless you request them in setting up your run.) If chi-square is invalid for a two-by-two table, there is an alternative test known as **Fisher's Exact Test,** which may be used in place of chi-square. Fortunately, most commercial chi-square computer programs also calculate Fisher's Exact Test, a real plus because Fisher's Exact Test is not an easy one to calculate by hand.

For larger than two-by-two tables, we must locate the offending expected frequencies and modify the table by collapsing or combining categories until all f_e's satisfy the size criteria. Figure 12.1 provides a flow chart for this decision-making process.

Suppose we are studying the relationship between religious affiliation and socioeconomic status (SES) and our study yields the following observed frequencies.

f_o	Religion				
SES	*Protestant*	*Catholic*	*Jewish*	*Other*	*Total*
High	25	10	5	0	40
Medium	20	10	5	5	40
Low	5	10	5	0	20
Total	50	30	15	5	100

For a 3×4 table we can tolerate two f_e's that are less than 5 but greater than 1. (Again note that we must examine *expected,* not *observed,* frequencies. There are two f_o's less than 5 above, but it does *not* mean that the f_e's will be less than 5.) We generate the expected frequencies to find:

f_e	Religion				
SES	*Protestant*	*Catholic*	*Jewish*	*Other*	*Total*
High	20	12	6	2	40
Medium	20	12	6	2	40
Low	10	6	3	1	20
Total	50	30	15	5	100

We see that there are four f_e's below 5: *Low, Jewish* and all three under the other religions category. Not only that, but the *Low, Other* f_e is 1. Thus chi-square is not valid as the table stands.

Note that most of the offending f_e's are in the other religion category. Because only 5 out of 100 people fell in that category, we could simply delete the Other group from the study and collapse the table into a 3×3 table.

f_o	Religion			
SES	*Protestant*	*Catholic*	*Jewish*	*Total*
High	25	10	5	40
Medium	20	10	5	35
Low	5	10	5	20
Total	50	30	15	95

The expected frequencies are now:

f_e		*Religion*		
SES	*Protestant*	*Catholic*	*Jewish*	*Total*
High	21.1	12.6	6.3	40.0
Medium	18.4	11.1	5.5	35.0
Low	10.5	6.3	3.2	20.0
Total	50.0	40.0	15.0	95.0

Now, all f_e's but one (*Low, Jewish*) are 5 or above, and the *Low, Jewish* f_e is greater than 1. Since in a 3×3 table we can tolerate that one f_e chi-square is a valid test for the modified table.

If we did not wish to delete the Other category, we could instead combine Other with one of the remaining categories as follows:

	(1) *Other*	*Catholic*	*Jewish*	(Other being Protestant plus the former Other)
or	(2) *Protestant*	*Other*	*Jewish*	(Other being Catholic plus the former Other)
or	(3) *Protestant*	*Catholic*	*Other*	(Other being Jewish plus the former Other)

Suppose we select the third alternative, perhaps because we want to compare Protestants and Catholics to non-Christians. (This assumes that those initially indicating Other were non-Christians. One has to be a bit wary, however, as both Mormons and Eastern Orthodox may have been included in the original Other category. Let us assume that this is not the case here.) We then have:

f_o	*Protestant*	*Catholic*	*Other*	*Total*
High	25	10	5	40
Medium	20	10	10	40
Low	5	10	5	20
Total	50	30	20	100

f_e	*Protestant*	*Catholic*	*Other*	*Total*
High	20	12	8	40
Medium	20	12	8	40
Low	10	6	4	20
Total	50	30	20	100

One f_e is below 5 (*Low, Other*), but it is larger than 1 and a 3×3 table can tolerate one exception. Thus Chi-square would be a valid test for this table and is calculated below:

H_0: In the population, there is no relationship between religion and SES.

H_1: In the population, religion and SES are related.

(Note: We could also have worded H_0 and H_1 using a measure like lambda: H_0: $\lambda_{pop}=0$.)

$f_o =$	$f_e =$	$f_o - f_e =$	$(f_o - f_e)^2 =$	$\frac{(f_o - f_e)^2}{f_e} =$
25	20	5	25	25/20 = 1.250
10	12	−2	4	4/12 = 0.333
5	8	−3	9	9/8 = 1.125
20	20	0	0	0/20 = 0.000
10	12	−2	4	4/12 = 0.333
10	8	2	4	4/8 = 0.500
5	10	−5	25	25/10 = 2.500
10	6	4	16	16/6 = 2.666
5	4	1	1	1/4 = 0.250

$$\chi^2 = \sum \frac{(f_o - f_e)^2}{f_e} = 8.957 \simeq 8.96$$

$df = (\#\text{rows} - 1) \times (\#\text{columns} - 1) = (3 - 1)(3 - 1) = (2)(2) = 4$

Examining Table 12.7, at $df = 4$,

$$\chi^2_{\text{critical}, .05} = 9.49 > 8.96$$

We cannot reject H_0. Chi-square is not statistically significant. We cannot conclude a relationship in the population between religion and socioeconomic status.

DIRECTIONAL ALTERNATIVE HYPOTHESES

Suppose we are investigating the possibility of a relationship between one's level of social activism and one's income. Although we could consider activism to be a function of income, let's reverse the order in this example and explore the possibility that a higher income could be a reward for social activism. We create a 3×3 table in which the independent variable, social activism, is listed as a three-category ordinal variable (high, medium, low) and the dependent variable, income, is also a three-category ordinal variable (high, medium, low). Since we have an ordinal-by-ordinal table we may use gamma to form our hypotheses.

$$H_0\colon \gamma_{\text{population}} = 0$$

$$H_1\colon \gamma_{\text{population}} \neq 0$$

A random sample of 50 respondents yields the following observed frequencies.

f_o	Social Activism			
Income	*High*	*Medium*	*Low*	*Total*
High	10	5	5	20
Medium	0	5	5	10
Low	5	10	5	20
Total	15	20	15	50

Calculating gamma from the observed frequencies, we obtain a value of .18, suggesting a small relationship. If H_0 is true, we must attribute the gamma of .18 to sampling error. If H_0 can be rejected, we may conclude that in the population there really is a relationship between one's social activism and one's income and we would estimate that population gamma with our sample gamma of .18. We test significance using chi-square. The following expected frequencies are generated.

f_e	*High*	*Medium*	*Low*	*Total*
High	6	8	6	20
Medium	3	4	3	10
Low	6	8	6	20
Total	15	20	15	50

(Ignore the fact that in this example too many f_e's are less than 5. This example intentionally utilizes small frequencies.)

$f_o =$	$f_e =$	$f_o - f_e =$	$(f_o - f_e) =$	$\frac{(f_o - f_e)^2}{f_e} =$
10	6	4	16	16/6 = 2.666
5	8	−3	9	9/8 = 1.125
5	6	−1	1	1/6 = 0.166
0	3	−3	9	9/3 = 3.000
5	4	1	1	1/4 = 0.250
5	3	2	4	4/3 = 1.333
5	6	−1	1	1/6 = 0.166
10	8	2	4	4/8 = 0.500
5	6	−1	1	1/6 = 0.166

$$\chi^2 = \sum \frac{(\text{size } f_o - f_e)^2}{f_e} = 9.372$$

$df = (\#\text{rows} - 1)(\#\text{columns} - 1) = (3 - 1)(3 - 1) = (2)(2) = 4$

Checking the table of critical values, we find

$$\chi^2_{\text{critical, .05 level, } 4\ df} = 9.49 > 9.372$$

We cannot reject H_0. $\chi^2_{\text{critical}} > \chi^2_{\text{obtained}}$.

Under the rules we have set so far, we can go no further. We could, however, redo the logic of the problem by making use of prior knowledge, which might enable us to reject the null hypothesis. We go back to "square one," prior to collecting our data, when the null hypothesis was formed.

$$H_0\text{: } \gamma_{\text{population}} = 0$$

Now if we reject H_0 and conclude H_1, we are saying $\gamma_{\text{population}} \neq 0$. Two possibilities are subsumed under such an inequality:

$$\text{either } \gamma_{\text{population}} > 0 \text{ (gamma is positive)}$$

$$\text{or } \gamma_{\text{population}} < 0 \text{ (gamma is negative)}$$

Before looking at our data we may have no reason to assume that one or the other condition is more appropriate. Suppose, though, that based on previous knowledge or experience, we have reason to believe that one of those possibilities is logically impossible (or at least very improbable). In this specific problem, suppose that based on previous research, there is reason to believe that a positive gamma is likely and that a negative gamma is highly unlikely. Thus it is logical to assume that in the population high activism levels go with high incomes and lower activism levels go with lower incomes. In our actual sample, this will turn out to be the case with gamma a positive .18, indicating clustering on the main diagonal.

Note however, that if the *reverse* were true, we might have come up with the table below.

f_o	*Social Activism*			
Income	*High*	*Medium*	*Low*	*Total*
High	5	5	10	20
Medium	5	5	0	10
Low	5	10	5	20
Total	15	20	15	50

Gamma for *this* table is – .18, but if we calculate chi-square we will still get a value of 9.372, the same as the one from the table where gamma was + .18. This means that for any given chi-square on the sampling distribution, half of all the tables generating that chi-square value reflect tables where gamma is positive and the other half reflect tables where gamma is negative. That is, half of the area under the chi-square sampling distribution reflects chi-squares from tables where gamma is positive and the other half reflects chi-squares from tables where gamma is negative.

Now if we may exclude in advance that either $\gamma_{\text{population}} > 0$ or $\gamma_{\text{population}} < 0$ is illogical, in effect we reduce by half all theoretically possible chi-square values in our sampling distribution. Thus the probability of obtaining any specific value of chi-square doubles. This means that the probability levels listed along the top of the table of critical values of chi-square may be cut in half. At 4 degrees of freedom, what was

	$p =$			
df	.10	.05	.01	.001
4	7.78	9.49	13.28	18.47

becomes

	$p =$			
df	.05	.025	.005	.0005
4	7.78	9.49	13.28	18.47

Each of the first probabilities is divided by 2 to produce the second probability: .10/2 = .05, .05/2 = .025, and so on.

This being the case, chi-square critical at the .05 level drops from the nondirectional 9.49 down to 7.78. Since the obtained chi-square value of 9.372 remains unchanged, our new conclusion is

$$\chi^2_{\text{critical, .05 level, 4 } df} = 7.78 < 9.372 \quad \text{Reject } H_0. \quad p < .05$$

But, H_1 is modified:

$$H_1: \gamma_{\text{population}} > 0.$$

By implication, though not stated, $\gamma_{\text{population}} < 0$ is excluded in advance as being illogical. In other situations, H_1 might have been

$$H_1: \gamma_{population} < 0$$

implying that $\gamma_{population} > 0$ is, based on prior knowledge, illogical. Recall that we followed the same logic with all our other tests of significance, with the exception of F, whose H_1 is always nondirectional.

TESTING SIGNIFICANCE OF ASSOCIATION MEASURES

In the previous examples, the chi-square test was used to determine statistical significance. In several instances, it is possible to test for significance by converting a measure of association into a z score and comparing the obtained z to the critical values developed earlier in this text.

In the case of gamma, an approximation of z can be determined with the following formula.

$$z = \gamma\sqrt{\frac{P_S + P_D}{n(1-\gamma^2)}}$$

Recall that the religion versus SES problem yielded a chi-square of 8.96, too low to reject H_0. If, in the same table, religion had been replaced by an ordinal variable such as the number of automobiles one owns—many, one or two, none—then gamma would be an appropriate measure for the table. If we calculated gamma for that table it would be:

$$\gamma = \frac{P_S - P_D}{P_S + P_D} = \frac{1100}{2450} = .449$$

Note:

$n = 100$
$P_S + P_D = 2450$
$\gamma^2 = (.449)^2 = .202$
$1 - \gamma^2 = 1 - .202 = .798$

Therefore:

$$z = \gamma\sqrt{\frac{P_S + P_D}{n(1-\gamma^2)}} = .449\sqrt{\frac{2450}{100(.798)}} = 2.487 = 2.49$$

Since 2.49 > 1.96, reject H_0, $p < .05$.

A good rule of thumb, therefore, is to first do the chi-square test, and if H_0 is not rejected with chi-square, try the z conversion. In doing so both P_S and P_D should exceed 100. An alternative for gamma and most other measures of association is to calculate what is called an **asymptotic standard error** or **ASE**. The conversion formula then becomes

$$z = \frac{\text{the measure of association}}{\text{ASE}}$$

Most measures of association have accompanying formulas for finding the ASE. Sometimes these formulas are complex. Fortunately, a few statistical computer programs calculate the ASEs for us. SAS, for instance, lists ASEs for gamma, lambda, and a number of other measures of association. They appear to the right of the association measure in the printout. (These were deleted in the exercises for Chapter 11 so as not to confuse you.)

ASSOCIATION VERSUS SIGNIFICANCE

Association and significance are related but separate concepts. Association measures show the strength of a relationship in a table, regardless of the size of the sample being studied. The significance level, by contrast, increases with increasing sample size. If the sample size is large enough, even tables whose association is small may be statistically significant. To illustrate this, let us calculate phi and chi-square for a series of tables all having the same amount of association in them but having differing sample sizes. Since this is an illustration, we will use two-by-two tables and calculate chi-squares without Yates' correction for continuity. We will also ignore the smallness of many of the expected frequencies.

Case 1, $n = 5$

f_o

2	1	3
1	1	2
3	2	5

f_e

1.8	1.2	3.0
1.2	0.8	2.0
3.0	2.0	5.0

$$\varphi = \frac{(2)(1) - (1)(1)}{\sqrt{(3)(2)(3)(2)}} = \frac{2 - 1}{\sqrt{36}} = \frac{1}{6} = .166$$

$f_o =$	$f_e =$	$f_o - f_e =$	$(f_o - f_e) =$	$\frac{(f_o - f_e)^2}{f_e} =$
2	1.8	.2	.04	.04/1.8 = 0.022
1	1.2	− .2	.04	.04/1.2 = 0.033
1	1.2	− .2	.04	.04/1.2 = 0.033
1	0.8	.2	.04	.04/0.8 = 0.050
				χ^2 = 0.138

df = (#rows – 1)(#columns – 1) = (2 – 1)(2 – 1) = (1)(1) = 1

At 1 df $\chi^2_{\text{critical, nondirectional, .05 level}} = 3.84$. Since 3.84 > 0.138 do not reject H_0. Chi-square is not significant and $\varphi = .166$ is quite low.

Case 2, n = 50

f_o

20	10	30
10	10	20
30	20	50

f_e

18	12	30
12	8	20
30	20	50

$$\varphi = \frac{(20)(10) - (10)(10)}{\sqrt{(30)(20)(30)(20)}} = \frac{200 - 100}{\sqrt{360{,}000}} = \frac{100}{600} = .166$$

$f_o =$	$f_e =$	$f_o - f_e =$	$(f_o - f_e) =$	$\frac{(f_o - f_e)^2}{f_e} =$
20	18	2	4	4/18 = 0.222
10	12	− 2	4	4/12 = 0.333
10	12	− 2	4	4/12 = 0.333
10	8	2	4	4/8 = 0.500
				χ^2 = 1.388

Chi-square, though larger than in Case no. 1, is *still* not significant; phi remains unchanged from Case no. 1.

Case 3, n = 500

f_o

200	100	300
100	100	200
300	200	500

f_e

180	120	300
120	80	200
300	200	500

$$\varphi = \frac{(200)(100) - (100)(100)}{\sqrt{(300)(200)(300)(200)}} = \frac{20{,}000 - 10{,}000}{\sqrt{3{,}600{,}000{,}000}} = \frac{10{,}000}{60{,}000} = .166$$

$f_o =$	$f_e =$	$f_o - f_e =$	$(f_o - f_e) =$	$\frac{(f_o - f_e)^2}{f_e} =$
200	180	20	400	400/180 = 2.222
100	120	– 20	400	400/120 = 3.333
100	120	– 20	400	400/120 = 3.333
100	80	20	400	400/80 = 5.000
				χ^2 = 13.888

Not only can we reject H_0, but since at 1 *df*, $\chi^2_{\text{critical, nondirectional, .001 level}} = 10.83 < 13.888$, $p < .001$. But while chi-square is significant and $p < .001$, the phi remains unchanged at a low .166.

Just because a table is statistically significant and we can reject the null hypothesis does not mean that the relationship actually existing in the population is large enough to have significance, in the sense of relevance or importance, to the researcher and reader. There are really two totally distinct ways in which the word *significance* is being used. To differentiate these we call the term as used in everyday life *research significance* to differentiate it from *statistical significance,* as redefined below.

- **Research Significance:** *Relevance; importance of a particular relationship or finding.*
- **Statistical Significance:** *The high probability that a relationship or finding based on a random sample is not the result of sampling error but reflects the characteristics of the population from which the sample was drawn.*

Whether a finding that is statistically significant also has research significance is up to the judgment of the researcher or reader. Think, though, how the term *significant* could be used to mislead the lay person who hears that there is "a significant difference between detergent A and detergent B," or "our group got significantly fewer colds by taking Vitamin C," or any context in which the term *significance,* meant to mean statistical significance, will be automatically interpreted to mean research significance or relevance. The latter conclusion must be based *not* on the size of the computed chi-square value but on the size of phi or some alternative measure of association.

In contrast, in Case 4 the phi is larger.

Case 4, $n = 5$

f_o

2	1	3
0	2	2
2	3	5

f_e

1.2	1.8	3.0
0.8	1.2	2.0
2.0	3.0	5.0

$$\varphi = \frac{(2)(2) - (1)(0)}{\sqrt{(3)(2)(2)(3)}} = \frac{4 - 0}{\sqrt{36}} = \frac{4}{6} = .666$$

$f_o =$	$f_e =$	$f_o - f_e =$	$(f_o - f_e) =$	$\frac{(f_o - f_e)^2}{f_e} =$
2	1.2	.8	.64	.64/1.2 = 0.533
1	1.8	−.8	.64	.64/1.8 = 0.355
0	0.8	−.8	.64	.64/0.8 = 0.800
2	1.2	.8	.64	.64/1.2 = 0.533
				χ^2 = 2.221

Since chi-square critical is 3.84, the difference is not significant, even though phi is relatively large.

Case 5, $n = 50$

f_o

20	10	30
0	20	20
20	30	50

f_e

12	18	30
8	12	20
20	30	50

$$\varphi = \frac{(20)(20) - (10)(0)}{\sqrt{(30)(20)(20)(30)}} = \frac{400 - 0}{\sqrt{360{,}000}} = \frac{400}{600} = .666$$

$f_o =$	$f_e =$	$f_o - f_e =$	$(f_o - f_e) =$	$\frac{(f_o - f_e)^2}{f_e} =$
20	12	8	64	64/12 = 5.333
10	18	−8	64	64/18 = 3.555
0	8	−8	64	64/8 = 8.000
20	12	8	64	64/12 = 5.333
				χ^2 = 22.221

Not only is the chi-square statistically significant, but $p < 001$.

As discussed earlier, significance is a function of both the amount of association and the sample size. So if you encounter a problem such as that in Case 4 with high association but no statistical significance, an increase in the magnitude of n will yield the same association *and* statistical significance.

Chi-Square and Phi

In the previous chapter we learned a formula for calculating phi for a two-by-two table and learned that we only use it on a two-by-two table.

In fact, however, phi may be calculated for a table of any size, by using the following formula in which the letter n indicates the grand total for the table.

$$\varphi = \sqrt{\frac{\chi^2}{n}}$$

While we may calculate phi and chi-square for any table, when the table is larger than a two-by-two table, the maximum possible value of phi may not be 1.0. It could be larger or smaller than 1.0. Thus a given phi value is difficult to interpret.

To correct for this, Pearson developed the **contingency coefficient,** C, calculated as follows:

$$C = \sqrt{\frac{\chi^2}{\chi^2 + n}}$$

While the maximum possible C cannot exceed 1.0, it still can be less than 1.0 (in a 2×2 table, it cannot exceed .71). To compensate for that problem, two other measures have been developed, **Tschuprow's *T*** (rarely used today) and **Cramer's *V*** (also Crämer's *V* or Craemer's *V*).

$$V = \sqrt{\frac{\chi^2}{mn}}$$

Where m (meaning minimum) is whichever is smallest: either #rows – 1 or #columns – 1.

For a two-by-two table, $\varphi = C = V$. For a larger than two-by-two table, the three may not all be equal. For obvious reasons, φ , V, and C are known as *chi-square based measures of association* and are appropriate at any levels of measurement. Though today they are often supplanted by the measures of association developed by Goodman and Kruskal (as well as by others) such as lambda and gamma, these measures are still in use and are commonly found on printouts where tables and their accompanying chi-squares are generated.

CONCLUSION

THE LIMITS OF STATISTICAL SIGNIFICANCE

Although tests of significance are useful adjuncts to research, their importance and results may be exaggerated. For one thing, whenever we reject a null hypothesis there is always the probability of a type I or alpha error: falsely rejecting a true null hypothesis. Also, these tests assume random sampling, whereas the actual selection of the sample might have resulted in some bias. Thus no test of significance can be as useful as the continued replication (repeating) of a study resulting in similar conclusions each time the study is done.

For example, in this and earlier chapters, we have been looking at the relationship between attitudes toward foreign drug intervention and attitudes toward the death penalty. For a random sample of 40 students at a selected college we found a statistically significant chi-square and concluded that the relationship in the table (phi = .50) reflected a relationship among all students at the college where the students were surveyed. If for some reason that college is atypical of other such institutions, the relationship will not hold elsewhere, and we would be wrong in concluding that the variables are related. The only way to find out is to run the study at other colleges or universities and see if the results are similar.

At some point we make a leap of faith and assume that a finding is true everywhere in the country, possibly everywhere on the continent, or even everywhere in the world. If cultures were similar, we might be right; what's true for Ohio and New York could also be true for Ontario and Nova Scotia, or Scotland, or New Zealand. Since cultures differ, however, differing results may easily occur. For example, suppose we find in the United States or Canada that among licensed physicians, the proportion of men predominates over that of women. The same could be true in many other settings but not necessarily everywhere: historically, in the former Soviet Union female physicians have outnumbered male physicians.

In these kinds of problems we are not dealing with random samples of a larger population but rather with self-selecting groups. For the people who attended the college we studied, the death penalty and intervention attitudes are related. That's all we can conclude. If, however, the 40 students we had studied had been *randomly* selected from the population of *all* the college students in the world, then our significant chi-square could allow us to conclude that the variables were related among all college students everywhere.

EXERCISES

Exercise 12.1

Using the sample from the *World Handbook* data from Chapter 6, the Civil Rights and Political Rights indices were cross-tabulated. Here, the low and very low categories were combined for both variables.

	Political Rights			
Civil Rights	*High*	*Medium*	*Low*	
High	7	1	0	8
Medium	1	2	0	3
Low	0	0	9	9
	8	3	9	20

Generate the expected frequencies. Even though these are too small for a valid chi-square, generate that measure anyway. If H_0 is rejected, find φ, *C,* and *V.* (Can you see why φ is sometimes hard to interpret in tables greater than 2×2?)

Exercise 12.2

Imagine that the same relationship in Exercise 12.1 had been determined for a sample $n = 200$.

	Political Rights			
Civil Rights	*High*	*Medium*	*Low*	
High	70	10	0	80
Medium	10	20	0	30
Low	0	0	90	90
	80	30	90	200

Find chi-square and if significant, find φ, *C,* and *V.* How do these four measures differ from those in the previous exercise?

Exercise 12.3

Since the table in Exercise 12.1 has too many low expected frequencies for a valid chi-square, reduce it to a 2×2 table by combining high and medium into a single category for each variable. Generate the f_e's for this 2×2 table. Although technically a chi-square is still invalid, generate chi-square uncorrected, and if H_0 is rejected, find φ, *C,* and *V.* Also calculate φ using the special formula for a 2×2 table presented in Chapter 11. Confirm that both formulas produce the same results except that the chi-square formula does not tell whether φ is positive or inverse. Examine the table to get the appropriate sign for φ.

EXERCISES

Exercise 12.4

Now redo the chi-square for the 2×2 table in Exercise 12.3 using Yates' correction. How does chi-square change? Does it change enough to alter a decision to reject H_0? Note that even though H_0 may still be rejected, we use the uncorrected chi-square to find measures of association. Thus we do *not* calculate those measures in this problem.

Exercise 12.5

	Percentage of Labor Force in Agriculture				
Per Capita GNP	*High*	*Medium*	*Low*	*Very Low*	
High	0	0	4	8	12
Medium	0	4	6	4	14
Low	6	6	2	0	14
	6	10	12	12	40

Generate the expected frequencies and begin combining rows and/or columns until chi-square is valid. (*Hint:* look for a 3×2 table.) Then generate and interpret chi-square, φ, *C,* and *V.*

Exercise 12.6

	Telephones per 1000 Population			
Per Capital GNP	*High*	*Medium*	*Low*	
High	12	3	0	15
Medium	6	6	3	15
Low	0	12	3	15
Very Low	0	0	15	15
	18	21	21	60

Generate expected frequencies. If necessary, combine rows until chi-square is valid. Calculate chi-square and its measures of association. Interpret the results.

Exercise 12.7

	High	Medium	Low	Very Low	
High	2	3	0	0	5
Medium	1	3	2	2	8
Low	0	1	2	3	6
	3	7	4	5	19

Generate expected frequencies. If necessary, combine or delete rows or columns. Determine the appropriate test of significance.

Exercise 12.8

Calculate chi-square for the following table.

9	6	15
7	8	15
9	11	20
25	25	50

Exercise 12.9

The number of protest demonstrations and the civil rights index have both been dichotomized in the table below.

Protest Demonstrations	*Civil Rights* High	*Civil Rights* Low	
High	10	4	14
Low	1	4	5
	11	8	19

What is the most appropriate test of significance for these data? Calculate both the uncorrected and corrected chi-squares. What are your conclusions? Calculate φ.

Exercise 12.10

The table in Exercise 12.9 has been redone using only the six Third World nations in the sample. Although an uncorrected chi-square would not be valid on this table (and even if valid would not be significant), calculate it anyway. Calculate φ. Compare this φ to the one in Exercise 12.9. Remember to inspect each table to see if φ is positive or negative. What are your conclusions? Explain them.

	For Third World Countries		
	Civil Rights		
Protest Demonstrations	*High*	*Low*	
High	0	2	2
Low	1	3	4
	1	5	6

KEY CONCEPTS

- Pearson's *r*/*r*/the Pearsonian product-moment correlation coefficient
- coefficient of determination/r^2
- regression equation
- correlation-regression analysis
- Cartesian coordinates
- origin of a graph
- *x*-axis
- *y*-axis
- ordered pair
- *x*-coordinate
- *y*-coordinate
- function
- linearity
- linear equation
- constant
- variable
- *y*-intercept
- slope
- positive versus inverse (negative) relationship
- curvilinear versus linear relationship
- linear regression
- scatter diagram/scattergram/scatter plot
- least squares method
- regression of *y* on *x*/regression of *x* on *y*
- coefficient of alienation/$1 - r^2$
- $\hat{y}$—the predicted value of *y*
- Intraclass correlation coefficient (r_I or r_i)
- Correlation ratio [E or η(eta)]

13 Correlation-Regression Analysis

INTRODUCTION

Correlation-regression analysis is a set of techniques that have gained widespread popularity throughout the social sciences, and in business and economics as well. It is a set of interrelated techniques designed for individual raw score data at the interval or ratio level of measurement. These techniques are quite versatile and relatively sophisticated.

We first generate a *correlation coefficient*, a measure designed to ascertain the strength of a relationship between two variables. (The measures of association presented in previous chapters were in fact, designed to do for data in tables what their forerunner, the correlation coefficient, does for interval or higher level data.)

If the correlation coefficient thus generated is large enough to demonstrate a meaningful relationship between the variables, we may then generate a *linear regression equation.* This is a mathematical equation designed to predict what a subject's score would be along the dependent variable if we had knowledge of the subject's score on the independent variable. Thus we go a step beyond measuring the strength of a relationship into the realm of making predictions. The larger the correlation coefficient, the more accurate our predictions will be.

THE SETTING

In Chapters 6 and 11, we examined relationships between grouped nominal or ordinal variables presented in cross-tabular form. We also measured association by means of several techniques. Most tables in this text present data at the nominal or ordinal level of measurement. From time to time, grouped interval level variables may appear in tabular format, but in general the grouping is done by the researcher for purposes of presenting the findings to the reader. Originally, the data were individual, interval level raw scores. In such a situation, we have at our disposal techniques of analysis that are more sophisticated and more useful than many of the techniques used to interpret cross-tabulations. Known collectively as correlation-regression analysis, these new techniques enable us not only to measure association, but also to make predictions of scores on the dependent variable. As researchers, we ideally want to collect as much data as possible at the interval level, because we can make use of these more sophisticated techniques of analysis to learn as much as possible about the variables under study. In this chapter, therefore, we will study relationships between interval level variables that are not in tables but in ungrouped, individual raw score format. See Box 13.1 for a summary.

Suppose we are doing a study of alleged political corruption. We have reason to believe that a locally elected district attorney is demanding kickbacks to his reelection campaign fund from those appointed to

BOX 13.1

When Is Correlation-Regression Analysis Used?

Given: Two individual (raw score) variables measured by interval scales.

Task: Measure the strength of the relationship between these two variables.

Task: If that relationship is sufficiently strong, describe the nature of the relationship between the two variables in such a way that it will be possible to predict a respondent's score on one variable if we know that person's score on the other variable.

patronage positions in his department. He, of course, denies this and claims that any contributions are strictly voluntary, reflect no coercion nor pattern, and certainly do not reflect a specific percentage of the employees' salaries.

From public sources, we acquire the names, salaries, and mean monthly contributions made by each employee to their boss's war chest. Following are the findings, with the employees' names discreetly replaced with numbers.

Employee	x = *Annual Income (in thousands of dollars)*	y = *Mean Monthly Contribution*
1	$ 80	$ 160
2	70	95
3	52	97
4	45	85

Each of the four people is measured along two variables by individual, interval level raw scores.

Our questions are: (a) Is there a relationship between one's income and one's monthly campaign contribution? (b) If there is, how strong is that relationship? (c) Is it possible to estimate (predict) someone's contribution if we know that person's income and base that prediction on the data for the original four individuals? (d) What is that prediction for the person in question?

Our first task is to determine if a relationship exists and if so how large it is. To do this, we calculate a measure known as a correlation coefficient. This measure resembles the measures of association for cross-tabs such as *Q*, gamma, and lambda discussed in previous chapters. *When data are not grouped in tables, but are individual raw scores, we refer to the measure of the strength of that relationship as a* **correlation coefficient** rather than a measure of association. When both variables are interval or ratio level of measurement, as is the case here, we use a coefficient formally known as the **Pearsonian product-moment correlation coefficient.** Since that is quite a mouthful, we usually refer to it as **Pearson's *r*** or just *r*.

Like gamma, Pearson's *r* is actually measuring how linear the relationship is (more on that later) and it ranges from 0 to 1 in value. It also carries a sign—positive (for a positive relationship) or negative (for an inverse relationship). So *r* reads like a gamma. Once we know *r*, we square that value to obtain what we call the coefficient of determination, another indicator of the strength of the relationship being studied. The

coefficient of determination, r^2, *will tell us the proportion of variation in the dependent variable (y) that can be explained by variation in the independent variable (x).*

If r and r^2 are low, we may conclude that going any further is useless. Even though we could develop a predictive model, it would not be precise enough a predictor to be of value. If r and r^2 are large enough to interest us (and this is for now strictly a personal judgment call), we may conclude that it would be useful for us to be able to predict y from x—in this case to predict someone's campaign contribution from that individual's income. *That predictive model, the mechanism for estimating a y score from the respective x score, is known as a* **regression equation.** Since these two steps are so closely interrelated, we call the procedure in its entirety **correlation-regression analysis.**

Before reaching the point of actually calculating and applying correlation coefficients and regression formulas, we need to discuss a number of preliminary topics. These topics—Cartesian coordinates, the concept of linearity, and linear equations—may be old material to many readers who have appropriate mathematical backgrounds. If so, treat these sections as necessary review material. However, do not worry if these topics are new to you; they are presented in enough detail for beginners. All readers should keep in mind the three topics that follow are *preparations* for correlation-regression analysis—not the analysis itself. The latter will follow in due course.

CARTESIAN COORDINATES

In analyzing data, a useful place to begin is with a pictorial representation of the relationship between the two variables under study. We can do this by means of **Cartesian coordinates,** *a method of graphing relationships* devised by the French mathematician René Descartes. Not only is this method of pictorial representation relatively simple, it is conceptually elegant and mathematically valuable. From it stems the field of mathematics known as analytic geometry, which links curves on a graph to algebraic equations. Many readers, in fact, will already be familiar with Cartesian coordinates.

We begin by constructing a graph composed of two perpendicular lines called *axes* (singular: *axis;* plural: *axes*). (See Figure 13.1.) Each axis is measured in scales resembling those on a ruler, with the point where the two axes intercept given a value of zero. *That point is the*

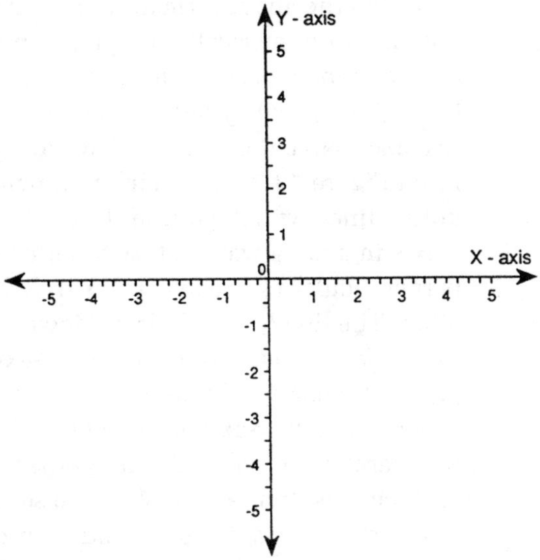

Figure 13.1

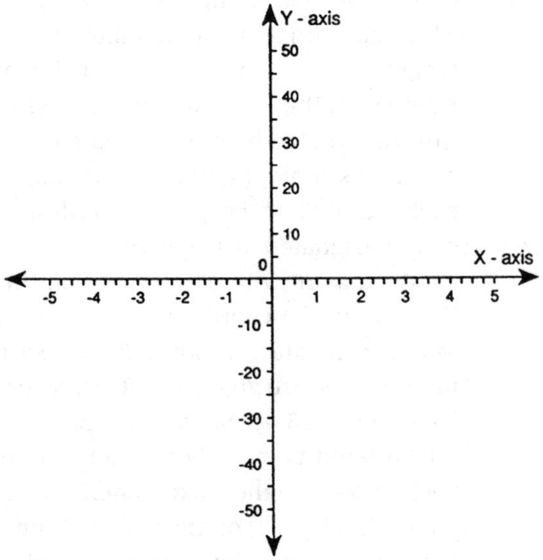

Figure 13.2

origin *of the graph. The axis that extends horizontally is the* **x-axis** *and the one that extends vertically is the* **y-axis.** On both axes, the measurement numbers increase in magnitude as one moves out from the origin. However, depending on the direction one is moving, we give to each measurement number a + or – sign. On the *x*-axis, the numbers to the right of the origin are positive (+) (we usually do not put + in front of each number) and those to the left are negative (–). On the *y*-axis, the numbers above the origin are positive and those below it are negative. On each axis, the distance between each unit is the same. For instance, if we lay a ruler on the *x*-axis we see that the distance from +1 to +2 is the same as the distance from +2 to +3 or from –4 to –5, and so on. Note, however, that it is not necessary that both axes be measured in units of the same size. That is, the distance between +1 and +2 on the *x*-axis need not be the same as the distance from +1 to +2 on the *y*-axis, as long as the distances between all unit intervals are the same on *each* axis.

In Figure 13.2, for instance, units of one on the *x*-axis are as far apart from each other as are units of ten on the *y*-axis (measure it yourself). The arrowheads at the ends of the axes are essentially reminders that the axes themselves are limited in length by the size of the pages on which they are printed. Mathematically, the page on which the graph is printed delineates a plane, a surface in space that can be extended infinitely beyond the edges of the page itself. Consequently, while in fact we run out of space in any real page, conceptually our axes continue out indefinitely. (Want to impress your friends? Say each axis approaches plus and minus infinity as its limit.)

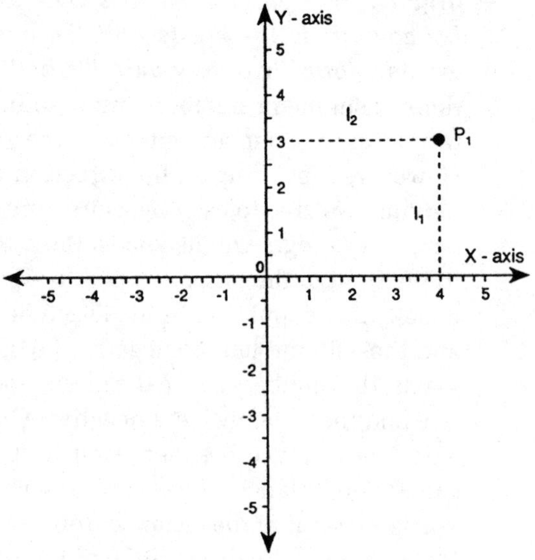

Figure 13.3

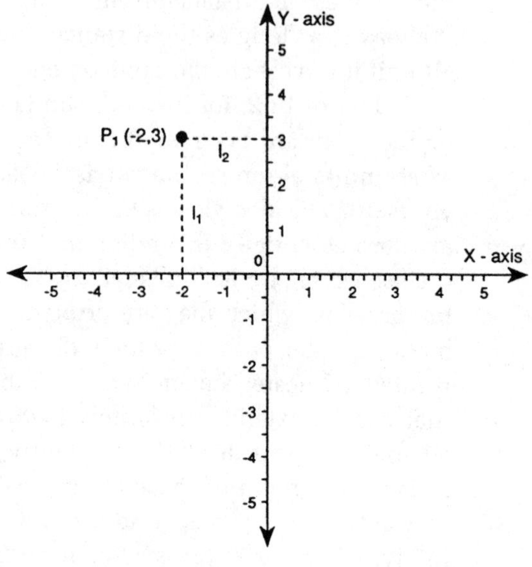

Figure 13.4

Given the above format, we note that it is possible to numerically designate any point on the graph's page (or its logical extension beyond the page) by relating that point to the two axes. Note the point we have designated P_1 on Figure 13.3. We begin by drawing two dotted lines, which go from P_1 to the *x*- and *y*-axis in such a way that each dotted line is perpendicular to one axis and parallel to the other. The line l_1, which drops down from P_1 to the *x*-axis, is parallel to the *y*-axis and perpendicular to the *x*-axis (the angle between l_1 and the *x*-axis is 90 degrees). The line l_2 is parallel to the *x*-axis and perpendicular to the *y*-axis. In the case of P_1, we see that l_1 intersects the *x*-axis at +4 and l_2 intersects the *y*-axis at +3. We use these two values to locate P_1. By convention, we designate the point by what is called an **ordered pair,** *a set of two numbers in parentheses separated by a comma. The first number is the* **x-coordinate** *and the second number is the* **y-coordinate.** The *x*-coordinate is the number on the *x*-axis where l_1 crosses it. The *y*-coordinate is the number on the *y*-axis where l_2 crosses it. P_1 is thus designated by the ordered pair (+4, +3) or, more simply, (4, 3) since we may assume each number to be positive unless specifically designated as negative.

We may now find any point on the graph if we know its coordinates. Suppose we are asked to locate the point P_2 whose coordinates (x_2, y_2) are given as (−2, 3), respectively (see Figure 13.4). Since the first number in our ordered pair is always the *x*-coordinate, we locate −2 on the *x*-axis. Then we locate the *y*-coordinate (+3) on the *y*-axis. From each of these points on the axes we construct our dotted lines (l_1 and l_2) perpendicular to each of the axes, and where the dotted lines intersect we have our point P_2.

The origin of the graph will always have the coordinates (0, 0). Note that any point exactly on the x-axis will have a y-coordinate of 0 and any point exactly on the y-axis will have an x-coordinate of 0.

THE CONCEPT OF LINEARITY

We use the Cartesian coordinate system to plot the relationships between variables we would like to explore. By convention we label the variable we presume to be the independent variable as x and the dependent variable, the one whose variation we are trying to explain, as y. Remember that in research, we begin with a number of cases or respondents and assign to each one a score or number related to each of the variables we are studying. Thus each respondent has both an x score and a y score. To be more specific, suppose that for a sample of six people we wish to explore the relationship between two characteristics that we know each of the respondents possesses. For each respondent we can measure both characteristics with at least an interval scale. We assume that one of the characteristics *(the dependent variable) is related to or can be explained by* the extent to which the respondent possesses the other characteristic *(the independent variable)*. Mathematically, we say that y is a **function** of x. In this case let us suppose that y, the dependent variable, is total savings and that x, the independent variable, is the respondent's level of education as measured by number of years of school he or she has completed. The hypothesis we wish to examine is that the more schooling one has, the more money one will save; savings is a function of education. If it is indeed true for everyone, we assume that it will be true for our group of six respondents. Consequently, we find out from each respondent his or her years of education (x) and amount of savings (y) and present the data in tabular form (see Table 13.1).

Notice that we can make use of Cartesian coordinates to graph each respondent in terms of scores. If we construct an x-axis to measure education and a y-axis to measure savings and for each respondent we have both an x and a y score, then each score may be treated as a coordinate in an ordered pair. Thus each person may be indicated by a point on the graph as determined by that person's education and savings (see Figure 13.5).

From the table and the graph, we immediately notice two things. First, in every one of the six cases, the greater the person's education,

Table 13.1

Respondents	*x = Education (years of schooling)*	*y = Total Savings (in thousands of dollars)*
Tom	0	0
Carol	4	2
Jim	8	4
Lucy	12	6
Jack	16	8
Jill	20	10

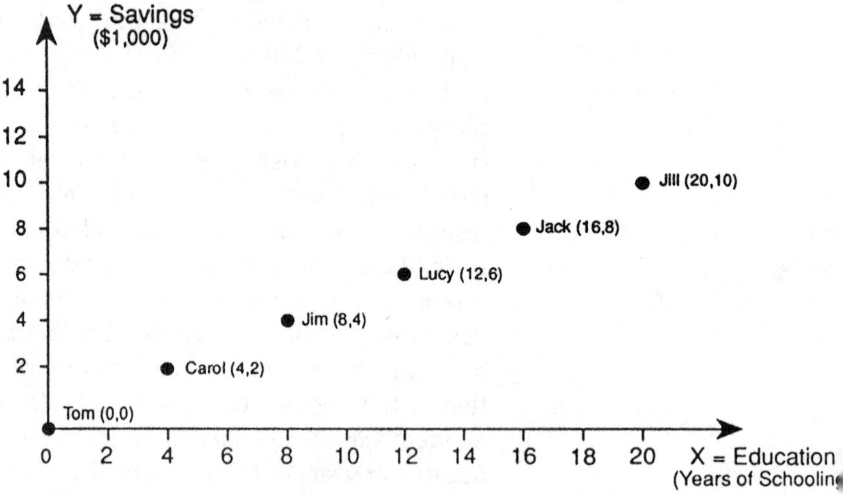

Figure 13.5

the greater the savings. On the basis of our hypothetical data at least, our hypothesis is confirmed. Second, we notice that the pattern of the points in Figure 13.5. is an exact straight line. We confirm the fact by laying a ruler over the points and drawing a straight line through them all, as we have done in Figure 13.6. *Since the relationship is an exact straight line we say that the two variables, education and savings, are* **linearly related.** The straight line in Figure 13.6 is an exact visual representation of that relationship.

The straight line in Figure 13.6 enables us to predict a person's score on one variable if we know that person's score on the other variable. If we assume that the straight line holds for everyone in the population

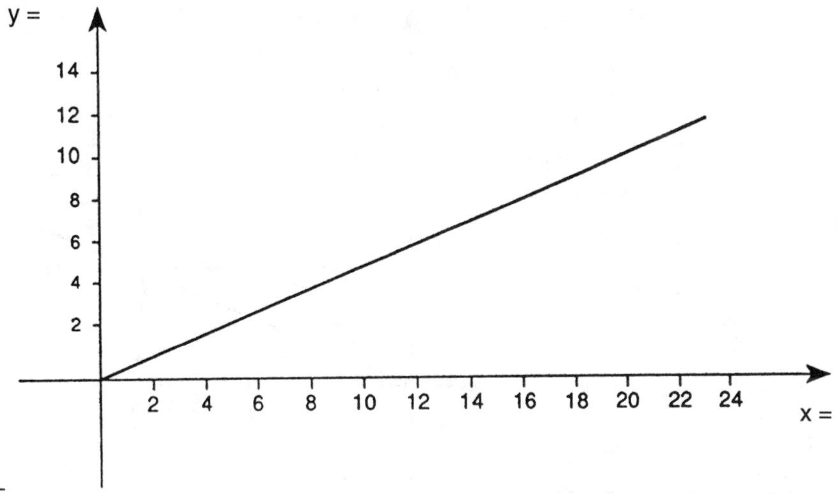

Figure 13.6

and another person comes along with a specific number of years of schooling, we can use the line to predict that person's savings.

For example, suppose Ann informs us that she has had a total of 24 years of schooling. To predict her savings we make use of the fact that if the straight line in Figure 13.6 reflects the relationship between education and income for everyone, then the ordered pair representing Ann's education and income must also fall on that line. We know that Ann's x-coordinate is 24, so if we designate her income as y_a (the subscript a stands for Ann), then the point $(24, y_a)$ will fall on that straight line. We locate Ann's education (24) on the x-axis and draw a perpendicular dotted line (l_1) up from that point until it intersects the line (see Figure 13.7). This will be the point $(24, y_a)$. To find y_a by visual inspection, we now construct dotted line l_2 that goes through $(24, y_a)$ and is parallel to the x-axis (perpendicular to the y-axis). The line l_2 crosses the y-axis at exactly y_a, at the spot where $y = 12$. Thus, $y_a = 12$. We predict, therefore, that Ann's savings will be $12,000.

Now, suppose yet another person, Frank, tell us he has completed 10 years of school. To predict Frank's savings, we find 10 on the x-axis and erect a perpendicular l_1, which crosses the line at the point $(10, y_f)$. Where l_1 crosses the line, we draw a new line l_2 parallel to the x-axis and see where l_2 crosses the y-axis. Since l_2 crosses the y-axis halfway between the 4 and the 6, we conclude that y_f is 5. We predict that Frank has saved $5,000.

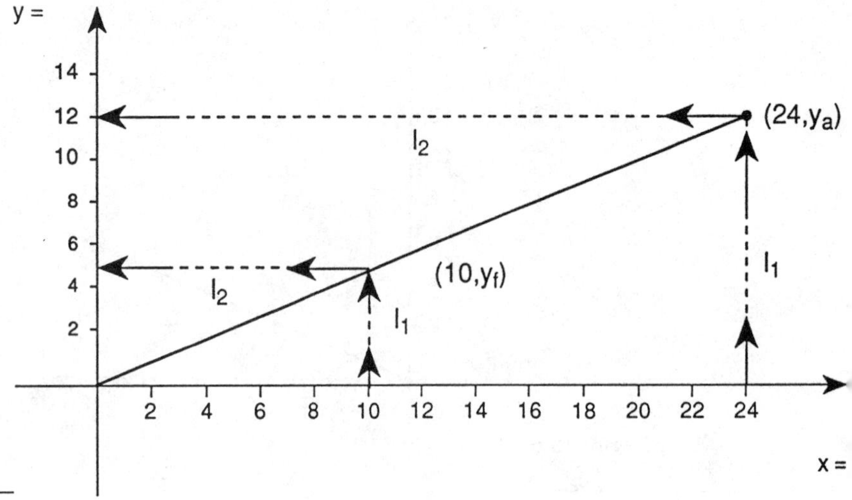

Figure 13.7

LINEAR EQUATIONS

The predictive process we have just been using may also be accomplished by means of simple algebra. If it is indeed true that income is linearly related to education as the data in Table 13.1 would suggest, then *it is possible to represent the straight line used in Figures 13.5 to 13.7 by an algebraic equation.* Specifically, *that equation will be of a form known as a* **linear equation,** *affirming that the points generated by the equation will graph as a straight line* rather than any other kind of graphic figure.

By way of review, we should recall what an equation really does. Essentially, an equation is a set of mathematical instructions—a program— that matches specified values of some variable x with specified values of some other variable y (it "maps" values of x into values of y). The equation tells us how to find the particular y value that matches it with the value of x that we have designated. We supply the value for x; the equation tells us the corresponding value of y. The line in Figures 13.5, 13.6, and 13.7 may be expressed algebraically by a linear equation so that instead of finding y graphically, we can calculate it by solving the equation for y. All linear equations can be written in the following form.

$$y = a + bx$$

In this equation, a and b are known as **constants.** This is because each straight line that we can draw on the graph will have a particular a-value and a particular b-value associated with it, and *for that specific line, a and b remain the same—constant*—no matter what value of x and y may be plugged into that equation. By contrast, *the x-value and the y-value are called* **variables,** because *any given straight line will have many different values of x and y associated with it; these values vary* from case to case. The constants, a and b, determine where on the graph the line is located and differentiate that line from all the others. The variables, x and y, relate to specific points on that line.

We are already familiar with the concept of a variable. Constants remain constant for a particular line (a particular relationship), although a and b do indeed vary from one line to the next. Suppose for a given relationship we know the constants a and b. We are left with x and y, the variables. But if x is specified, we then know a, b, and x and can easily solve for y, the remaining unknown. Solving for y—predicting y for a given x—is the major goal of this process.

The constants are not merely numbers drawn like rabbits from a magician's hat; both a and b can be defined in terms of the characteristics of the particular line under study. The first of these, a, is known as the **y-intercept** of the line. It *is the value of y at that point where the line crosses the y-axis.* Note that because we have indicated that the axes extend out indefinitely, we can specify that in every case other than where a line is parallel to or on the y-axis, a line will eventually cross the y-axis and thus will have a y-intercept. This can be seen by placing a ruler's edge on a graph, moving it and rotating it around. The line it delineates will either cross the y-axis on the graph itself or on some extension of the line and the y-axis (see Figure 13.8). *The y-intercept is the value of y where* $x = 0$. Note that in our savings/income problem, the line crosses the y-axis exactly at the origin of the graph (0, 0). In that case, when $x = 0$, $y = 0$ also, and the y-intercept (a) of that particular line is 0.

The second constant in the equation of the line is designated b and is called the **slope** of the line. It is defined as *the change in y per unit change in x.* Imagine two points on the line, P_1 with coordinates (x_1, y_1) and P_2 with coordinates (x_2, y_2). The slope of the line is the ratio of the change in y to the change in x, that is, the difference between the two y values divided by the difference between the two x values.

$$b = \frac{y_2 - y_1}{x_2 - x_1}$$

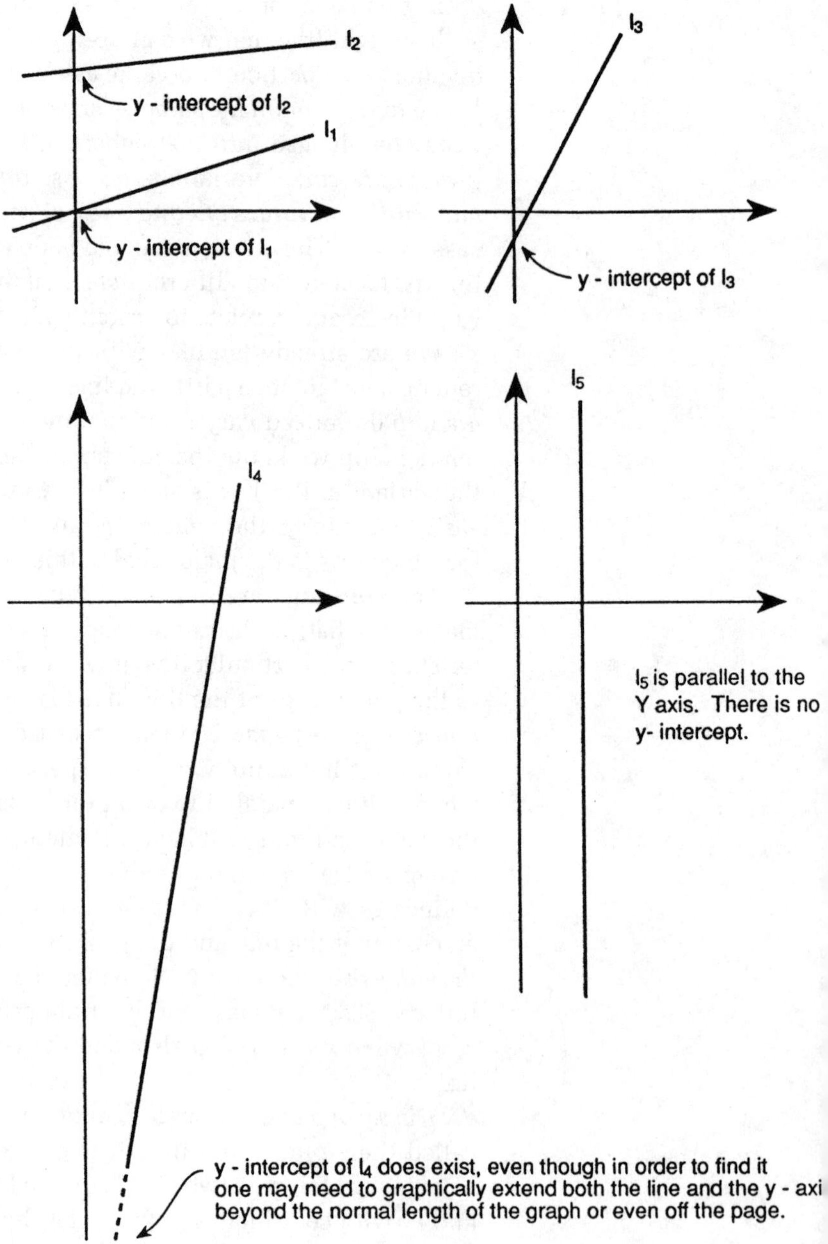
l2
y - intercept of l2
l1
y - intercept of l1
l3
y - intercept of l3
l5
l4
l5 is parallel to the Y axis. There is no y- intercept.
y - intercept of l4 does exist, even though in order to find it one may need to graphically extend both the line and the y - axi beyond the normal length of the graph or even off the page.

Figure 13.8

P_1 (x_1, y_1) and P_2 (x_2, y_2) may be any two points on the line; it does not matter which ones. In our example, for instance, Jim's two score have been graphed as a point on the line designated by the ordered pair (8, 4) (see Figure 13.5). Let that be P_1. Thus, $x_1 = 8$ and $y_1 = 4$. Jack's scores are designated by the ordered pair (16, 8). If we let this be P_2, then $x_2 = 16$ and $y_2 = 8$. We can now find b.

$$b = \frac{y_2 - y_1}{x_2 - x_1} = \frac{8 - 4}{16 - 8} = \frac{4}{8} = \frac{1}{2} = \frac{.5}{1} = .5$$

By inspection of Figure 13.6, we see that every time we move a point up the line far enough to change x by one unit, the corresponding change in y is 1/2 or .5 units.

To confirm our conclusion, we may designate another two points on the line as P_1 and P_2 to see if we still get b equal to .5. Let us call Jill's score (20, 10) P_1 and Carol's score (4, 2) P_2.

$$b = \frac{y_2 - y_1}{x_2 - x_1} = \frac{2 - 10}{4 - 20} = \frac{-8}{-16} = \frac{-1}{-2} = .5$$

We can now indicate the equation of the line in our example. Remember that the standard form of a linear equation is

$$y = a + bx$$

We know by visual inspection that a is 0, and we have calculated a b of .5; thus

$$y = 0 + .5x$$

To simplify the equation, when $a = 0$ we drop it altogether. Our equation thus becomes

$$y = .5x$$

Earlier we found the y values for Ann and Frank by visual inspection. We now find them algebraically by plugging Ann's and Frank's x-values into the formula and solving for y.

For Ann:

$$y = .5(24) = 12$$

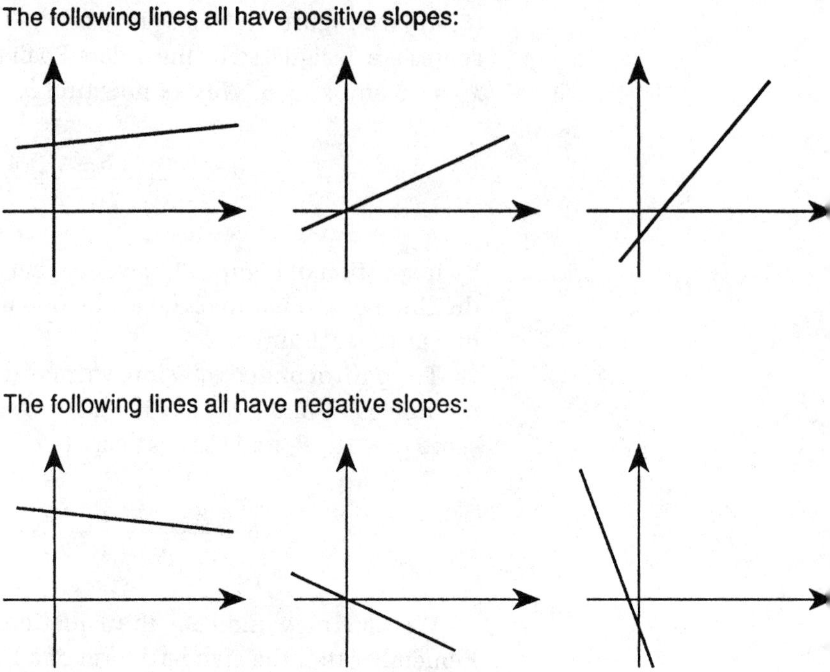

The following lines all have positive slopes:
The following lines all have negative slopes:

Figure 13.9
Positive and Negative Slopes

For Frank:

$$y = .5(10) = 5$$

Since y was measured in thousands of dollars, we now know that Ann and Frank saved $12,000 and $5,000, respectively.

A few more points are in order about b, the slope of the line. In our sample problem, where $b = .5$, the slope is a positive number. However, in other cases the slope may be negative. The sign of the slope gives us an indication of the direction of the line and the nature of the relationship that the line reflects. When the slope of a line is positive, the line will slope upward from the lower left-hand side of the graph to the upper right-hand side (see Figure 13.9). If the slope of the line is negative, the line will slope from the upper left-hand side to the lower right-hand side (Figure 13.9). *When the slope of the line is positive, x and y are* **positively related.** *As x increases in magnitude, so will y; as x decreases in magnitude, so will y.* In our example problem, the greater the amount of one's schooling, the greater one's savings; conversely, the lower one's schooling, the less one's savings. If the b in the equation were negative, we would have two variables that were **inversely (negatively) related:** *as one variable increased in magnitude, the other variable would decrease.* Later, when we are working with the slope of a regression equation, we will see that the slope has the same sign as *r*.

To visualize the difference between positive and negative relationships, imagine a room with an electric heater in it. The more power put into the heater, the greater its output and the warmer the room becomes. Power input and temperature are positively related: the more power, the more heat. If, however, we replaced the heater with an air conditioner the relationship between power and heat would be negative, or inverse: the more power going into the unit, the *less* the heat in the room. Power input and room temperature are still related to each other, but the nature of the relationship that was positive in the first case is negative in the second. Therefore we will need to be aware of the sign of the slope whenever we undertake this kind of analysis.

LINEAR REGRESSION

Up until now the problems discussed have been graphed as straight lines and a linear equation could easily be derived from the data. It would be rare indeed, however, to find actual social data that could automatically be graphed as a straight line. Several reasons account for this fact.

The first and most obvious reason the data may not graph as a straight line is that the relationship may not be linear. Some other kind of curve is best for describing such a relationship. Figure 13.10, for instance, presents a somewhat refined picture of the relationship between respondents' ages and their levels of participation in athletic activities. For instance, *y* might be the number of athletic events held in the previous 12 months in which the respondent actually played. The respondent's age is treated as the independent variable. There is a relationship between the two variables, but it is **curvilinear** not linear.

In a case like this, the equation best fitting the curve would not be linear but would be more mathematically complicated. Finding such an equation is beyond the scope of this text, but we can easily make the observation that there is a relationship that is not linear by graphing the relationship and inspecting the graph. It is an all-important step that is often left out of research and yet could be of crucial significance to the social scientist. For our purpose here, in the remaining procedures

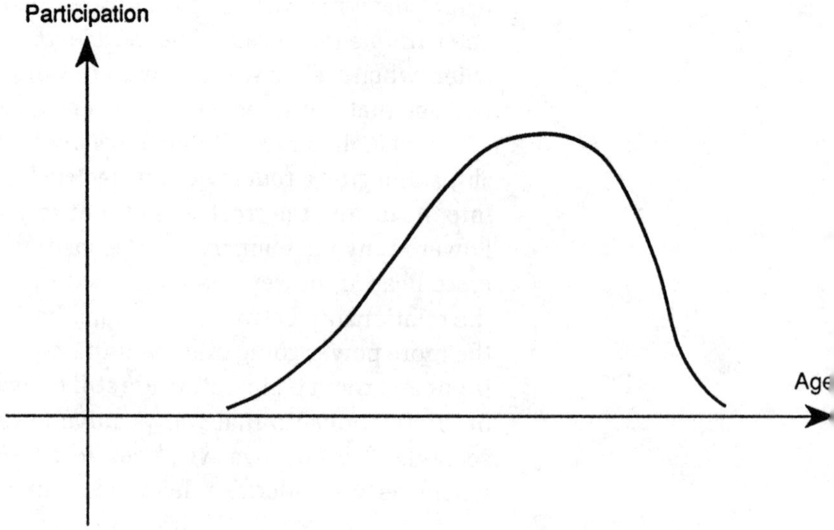

Figure 13.10
A Curvilinear Relationship

presented in this chapter we will assume that the underlying relation ship we are studying is linear. If the data do not graph exactly as straight line, we will assume it is due to other factors.

Of those other factors, measurement error or procedural deviation can be major reasons for deviation from linearity. For example, in th annual income versus campaign contribution problem, we might hav been relying on the people to report their incomes accurately. If the estimate or even falsify these figures, the data we have will only be a inexact estimate of actual income. How is the value of their contribu tions determined? Tax returns? Verbal estimates?

In addition, other factors can keep the relationships in the socia sciences from being strongly linear. Variables other than the two variable we are studying may be involved. For example, in the income/contributio problem, people appointed to patronage positions may not all have bee the D.A.'s cronies. Some may have been appointed because they pos sessed specific expertise not available elsewhere. Perhaps the one an only arson specialist in the county was hired for her expertise alone an thus felt less pressure to contribute to the boss. Or, perhaps a stron partisan contributed far more than expected out of ideological convic tion alone. Such factors would obviously prevent a perfect "real world linear relationship between income and campaign contribution.

Some combination of error in measurement or other factors coul lead to a distortion of the overall data that when graphed does not resul in an exact straight line. The set of points may show a pattern that looks somewhat like a straight line but is not definitely nonlinear. In such cases, we may study the relationship by means of a technique known as **linear regression** *which finds a line that "fits" the scatter of data points in such a way as to provide for any given value of* x *the best estimate of the corresponding value of* y. We are then able to predict y from x the same way that we did for perfect linear relationships.

Suppose we are studying the relationship between social alienation and religiosity. We develop a scale, ranging from 0 to 100, to measure the extent to which an individual may feel alienated from the nonpolitical structures of society—friends, family, church, and other reference groups or their values. The higher the score, the more the alienation felt. Religiosity is measured by another scale, ranging from 0 (least religious) to 10. Our hypothesis is that the higher one's level of social alienation, the lower will be that person's intensity of religious belief.

x = *Social Alienation*	y = *Religiosity*
25	10
30	9
35	8
40	8
40	7

The graph of this relationship, shown in Figure 13.11, is known as a **scatter diagram** (or **scattergram** or **scatter plot**) *because the data points are scattered on the graph and not a perfect line.*

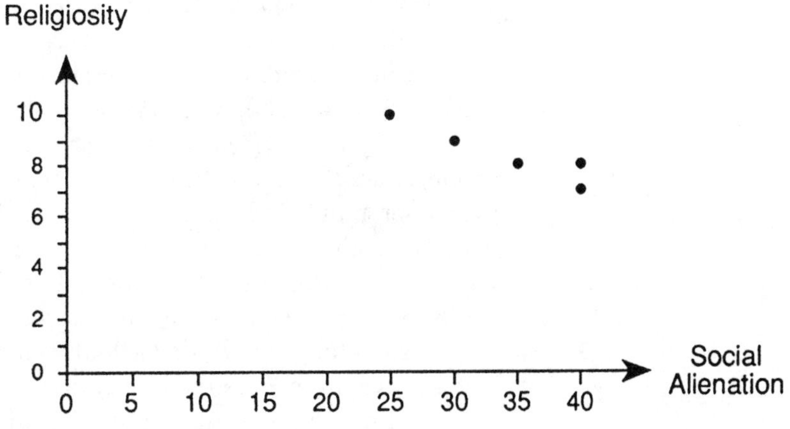

Figure 13.11

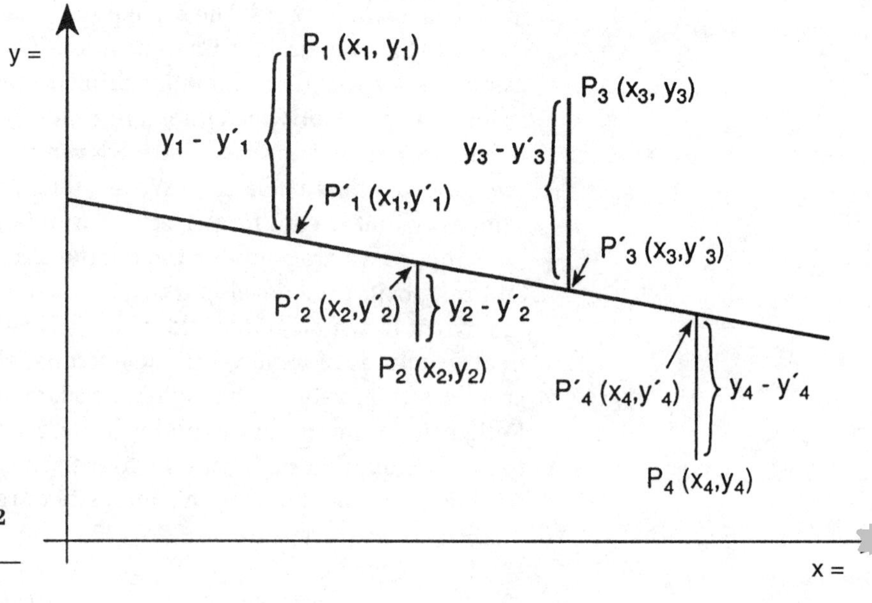

Figure 13.12

The scatter diagram indicates a relationship near enough to linearity that we can imagine that had all other factors (such as the influence of other variables on the relationship) and all measurement errors been eliminated, we would indeed have a straight line. *We can use a technique called the* **least squares method** of linear regression to find the equation of the line that best fits these points and that we will call **the regression of y on x.** *This phrase informs us that the line we generate will be the line that enables us to most accurately predict y from x.*

To understand what the least squares method actually does, look at the scatter diagram in Figure 13.12, where the regression line has already been drawn in between the points of some hypothetical relationship. The actual observations are P_1, P_2, P_3, and P_4. The line running between these points, the regression line, has a set of points on it—P'_1, P'_2, P'_3, P'_4— which have the same x values as their corresponding points P_1, P_2, P_3, and P_4. (The apostrophe is read "prime." If P_1 is read "P-sub-one," then P'_1 is read "P-sub-one-prime.") We use P'_1 instead of a P with another numbered subscript because P_1 and P'_1 are related; in this instance, they share a common x coordinate.

What distinguishes P_1 from P'_1 is that the y-value for the actual observation P_1 is different (in this case, larger) than the y-value of the corresponding point P'_1 that actually falls on the line. In fact, the shortest y distance from P_1 to the line is $y_1 - y'_1$. For P_2, the shortest y distance to the line is $y_2 - y'_2$. (It will be a negative number because the point is below the line; hence y'_2 is larger than y_2. But the absolute value of the distance is still $| y_2 - y'_2 |$.)

The dotted lines in Figure 13.12 show each of these distances, indicating the deviation of these points from the regression line. The regression line is the line (found by means of the calculus) such that the sum of the square of all the y deviations from it is less than it would be for any other line that one might construct between the points. Actually, we state this characteristic with the following expression.

$$\sum (y_i - y'_i)^2 = \text{a minimum}$$

$y_i - y'_i$ is the y distance from the point to the line. We square each such distance, or deviation, to get rid of negative numbers. Then we add up all the squared deviations. The number we get will be smaller than it would be for any other line we might have constructed to pass between the data points. We will never have to derive the calculus part of this problem ourselves; it has been done for us already. But it is important to understand what this regression of y on x actually does. We have generated a line that minimizes the y distances from the actual data points to the line. Thus, if we predict y from x using this regression line, our prediction of y should be closer to the true value of y than any other prediction we might have made using any other line drawn between the data points.

To see this effect, examine Figures 13.13 and 13.14. In Figure 13.13, the least squares regression for predicting y has been generated from three initial data points: P_1 (1, 3), P_2 (3, 0), and P_3 (6, 2). The distance from each point to the line $(y - y')$ is then determined to be 1.03, −1.71, and .68 for P_1, P_2, and P_3, respectively. Squaring each distance and adding up the squared distances yields a $\Sigma(y - y')^2 = 4.44$. Any other line drawn through the three points will yield a $\Sigma(y - y')^2$ greater than 4.44.

Figure 13.14 illustrates this by showing another line. In this instance, $\Sigma(y - y')^2 = 7.46$, which of course is larger than 4.44. The least squares line's $\Sigma(y - y')^2 = 4.44$ is less than the one for the line in Figure 13.14 or any other line through these points.

We must point out that the line used to predict y from x, the regression of y on x, is *not* the line that would give us the best prediction of x from y. This latter line, the **regression of *x* on *y*,** *is obtained by minimizing the sum of the squared x deviations* instead of minimizing the y distances.

Figure 13.15 gives a visual portrayal of what we mean by minimizing the x distances. The line we have drawn in that figure is the regression of x on y, the line such that:

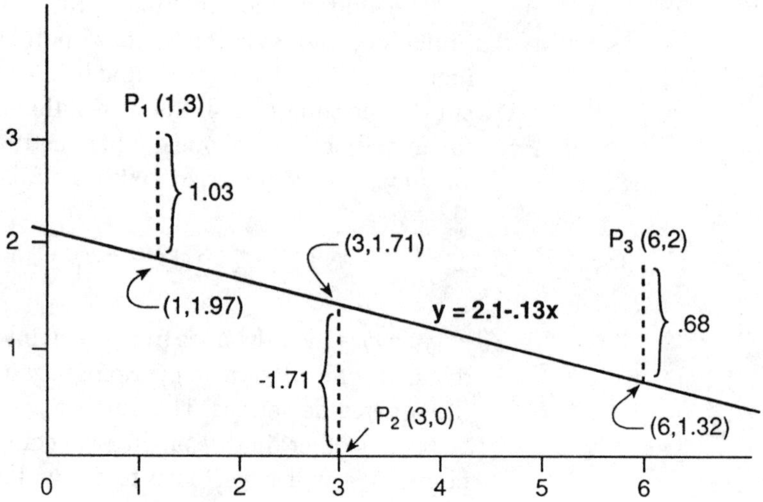

$y = 2.1 - 1.3x$ The least squares regression.

$$\Sigma(y - y')^2 = (1.03)^2 + (-1.71)^2 + (.68)^2$$
$$= 1.06 + 2.92 + .46$$
$$= \underline{4.44}$$

There is ***no*** line that can be drawn through P_1, P_2, and P_3 whose $\Sigma(y - y')^2$ will be less than 4.44.

Figure 13.13

$$\sum (x_i - x'_i)^2 = \text{a minimum}$$

Usually we will not need to know the regression of *x* on *y* since we usually want to predict the dependent variable *y* from given knowledge of the independent variable *x*, and not vice versa. Still, the differences between the regression of *y* on *x* and the regression of *x* on *y* will be conceptually important to us. Only when we begin with a perfect linear relationship will the two regression lines be identical to each other.

In any event, our eventual need is to find the line for the regression of *y* on *x* from our original data. To clarify this, we first make a small notational addition to the general form for a linear equation.

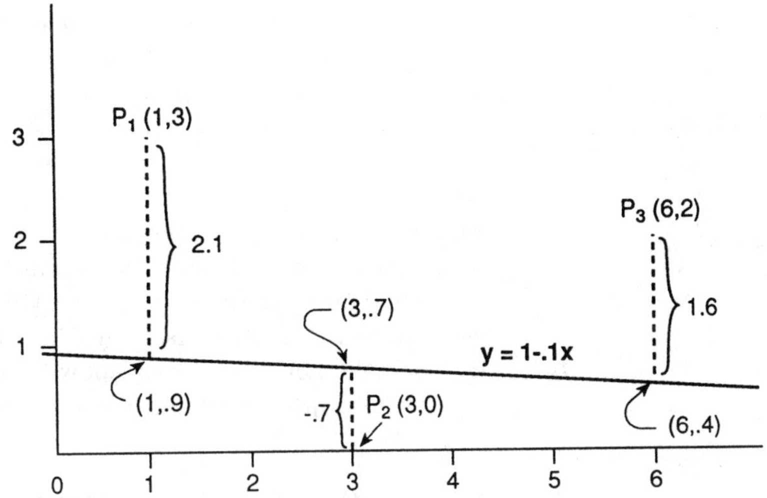

$y = 1 - .1x$ is ***not*** the least squares regression.

$\Sigma(y - y')^2 = (2.1)^2 + (-.7)^2 + (1.6)^2$

$= 4.41 + .49 + 2.56$

$= 7.46$

This, or any line other than the regression, has a $\Sigma(y - y')^2$ ***larger*** than that of the regression line's 4.44.

Figure 13.14

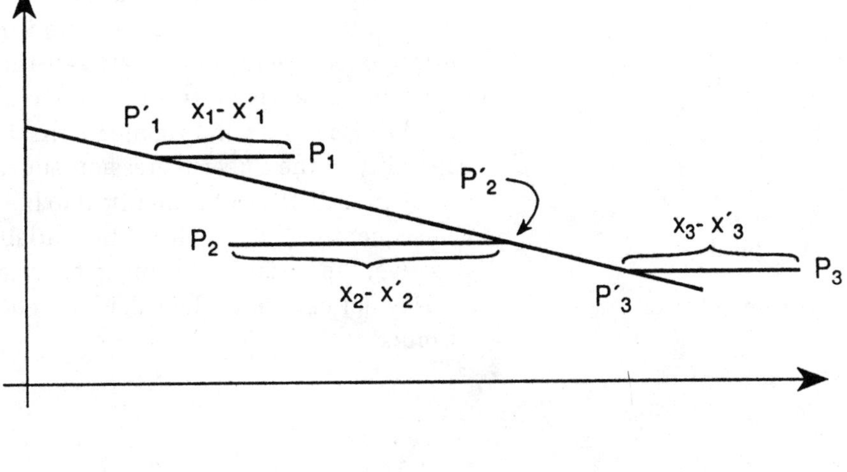

Figure 13.15

$$y = a_{yx} + b_{yx}x$$

The subscripts on a_{yx} and b_{yx} tell us that this line will be the least squares regression of y on x. If we want to predict x from y, we we would use the following notation instead.

$$x = a_{xy} + b_{xy}y$$

And of course a_{yx} does not equal a_{xy} and b_{yx} does not equal b_{xy} except where the original relationship is perfectly linear.

To find the regression of y on x, we will need to find a_{yx} and b_{yx}. Once we know these two numbers, we will have the equation. The formulas that will yield the slope and y-intercept will be presented a bit later. But first we need to return to the topic of the correlation coefficient.

THE CORRELATION COEFFICIENT

Before we plunge into calculating b_{yx} and a_{yx}, let us determine their value to us in predicting y from x. We do this by first finding the correlation coefficient r and its square, the coefficient of determination.

First let us understand the purpose of calculating r. Pearson's r is actually a measure of how close the point distribution comes to linearity. When r is 0, it usually means that the points are randomly distributed throughout the scatter diagram. Thus the regression of y on x is a horizontal line with slope equal to 0 and knowing the regression does not improve our ability to predict y from x. In some instances, an r of 0 means the data are related in a manner other than a linear one. In short an r equal to 0 means the variables are not linearly related (see Figure 13.16) and so a linear regression is of no predictive use.

By contrast, an r whose absolute value is 1.0 tells us that there is a perfect linear relationship between the two variables, and the linear regression gives us perfect predictive capability. The higher the value of r, the closer the points come to linearity and the better the predictive capability of the linear regression equation. Moreover, the sign on the correlation coefficient is identical to the sign of the regression slope b_{yx}. Consequently, if r is positive, the variables are positively related; if r is negative, the variables are inversely related.

Now let us turn to the calculation of r. Pearson's r has a formidable formula:

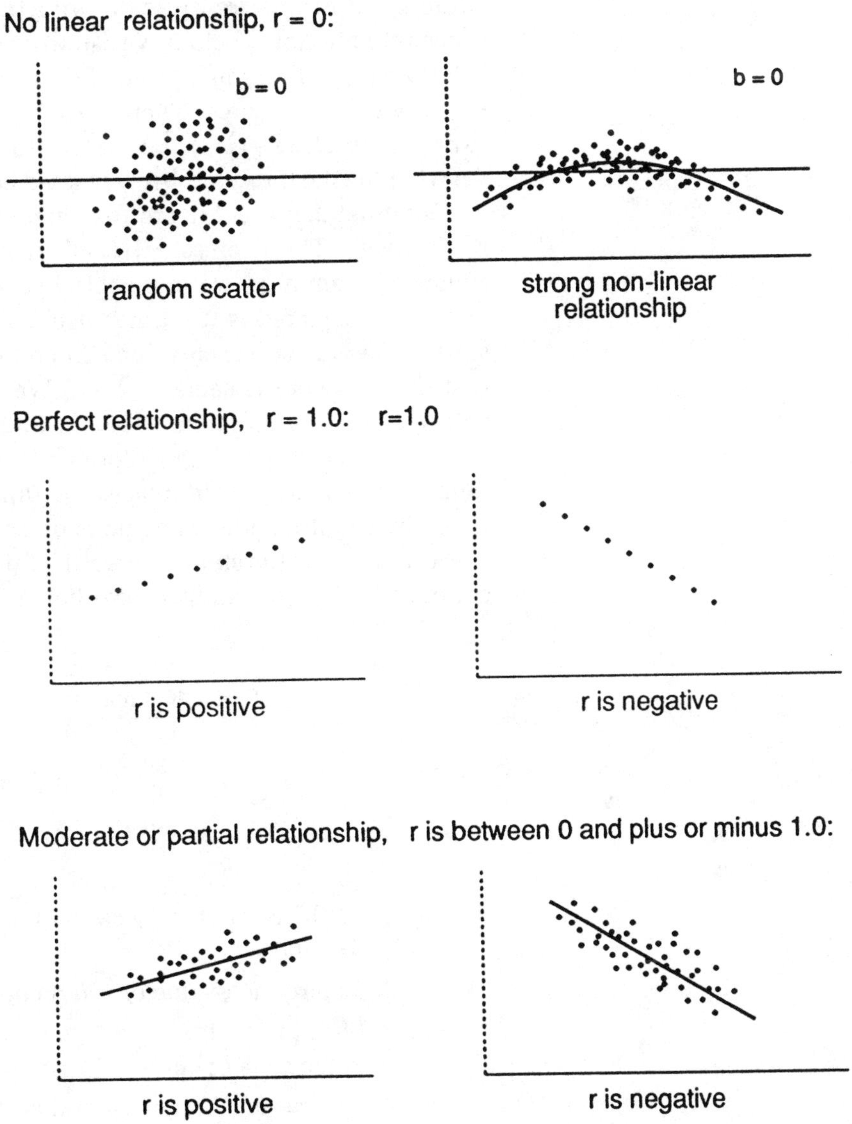

Figure 13.16

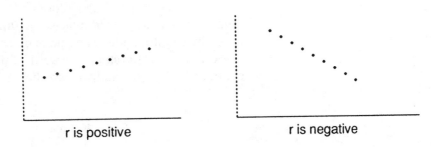

$$r = \frac{n\sum xy - \left(\sum x\right)\left(\sum y\right)}{\sqrt{\left[n\sum x^2 - \left(\sum x\right)^2\right]\left[n\sum y^2 - \left(\sum y\right)^2\right]}}$$

Although this appears difficult, with the aid of a calculator and the procedure presented below, we can work it without too much difficulty. What we must do is generate all of the components of the equation from the original raw scores. Then we will separately calculate the three expressions $n\sum xy - (\sum x)(\sum y)$, $n\sum x^2 - (\sum x)^2$, and $n\sum y^2 - (\sum y)^2$. Then we will plug the three numbers obtained back into the formula to find r.

Note that $\sum x$ and $\sum y$ are the sums of the original x and y scores, respectively. The expression $(\sum x)^2$, read, "summation of x, quantity squared," is simply $\sum x$ times itself. Likewise, $(\sum y)^2$, the summation of y, quantity squared is $\sum y$ times itself. To find $\sum x^2$, "summation of x squared," we square each value of x and then add the values up. (Note that $(\sum x)^2$ is *not* the same as $\sum x^2$.) We will also need to square all y values and add them up to get $\sum y^2$. To find $\sum xy$, we multiply each x score by its corresponding y score and add the products together. The number of cases (people, places, or things) actually measured, n, is found by counting how many pairs of x and y scores we have.

Following are the steps elaborating this procedure. The original data columns labeled $x =$ and $y =$ are shown first.

$x =$ *Social Alienation*	$y =$ *Religiosity*
25	10
30	9
35	8
40	8
40	7

Step 1. To the right of the y column, add three new columns: $x^2 =$, $xy =$, and $y^2 =$.

Step 2. Square each value of x and enter it in the $x^2 =$ column.

Step 3. Multiply each x score times its respective y score and enter it in the $xy =$ column.

Step 4. Square each value of y and enter it in the $y^2 =$ column.
Your data should now look like this:

$x =$	$y =$	$x^2 =$	$xy =$	$y^2 =$
25	10	625	250	100
30	9	900	270	81
35	8	1225	280	64
40	8	1600	320	64
40	7	1600	280	49

Step 5. Count up n, the number of cases (pairs of scores) and enter it as $n =$ __ (in this case $n = 5$) to the left of the bottommost value of x.

Step 6. Add each column up to obtain Σx, Σy, Σx^2, Σxy, and Σy^2. Put these totals under a totals line below the bottommost score in each column, and label each total as indicated below.

Step 7. Below the values for Σx and Σy, square those numbers to obtain $(\Sigma x)^2$ and $(\Sigma y)^2$ and enter these two numbers as labeled below.

You should now have:

	$x =$	$y =$	$x^2 =$	$xy =$	$y^2 =$
	25	10	625	250	100
	30	9	900	270	81
	35	8	1225	280	64
	40	8	1600	320	64
	40	7	1600	280	49
$n = 5$	$\Sigma x = 170$	$\Sigma y = 42$	$\Sigma x^2 = 5950$	$\Sigma xy = 1400$	$\Sigma y^2 = 358$

$(\Sigma x)^2 = (170)^2 = 28{,}900$

$(\Sigma y)^2 = (42)^2 = 1764$

Step 8. Using the figures calculated above, find $n\Sigma xy - (\Sigma x)(\Sigma y)$.

Step 9. Calculate $n\Sigma x^2 - (\Sigma x)^2$.

Step 10. Calculate $n\Sigma y^2 - (\Sigma y)^2$.

You should obtain the following:

$$n\Sigma xy - (\Sigma x)(\Sigma y) = 5(1400) - (170)(42)$$
$$= 7000 - 7140$$
$$= -140$$

$$n\Sigma x^2 - (\Sigma x)^2 = 5(5950) - 28{,}900$$
$$= 29{,}750 - 28{,}900$$
$$= 850$$

$$n\Sigma y^2 - (\Sigma y)^2 = 5(358) - 1764$$
$$= 1790 - 1764$$
$$= 26$$

Step 11. Calculate Pearson's r from the formula.

$$r = \frac{n\sum xy - \left(\sum x\right)\left(\sum y\right)}{\sqrt{\left[n\sum x^2 - \left(\sum x\right)^2\right]\left[n\sum y^2 - \left(\sum y\right)^2\right]}} = \frac{-140}{\sqrt{(850)(26)}}$$

$$= \frac{-140}{\sqrt{22{,}100}} = \frac{-140}{148.66} = -.9417$$

Step 12. Square r (before rounding) to obtain the coefficient of determination.

$$r^2 = (-.9417)^2 = .8867$$

Let us round off and summarize.

$r = -.94$	a large inverse relationship.
$r^2 = .89$	.89 proportion (89%) of the variation in religiosity can be accounted for by the variation in social alienation.

This relationship is large enough that few would dispute the usefulness of moving on to find the regression formula, which we will do shortly. First, however, a bit more discussion of r^2 is in order.

THE COEFFICIENT OF DETERMINATION

For now, perhaps it is best to say that the concept of the proportion of variation explained by the independent variable is a way of specifying the impact of the independent variable in explaining changes in the dependent variable. There is a very specific explanation for the meaning of r^2, which will be discussed in Chapter 14. Explained briefly here, variation is the sum of all the squared distances of the points' y values from the mean value of y. The distances are squared to get rid of negative numbers. This sum of squared deviations is then broken into two components: (a) the squared distances from the mean y value to the regression line (explained variation) and (b) the squared distances from the regression line to the point (unexplained variation). The closer the points come to the regression line, the

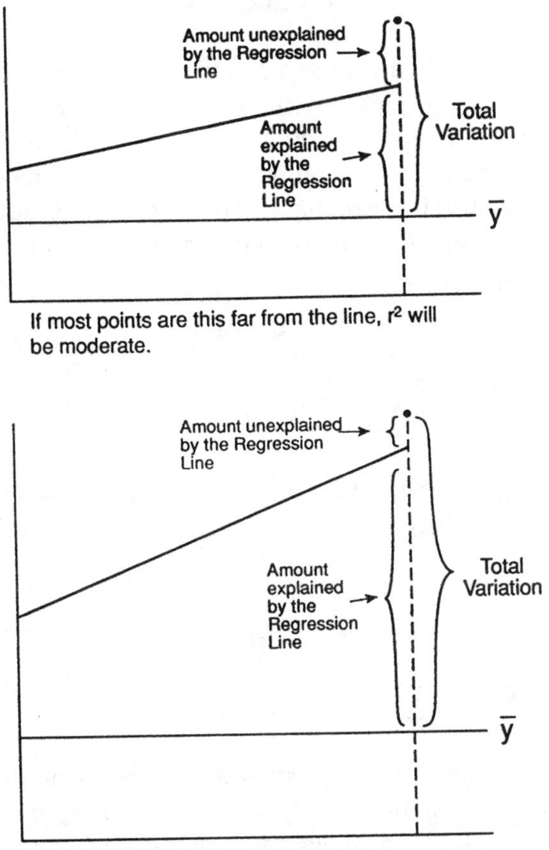

Figure 13.17

greater the amount of the total variation that can be attributed to the line and the less left unexplained (see Figure 13.17).

In some ways, the coefficient of determination is a kind of police officer, keeping us honest when we may feel the urge to inflate the importance of our findings. Relatively few r's in research are .7, .8, or .9; rather they tend to be in the .2 to .5 range. To see what this means in terms of r^2, look below where r's and their r^2's are compared as r shrinks from 1 to 0.

$r =$	$r^2 =$
1.00	1.00
.90	.81
.80	.64
.70	.49
.60	.36
.50	.25
.40	.16
.30	.09
.20	.04
.10	.01
0.00	0.00

By the time r drops to .7, r^2 is only .49; only .49 proportion (49%—less than one half) of the variation is explained and .51 is left unexplained by the independent variable. *The proportion left unexplained,* $1 - r^2$, is referred to as the **coefficient of alienation.** By the time $r = .40$, only .16 of the variation is explained and the coefficient of alienation is .84; 84% is left unexplained.

At the same time, recall the earlier discussion of the many real-world factors that mitigate linearity in the social sciences. It would be rare indeed to see coefficients of determination approaching 1.0 (or even .9 or .8) in social or behavioral research. For this reason, while r^2 is a useful concept in statistics, it is less useful in the real world of data analysis in determining what is or is not a "good" linear relationship. Thus statistical significance is more often used to demonstrate "good" relationships. This will be illustrated in Chapter 14.

FINDING THE REGRESSION EQUATION

Let us return to our social alienation versus religiosity example. With $r = -.94$ and r^2 explaining 89% of the variation, it is worth generating the regression formula for predicting religiosity from alienation. If you found the calculation of Pearson's r to be tedious, here is joyous news: In calculating r we also obtained the essential components for finding b_{yx}.

According to the least squares principle, the following formula will yield the regression slope.

$$b_{yx} = \frac{n\sum xy - \left(\sum x\right)\left(\sum y\right)}{n\sum x^2 - \left(\sum x\right)^2}$$

In calculating r, we already found that $n\sum xy - (\sum x)(\sum y) = -140$ and that $n\sum x^2 - (\sum x)^2 = 850$. Plugging these numbers into the formula we obtain:

$$b_{yx} = \frac{n\sum xy - \left(\sum x\right)\left(\sum y\right)}{n\sum x^2 - \left(\sum x\right)^2} = \frac{-140}{850} = -.1647058$$

The b_{yx}, just like r, is negative, indicating an inverse relationship. Later we will round off the slope to $-.17$, but for now, keep b_{yx} to as many decimals as your calculator allows. We need b_{yx} to find a_{yx}, and even a slight rounding of b_{yx} can have a large impact on a_{yx}.

There are two formulas that may be used to find a_{yx}. If the means for x and y have already been calculated, we may use the following formula.

$$a_{yx} = \bar{y} - b_{yx}\bar{x}$$

If no means are available, we may use the following equation, which uses information available from the initial calculation of Pearson's r.

$$a_{yx} = \frac{\sum y - b_{yx}\sum x}{n}$$

For our problem:

$$a_{yx} = \frac{\sum y - b_{yx}\sum x}{n} = \frac{42 - (-.1647058)(170)}{5} = \frac{42 - (-27.999986)}{5}$$

$$= \frac{42 + 27.999986}{5} = \frac{69.999986}{5} = 13.999997$$

rounding off: $a_{yx} = 14.00$. Thus $y = 14 + (-.17)x$, which simplifies to our final equation.

$$y = 14 - .17x$$

This is our best model for predicting level of religiosity (y) from the social alienation score (x). Before we complete this procedure by completing the scatter diagram, let us make use of our newly generated equation. Remember that all of our work up to now has been to (a) establish the existence of a viable linear relationship by finding r and r^2 and (b) develop the regression formula for predicting y from x. Now that we have the equation, let us put it to use.

Three things before we start. First, note that these are **predictions of y,** not necessarily the actual y values from which we generated the regression formula. To stress this point, we often place a circumflex above the y ($\hat{y}$ read "y-hat" because the circumflex resembles a hat) to indicate an *estimate* of some value.

Second, even though the original religiosity scores ranged from 0 to 10, our estimates may be above 10 or below 0. We are fitting actual data to a mathematical model. If the estimate is higher than 10, it simply means that were the index larger, this person would score even higher than someone with a predicted score of exactly 10.

Third, the actual scores, based on only 10 items, produce only whole numbers—10, 9, 8, and so on—but our estimate is a continuous variable that can take on values between whole numbers: 9.8, 6.4, and so forth. If the scale had more items, say 100 items worth 1/10 of a point each, then the respondent could be expected to achieve the score estimated by the regression. If the scale only has 10 items, we could choose to round our estimate to the nearest whole number between 0 and 10. In our example, however, we will leave the estimates as the regression equation predicts them.

Now, to make an estimate using the regression equation, we simply replace the x in the equation with the actual value specified for a person and solve for y.

Suppose $x = 0$, no alienation:

$$y = 14 - .17x = 14 - .17\ (0) = 14 - 0 = 14$$

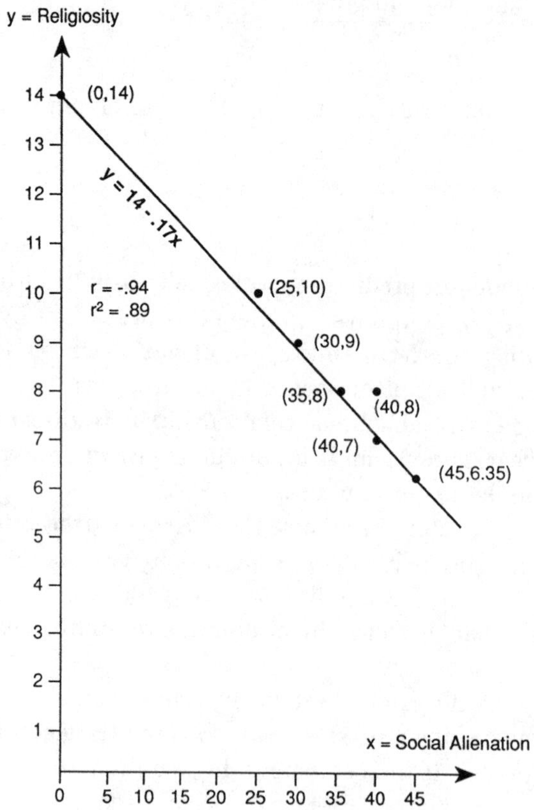

Figure 13.18
Predicted Religiosity Score, by Social Alienation

We estimate 14 even though 10 is the highest possible score. Remember that the definition of a_{yx}, the value of y where the line crosses the y-axis, is the y value where $x = 0$. Thus $(0, a_{yx})$ is always a point on the regression line. In this case, (0, 14) will be on the regression line—a useful thing to remember when we add the regression line to the scatter diagram.

Now suppose $x = 5$:

$$y = 14 - .17\,(5) = 14 - .85 = 13.15$$

Thus the ordered pair (5, 13.15) will also be a point on the regression line.

For $x = 45$:

$$y = 14 - .17\,(45) = 14 - 7.65 = 6.35$$

and so (45, 6.35) will also be on the regression line.

As an exercise, plug into the formula each of the five original x values for our social alienation versus religiosity problem to get the corresponding estimates of y. Compare these estimates to the actual y score for each person.

We now want to add the regression line to the scatter diagram originally presented in Figure 13.11. We do this in Figure 13.18, which has the same data as Figure 13.11. Remember that any two points determine a straight line, Thus once we know two points on the line we can plot them on the graph, lay a ruler between them, and draw in the line. We already know three points on the line: (0, 14), (5, 13.15), and (45, 6.35). We plot the first point (0, 14), (indicated with a small circle on Figure 13.18) and then another point within a ruler's distance of the first. In this case we will use (45, 6.35).

We then draw in the line, label our axes, add a title to the diagram, and either along the line or at some uncluttered place on the graph, write the equation, r, and r^2. We have now completed the entire correlation-regression process.

For further practice, let us return to the political corruption problem presented earlier in this chapter, where we predicted monthly campaign contributions from annual income. The following computations yield an $r = .81$ and an $r^2 = .65$. Although .65 is less than the r^2 of the previously worked problem, it is still quite satisfactory to justify finding the regression formula. (Even though the findings for r and r^2 might not be conclusive enough to convince the electorate [let alone a jury] that the district attorney in question enforced an organized kickback scheme, .65 proportion of the variation in contribution can be explained by income.)

Employee	x = Annual Income (in thousands of dollars)	y = Mean Monthly Contribution	x^2 =	xy =	y^2 =
1	80	160	6,400	12,800	25,600
2	70	95	4,900	6,650	9,025
3	52	97	2,704	5,044	9,409
4	45	85	2,025	3,825	7,225
$n = 4$	$\sum x = 247$	$\sum y = 437$	$\sum x^2 = 16{,}029$	$\sum xy = 28{,}319$	$\sum y^2 = 51{,}259$

$$(\sum x)^2 = (247)^2 = 61{,}009 \qquad (\sum y)^2 = (437)^2 = 190{,}969$$

$$\begin{aligned} n\sum xy - (\sum x)(\sum y) &= 4(28{,}319) - (247)(437) \\ &= 113{,}276 - 107{,}939 \\ &= 5337 \end{aligned}$$

$$\begin{aligned} n\sum x^2 - (\sum x)^2 &= 4(16{,}029) - 61{,}009 \\ &= 64{,}116 - 61{,}009 \\ &= 3107 \end{aligned}$$

$$\begin{aligned} n\sum y^2 - (\sum y)^2 &= 4(51{,}259) - 190{,}969 \\ &= 205{,}036 - 190{,}969 \\ &= 14{,}067 \end{aligned}$$

$$r = \frac{n\sum xy - (\sum x)(\sum y)}{\sqrt{\left[n\sum x^2 - (\sum x)^2\right]\left[n\sum y^2 - (\sum x)^2\right]}} = \frac{5337}{\sqrt{(3107)(14{,}067)}}$$

$$= \frac{5337}{\sqrt{43{,}706{,}169}} = \frac{5337}{6611.06} = +.807$$

$r^2 = (.807)^2 = .651$

Thus $r = .81$
$r^2 = .65$

$$b_{yx} = \frac{n\sum xy - \left(\sum x\right)\left(\sum y\right)}{n\sum x^2 - \left(\sum x\right)^2} = \frac{5337}{3107} = 1.7177341$$

$$a_{yx} = \frac{\sum y - b_{yx}\sum x}{n} = \frac{437 - (1.7177341)(247)}{4} = \frac{437 - 424.28032}{5}$$

$$= \frac{12.71968}{4} = 3.17992$$

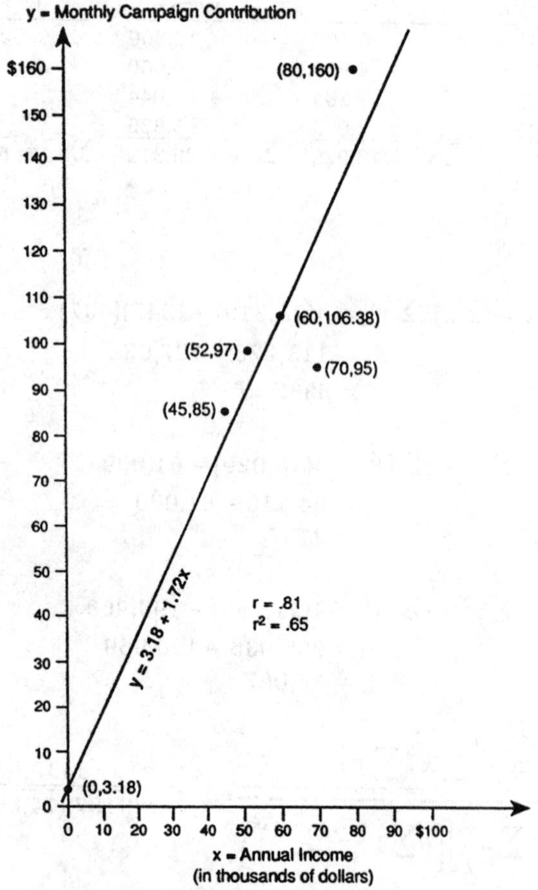

Figure 13.19

After rounding to two decimal places, we have our equation:

$$y = 3.18 + 1.72x$$

The scatter diagram is presented in Figure 13.19, where the ordered pairs for each data point are indicated. (If n were very large, we would probably not label the points at all.) Since $a_{yx} = 3.18$, we locate that point on the y-axis as one known point on the regression line. To find a second point, we select a value of x and solve for y. In this case, we chose $x = 60$.

$$y = 3.18 + 1.72\ (60) = 3.18 + 103.20$$
$$= 106.38$$

Thus the point (60, 106.38) is plotted on the scatter diagram and the regression line is drawn in.

CORRELATION MEASURES FOR ANALYSIS OF VARIANCE

Once we have rejected the null hypothesis in an ANOVA problem, we are concluding a difference in the population between at least two of our category means. In the nonexperimental context, that observation is tantamount to concluding the existence of a relationship between the variables. Once H_0 is rejected, we may measure the amount of relationship in the data and use that measure as an estimate of the relationship in the population.

There are two commonly used measures of association: the intraclass correlation and the correlation ratio. The **intraclass correlation coefficient,** designated r_i or sometimes r_I, was once commonly used by researchers. The simplest formula for calculating r_i is

$$r_i = \frac{F - 1}{F + (\overline{n} - 1)}$$

$\overline{n}$ is the mean category size.

In our area versus pro-life score problem presented in Chapter 10, we had two categories of four people each. Since both n's are the same, $\overline{n} = 4$.

$$\overline{n} = \frac{n_{\text{rural}} + n_{\text{urban}}}{2} = \frac{4+4}{2} = \frac{8}{2} = 4$$

We review the source table to find that $F = 10.80$.

Source	*SS*	*df*	*MS*	*F*	*p*
Total	2800				
Between	1800	1	1800	10.80	< .05
Within	1000	6	166.66		

Thus

$$r_i = \frac{F - 1}{F + (\overline{n} - 1)} = \frac{10.80 - 1}{10.80 + (4 - 1)} = \frac{10.80 - 1}{10.80 + 3} = \frac{9.80}{13.80} = .710$$

There are several problems in interpreting r_i. For one thing, a negative r_i does not necessarily indicate an inverse relationship in the data. Also, when category n's differ greatly a more complex formula must be used to find n. Finally, although based on Pearson's r, r_i is more like a stand-in for r^2, not r.

More commonly found than r_i is a measure known as the **correlation ratio,** designated by either the capital letter E or by the Greek letter *eta*, η. We always square the correlation ratio to make a generalization about proportion of variation explained, thus we really want to find E^2 or eta-squared (η^2).

E^2 is readily calculated from information in the source table.

$$E^2 = \frac{SS_B}{SS_T}$$

For our pro-life problem, SS_B and SS_T are 1800 and 2800, respectively.

$$E^2 = \frac{SS_B}{SS_T} = \frac{1800}{2800} = .643$$

Thus about .64 proportion (64%) of the total variation (total sum of squares) can be explained by the categories of the independent variable.

The square of the correlation ratio is also often designated R^2. (This is similar to the R^2 known as the coefficient of multiple determination, to be presented in Chapter 14. For now, simply interpret R^2 or E^2 as if it were r^2.) For instance, on the SAS Source Table printout (see Table 10.2), to the right of PR > F you will see R-SQUARE and below it the number 0.244660. If we calculate E^2 from the same source table, remembering that MODEL SS means between SS in SAS usage,

$$E^2 = \frac{SS_B}{SS_T} = \frac{12.6}{51.5} = .244660$$

So E^2 is the same as R-SQUARE on the printout. In that problem, about 24.4% of the variation in course satisfaction can be explained by the time of the classes: day, night, or Saturday.

CONCLUSION

Correlation-regression techniques have become widely used throughout all of the social sciences. Business and economics research also relies heavily on this form of analysis. This is largely due to the predictive capability of regression. When r is large enough, we can rely on regression models for predicting scores on such diverse dependent variables as electoral outcomes, voting in legislative bodies, arms acquisitions, economic growth, unemployment, health expenditures, academic performance, and crime rates.

Our predictions are enhanced when we turn our attention to an extension of linear regression known as multiple regression, a topic to be encountered in Chapter 14. With multiple regression, we are predicting a score on a dependent variable from several independent variables at once. As more independent variables are added to the model, its predictive capability usually rises above the capability of a one-independent-variable model such as we have been using here.

EXERCISES

Use the following study to work exercises 13.1–13.6 in this section.

> A school psychologist has developed a series of inventories to measure student interest in studying selected academic subjects. These inventories range from 0 to 100; the higher a student scores, the greater that student's interest in studying the particular subject. The academic subjects being inventoried are health, physics, athletics, algebra, literature, geometry, drama, and chemistry.

Exercise 13.1

For all 10th graders, the correlation between student interest in drama and student interest in chemistry is –.90664. Below are the scores of the five 10th-grade representatives elected to the student council.

Student	*DRAMA*	*CHEM*
Emile	14	93
Dimitri	8	93
Adam	7	100
Laura	0	85
Deborah	92	29

1. Calculate r for these five council representatives and compare it to r for all 10th graders. Does the relationship hold for these five students? In case your calculator overloads:

$$\sqrt{\left[n\sum x^2-\left(\sum x\right)^2\right]\left[n\sum y^2-\left(\sum y\right)^2\right]}=\sqrt{(29{,}224)(16{,}820)}$$

$$=\sqrt{491{,}547{,}680}=22{,}170.875$$

2. For the council representatives, what proportion of variation in CHEM scores can be explained by the DRAMA interest scores?

Exercise 13.2

1. Complete the regression analysis for the data in Exercise 13.1. Find *b* and *a*. Assemble the regression formula.
2. Do a scatter diagram with the regression line included.
3. Compare the predicted CHEM scores with the actual scores of the five council representatives.

EXERCISES

Exercise 13.3

Here is a matrix of Pearson's *r* correlation coefficients for the interest inventory scores for all 11th graders in the same school. Review the coefficients.

	HEALTH	PHYSICS	ATHLET	ALGEBRA	LITER	GEOM	DRAMA	CHEM
HEALTH	1.00							
PHYSICS	.33	1.00						
ATHLET	.45	−.02	1.00					
ALGEBRA	.20	.87	−.14	1.00				
LITER	.12	−.83	.22	−.83	1.00			
GEOM	.09	.90	−.07	.86	−.95	1.00		
DRAMA	−.01	−.81	.01	−.72	.90	−.92	1.00	
CHEM	−.08	.77	−.12	.70	−.89	.89	−.92	1.00

1. Which two variables have relatively little correlation with most of the other variables? (*Hint:* these two variables are *moderately* correlated with each other.)
2. Assuming that all the others tap a verbal versus quantitative interest continuum, which variables are positively associated with interest in quantitatively oriented courses?
3. Which are positively associated with verbally oriented courses?
4. What are the highest positive and highest inverse correlations? (Ignore the *r* of 1.00 between a variable and itself.)

Exercise 13.4

Here are the regression formulas for predicting literature interest scores from several of the other indicators.

LITER = 98.94 – .90 ALGEBRA	$r = -.83$
LITER = 87.34 – .91 GEOM	$r = -.95$
LITER = 117.22 – 1.27 CHEM	$r = -.89$
LITER = –13.11 + .99 DRAMA	$r = +.90$

1. Predict the literature score for 11th graders whose algebra interest is 0; 30; 80; 100.
2. One student has an actual LITER score of 5. His scores for each of the independent variables are:

   ALGEBRA = 81
   GEOM = 100
   CHEM = 75
   DRAMA = 23

   Using the above regression formulas, predict his LITER score. In this case, what index comes closest to predicting his actual score? Which is least close?
3. Suppose that Joan and David tied for the highest LITER score in the 11th grade (100). Joan's other scores are

   ALGEBRA = 8
   GEOM = 0
   CHEM = 36
   DRAMA = 100

   Predict her LITER score from each of the regression formulas. Which is the closest predictor and which is the least close?
4. David has the same ALGEBRA, GEOM, and CHEM scores as Joan, but his DRAMA score is only 86. Predict his LITER score from the DRAMA score. Which independent variables are the best and worst predictors, respectively, for David?

Exercise 13.5

When regression is performed by a computer program, so much information is provided that one must hunt for what one needs. Following is a copy of an SAS printout for predicting an 11th grader's physics score from that individual's geometry score. Be-

Source	DF	Sum of Squares	Mean Square	F Value	Prob > F
Model	1	16437.71936	16437.71936	371.338	0.0001
Error	87	3851.15704	44.26617		
C Total	88	20288.87640			

Root MSE	6.65328	R-square	0.8102
Dep Mean	57.11236	Adj R-sq	0.8080
C. V.	11.64946		

Parameter Estimates

Variable	DF	Parameter Estimate	Standard Error	T for HO: Parameter=0	Prob > |T|
Intercept	1	41.122668	1.08898266	37.762	0.0001
GEOM	1	0.375980	0.01951099	19.270	0.0001

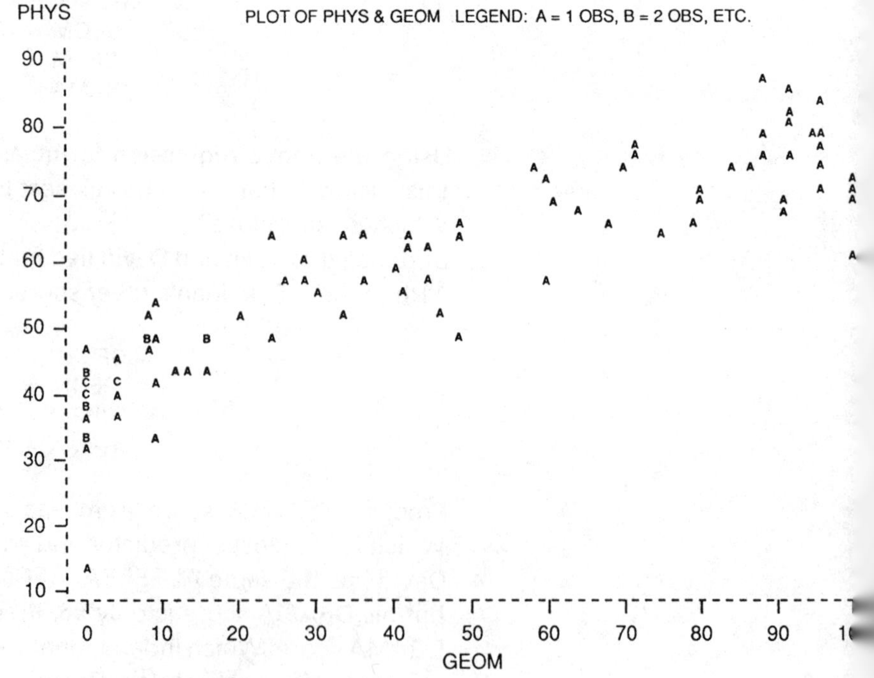

Figure 13.20

fore you examine the full printout and scatter diagram, note that some of what you see, such as Analysis of Variance as applied to regression, will be covered in the next chapter. What you need for now is found in the lower portion of the printout, under the heading PARAMETER ESTIMATE (see the partial printout below). The first number in that column, adjacent to INTERCEP, is the *y*-intercept, *a*.

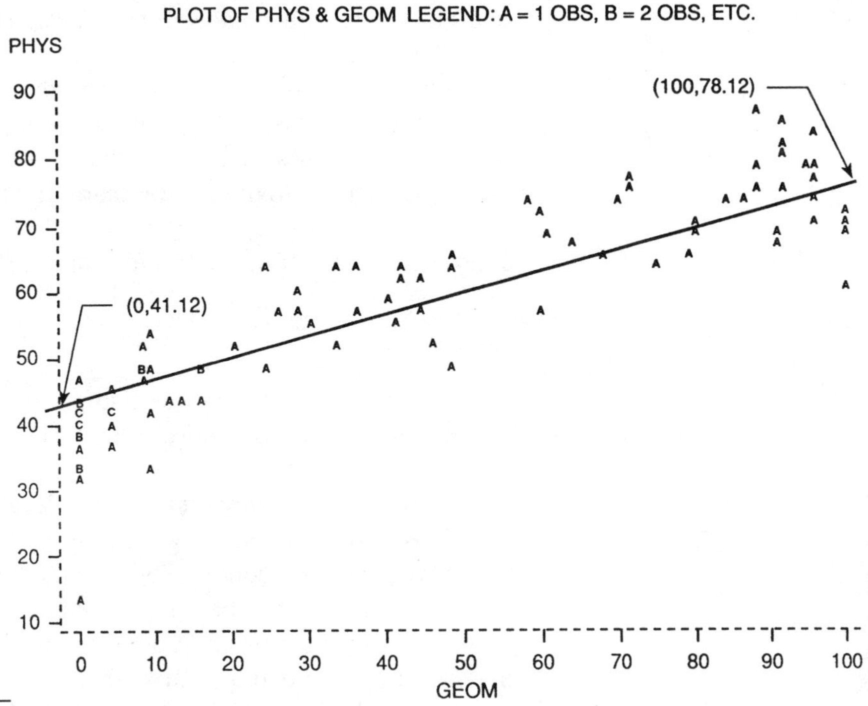

Figure 13.21

Below that number, adjacent to the name of the independent variable, GEOM, is the slope, *b*. (Ignore the two number ones under DF.)

VARIABLE	DF	PARAMETER ESTIMATE	
INTERCEP	1	41.122668	____*a*
GEOM	1	0.375980	____*b*

See Figure 13.20 for the complete printout.

In the scatter diagram, a pair of scores is indicated by the letter A instead of a dot. If a pair of scores occurs twice, SAS prints a B. If three times, a C. (In SPSSX the number of occurrences, 1, 2, 3, and so on replaces the dot.) The actual regression line is not printed here, though it is possible for the program to place a line of letters approximately where the regression line goes. As an alternative, we could add the line by hand by finding two points on the line using the regression formula, exactly as we have been doing.

The modified scatter diagram appears in Figure 13.21.

1. From the printout, form the regression equation for predicting PHYSICS from GEOM (use only the first two decimal places).
2. Use the formula to confirm that the two points used to find the regression line in the scatter diagram are two points on the regression line: (0, 41.12) and (100, 78.12).
3. Predict the PHYSICS score for someone with a GEOM score of 55.

Exercise 13.6

See Figure 13.22 for the regression and scatter diagram printout for predicting LITER score from the GEOM score.

1. Put together the regression formula after finding *a* and *b* on the printout. (Again, use only the first two decimal places.)
2. Identify from the equation two points that could be used to find the regression line on the scatter diagram. (Do not actually draw the line, except on a photocopy of the graph.)
3. You know from the matrix in Exercise 13.3 that the correlation between LITER and GEOM is –.95. What proportion of the LITER variance can be explained by GEOM? Should the regression be a good predictor?
4. Following are the actual LITER scores for three students. Predict their LITER scores using the regression formula.

	GEOM	*LITER*	*LITER*
Student	*Actual*	*Predicted*	*Actual*
Julie	100	?	0
Jack	48	?	35
Jill	0	?	95

How close were your predictors?

Exercise 13.7

Calculate and interpret r_i and E^2 for the ANOVAs in Exercises 10.2, 10.3, and 10.4.

Source	DF	Sum of Squares	Mean Square	F Value	Prob > F
Model	1	96488.48532	96488.48532	901.572	0.0001
Error	87	9310.95289	107.02245		
C Total	88	105799.43820			

Root MSE	10.34517	R-square	0.9120
Dep Mean	48.59551	Adj R-sq	0.9110
C. V.	21.28832		

Parameter Estimates

Variable	DF	Parameter Estimate	Standard Error	T for HO: Parameter=0	Prob > |T|
Intercept	1	87.335276	1.69325518	51.578	0.0001
GEOM	1	-0.910922	0.03033757	-30.026	0.0001

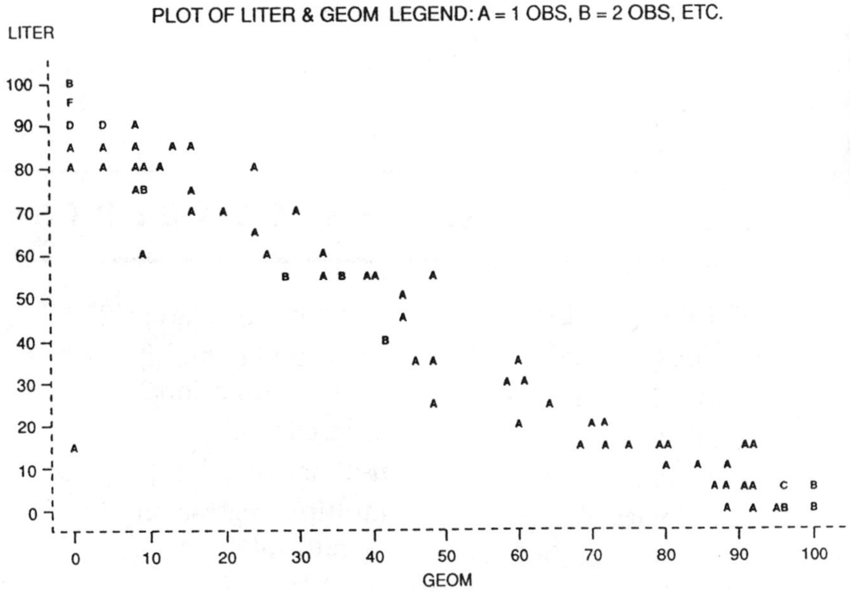

Figure 13.22

KEY CONCEPTS

critical value of r/$r_{critical}$
spurious correlation
intervening variable
causal modeling
partial correlation
control variable
zero-order correlation
first-order partial correlation
second-order (etc.) partial correlation
multiple correlation/R
coefficient of multiple determination/R^2
ordered triplet
n-dimensional hyperplane
multiple regression/multiple linear regression
partial regression slope
zero-order regression slope
first-order (etc.) partial regression slope
mnemonics
standardized partial regression slope/beta coefficient/beta weight/β
adjusted R^2/R^2 adjusted
stepwise multiple regression

14 Additional Aspects of Correlation-Regression Analysis

INTRODUCTION

In this, the final chapter of this book, we continue with the topic of correlation-regression analysis and expand its techniques beyond two variables. In addition, a number of loose ends will be tied up. We will first seek to tie together the two branches of statistics—descriptive and inferential—showing how correlation coefficients and regression slopes may be tested for statistical significance. We will pay particular attention to the analysis of variance procedure as it is applied to a regression, as is commonly found on a computer printout. We will also explain the origin of the critical values of a correlation coefficient table.

The section on partial correlations and causal models presents a new way of studying the impact of a third variable (the control variable) on the relationship between two other variables. We first encountered this problem in Chapter 6, when the use of partial tables was discussed. Now we will make use of partial correlations to extend our analytical capabilities.

In the latter part of the chapter we discuss multiple regression, the extension of linear regression beyond a single independent variable. We will make use of this and related concepts such as the coefficient of multiple determination to develop and evaluate predictive models. These are regression models aimed at enabling us to predict a score on a dependent variable from the scores of several independent variables working together.

We will see how such techniques may be used to verify theory and even to help us develop further theory.

STATISTICAL SIGNIFICANCE FOR *r* AND *b*

When linear regression is performed by most library computer pro grams, an ANOVA source table also appears and an F is generated. Here analysis of variance is being used to test a null hypothesis that $b = 0$ Because where $b = 0$, $r = 0$ also, the null hypothesis is stating that i the population, $b = 0$ and $r = 0$. If H_0 is true, any regression slop generated from the sample and any Pearson's r generated from th sample will differ from 0 only as the result of sampling error. If however, the F_{obtained} in the analysis of variance is statistically signifi cant and we can reject H_0, we conclude that in the population $b \neq 0$ an $r \neq 0$ and we use the sample regression/correlation data to estimate thei respective population parameters.

The conversion factors between the components of the ANOVA sourc table and the original raw score data and r^2 are summarized below.

Source	*SS*	*df*	*MS*	*F*
Total	$\Sigma(y-\bar{y})^2$			
Explained (Between)	$r^2 \cdot \Sigma(y-\bar{y})^2$	1	$\dfrac{r^2 \cdot \Sigma(y-\bar{y})^2}{1}$	$\dfrac{r^2(n-2}{1-r^2}$
Unexplained (Within)	$(1-r^2) \cdot \Sigma(y-\bar{y})^2$	$n-2$	$\dfrac{(1-r^2) \cdot \Sigma(y-\bar{y})^2}{n-2}$	

So from the regression, F is calculated as follows:

$$F = \frac{r^2(n-2)}{1-r^2} \qquad df = 1 \quad \text{and} \quad n-2$$

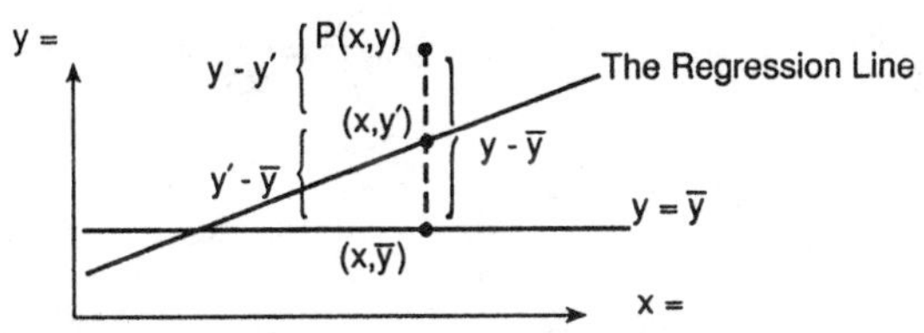

Figure 14.1

To see what is happening, observe the deviation in the y direction of some point on the scatter diagram presented in Figure 14.1. In the y direction, the distance from point $P(x, y)$ to the mean value of y, $\bar{y}$, is $y - \bar{y}$. The distance from any point to the mean value of y will be that point's y coordinate minus $\bar{y}$. Note that the distance from y to $\bar{y}$ can be broken into two components. The first of these is the distance, in the y direction, from the point to the regression line, which is $y - y'$, and the second is the distance from the regression line to the mean value of y, which is $y' - \bar{y}$. Adding the two components back together yields the original distance: $(y - y') + (y' - \bar{y}) = y - y' + y' - \bar{y} = y - \bar{y}$.

Of the total deviation $y - \bar{y}$, the component $y' - \bar{y}$ is the part *explained* by the regression line and the remaining component, $y - y'$, is that part *unexplained* by the regression line. If we square these deviations to eliminate negative signs and add up all the squared deviations, we get $\Sigma(y - \bar{y})^2$, which is the same as the total sum of squares (in the y direction). We do the same for the two components of $y - \bar{y}$ and the following emerges.

$$\Sigma (y - \bar{y})^2 = \Sigma (y' - \bar{y})^2 + \Sigma (y - y')^2$$

This is the same as saying

$$SS_{\text{Total}} = \begin{array}{c}\text{The portion of } SS_T \\ \text{explained by the} \\ \text{regression line.}\end{array} + \begin{array}{c}\text{The portion of } SS_T \\ \text{unexplained by the} \\ \text{regression line.}\end{array}$$

and is analogous to

$$SS_{\text{Total}} = SS_{\text{Between}} + SS_{\text{Within}}$$

except that "Between" means explained by the regression and "Within" is the error SS—the *SS* that the regression line does *not* account for.

When $b = 0$ and $r = 0$, the regression line is actually the line parallel to the x-axis where $y = \bar{y}$. In effect, the y's are at $\bar{y}$, each $y' - \bar{y} = 0$, and $\Sigma(y' - \bar{y})^2$ (the explained *SS*) drops to 0. (See Figure 14.2.)

Since $y' - \bar{y} = 0$, $\Sigma(y' - \bar{y})^2 = 0$ and there is no *SS* explained by the line. Since $SS_{\text{Explained}}$ is the same as SS_B, it stands to reason that

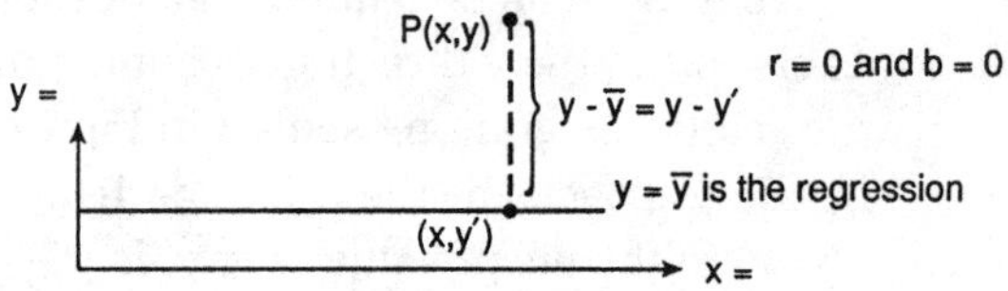

Figure 14.2

$$MS_B = \frac{SS_B}{df_B} = \frac{0}{df_B} = 0$$

and since

$$F = \frac{MS_B}{MS_W} = \frac{0}{MS_W} = 0$$

$F = 0$ when $r = 0$ and $b = 0$.

Let's take a concrete example and do both the correlation/regression part and the ANOVA. This time, though, we will use the definitional formulas instead of the computational formulas to calculate F. Assume the six data points shown in Figure 14.3.

	$x =$	$y =$	$x^2 =$	$xy =$	$y^2 =$
	1	2	1	2	4
	2	2	4	4	4
	3	2	9	6	4
	1	4	1	4	16
	2	4	4	8	16
$n = 6$	3	4	9	12	16
	$\Sigma x = 12$	$\Sigma y = 18$	$\Sigma x^2 = 28$	$\Sigma xy = 36$	$\Sigma y^2 = 60$
	$\bar{x} = 2$	$\bar{y} = 3$			

$$n\Sigma xy - (\Sigma x)(\Sigma y) = 6(36) - (12)(18) = 216 - 216 = 0$$

$$n\Sigma x^2 - (\Sigma x)^2 = 6(28) - (12)^2 = 168 - 144 = 24$$

$$n\Sigma y^2 - (\Sigma y)^2 = 6(60) - (18)^2 = 360 - 324 = 36$$

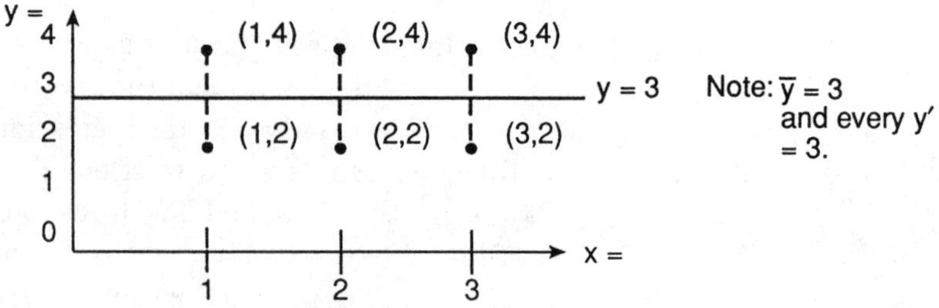

Figure 14.3

$$r = \frac{n\sum xy - \left(\sum x\right)\left(\sum y\right)}{\sqrt{\left[n\sum x^2 - \left(\sum x\right)^2\right]\left[n\sum y^2 - \left(\sum y\right)^2\right]}} = \frac{0}{\sqrt{(24)(36)}} = \frac{0}{\sqrt{864}}$$

$$= \frac{0}{29.39} = 0$$

$$b = \frac{n\sum xy - \left(\sum x\right)\left(\sum y\right)}{n\sum x^2 - \left(\sum x\right)^2} = \frac{0}{24} = 0$$

$$a = \frac{\sum y - b\sum x}{n} = \frac{18 - (0)(12)}{6} = \frac{18 - 0}{6} = \frac{18}{6} = 3$$

$r = 0$, $b = 0$, $a = 3$

The regression line is $y = a + bx = 3 + (0)x = 3 + 0$. Thus $y = 3$. So here is a case reflecting a null hypothesis: $b = 0$ or $r = 0$.

Now we will do the ANOVA using the definitional formulas. Remember: $SS_{\text{Total}} = \sum (y - \bar{y})^2$, $SS_{\text{Explained}} = \sum (y' - \bar{y})^2$, and $SS_{\text{Unexplained}} = \sum (y - y')^2$. And in this problem, $\bar{y} = 3$ and every $y' = 3$.

	$x =$	$y =$
	1	2
	2	2
	3	2
	1	4
	2	4
n = 6	3	4
		$\sum y = 18$

$\bar{y} = 18/6 = 3$

$y - \bar{y} =$	$(y - \bar{y})^2 =$	$y' - \bar{y} =$	$(y' - \bar{y})^2 =$	$y - y' =$	$(y - \bar{y}')^2 =$
−1	1	0	0	−1	1
−1	1	0	0	−1	1
−1	1	0	0	−1	1
1	1	0	0	1	1
1	1	0	0	1	1
1	1	0	0	1	1
	$\sum (y - \bar{y})^2 = 6$ $= SS_{\text{Total}}$		$\sum (y' - \bar{y})^2 = 0$ $= SS_{\text{Explained}}$		$\sum (y - y')^2 = 6$ $= SS_{\text{Unexplained}}$

Confirming:

$$SS_{\text{Total}} = SS_{\text{Explained}} + SS_{\text{Unexplained}}$$

$$\sum (y - \bar{y})^2 = \sum (y' - \bar{y})^2 + \sum (y - y')^2$$

$$6 = 0 + 6$$

$$6 = 6$$

To get $MS_{\text{Explained}}$, we divide $SS_{\text{Explained}}$ by $df_{\text{Explained}}$, which is alway 1. Think of the categories as (a) Explained by the regression line and (b) Unexplained by the regression line. Then (no. of categories − 1) 2 − 1 = 1.

$$MS_{\text{Explained}} = \frac{SS_{\text{Explained}}}{df_{\text{Explained}}} = \frac{0}{1} = 0$$

$$MS_{\text{Unexplained}} = \frac{SS_{\text{Unexplained}}}{df_{\text{Unexplained}}}$$

$$df_{\text{Unexplained}} = (n_{\text{Total}} - 1) - (\text{no. of categories} - 1)$$

$$= (6 - 1) - (2 - 1) = 5 - 1 = 4$$

or more simply,

$$df_{\text{Unexplained}} = n - 2$$

So,

$$MS_{\text{Unexplained}} = \frac{SS_{\text{Unexplained}}}{df_{\text{Unexplained}}} = \frac{6}{6 - 2} = \frac{6}{4} = 1.5$$

$$F = \frac{MS_{\text{Explained}}}{MS_{\text{Unexplained}}} = \frac{0}{1.5} = 0$$

By contrast, in Figure 14.4, there is a line where $r \neq 0$, $b \neq 0$, and $F \neq$ (

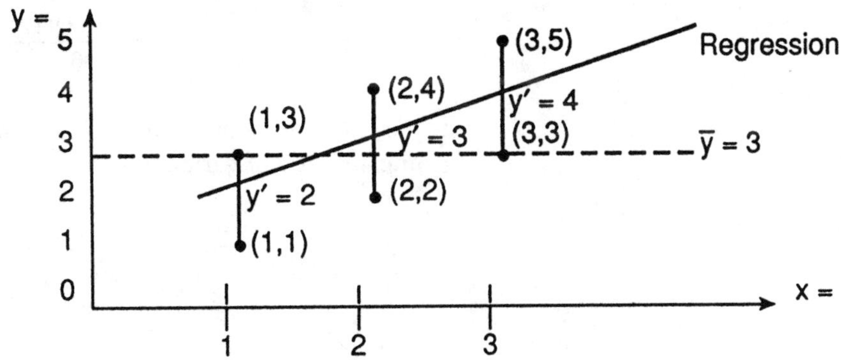

Figure 14.4

	$x =$	$y =$	$x^2 =$	$xy =$	$y^2 =$
	1	1	1	1	1
	2	2	4	4	4
	3	3	9	9	9
	1	3	1	3	9
	2	4	4	8	16
$n = 6$	3	5	9	15	25
	$\sum x = 12$	$\sum y = 18$	$\sum x^2 = 28$	$\sum xy = 40$	$\sum y^2 = 64$

$$n\sum xy - (\sum x)(\sum y) = 6(40) - (12)(18) = 240 - 216 = 24$$

$$n\sum x^2 - (\sum x)^2 = 6(28) - (12)^2 = 168 - 144 = 24$$

$$n\sum y^2 - (\sum y)^2 = 6(64) - (18)^2 = 384 - 324 = 60$$

$$r = \frac{n\sum xy - \left(\sum x\right)\left(\sum y\right)}{\sqrt{\left[n\sum x^2 - \left(\sum x\right)^2\right]\left[n\sum y^2 - \left(\sum y\right)^2\right]}} = \frac{24}{\sqrt{(24)(60)}} = \frac{24}{\sqrt{1440}}$$

$$= \frac{24}{37.947} = .63246 = .63$$

$$r^2 = (.63246)^2 = .40$$

$$b = \frac{n\sum xy - \left(\sum x\right)\left(\sum y\right)}{n\sum x^2 - \left(\sum x\right)^2} = \frac{24}{24} = 1.0$$

$$a = \frac{\sum y - b\sum x}{n} = \frac{18-(1)(12)}{6} = \frac{18-12}{6} = \frac{6}{6} = 1.0$$

Thus $y = 1.0 + 1.0\,x$ or simply $y = 1 + x$.

Doing analysis of variance:

	$x =$	$y =$
	1	1
	2	2
	3	3
	1	3
	2	4
$n = 6$	3	5
		$\sum y = 18$

$\bar{y} = 18/6 = 3$

$y - \bar{y} =$	$(y - \bar{y})^2 =$	$y' =$	$y' - \bar{y} =$	$(y' - \bar{y})^2 =$	$y - y' =$	$(y - \bar{y}')^2 =$
−2	4	2	−1	1	−1	1
−1	1	3	0	0	−1	1
0	0	4	1	1	−1	1
0	0	2	−1	1	1	1
1	1	3	0	0	1	1
2	4	4	1	1	1	1
	$\sum (y - \bar{y})^2 = 10$ $= SS_{\text{Total}}$			$\sum (y' - \bar{y})^2 = 4$ $= SS_{\text{Explained}}$		$\sum (y - y')^2 = 6$ $= SS_{\text{Unexplained}}$

Confirming:

$$SS_{\text{Total}} = SS_{\text{Explained}} + SS_{\text{Unexplained}}$$

$$10 = 4 + 6$$

$$10 = 10$$

$$MS_{\text{Explained}} = \frac{SS_{\text{Explained}}}{df_{\text{Explained}}} = \frac{4}{1} = 4$$

$$MS_{\text{Unexplained}} = \frac{SS_{\text{Unexplained}}}{df_{\text{Unexplained}}} = \frac{6}{n-2} = \frac{6}{6-2} = \frac{6}{4} = 1.5$$

$$F = \frac{MS_{\text{Explained}}}{MS_{\text{Unexplained}}} = \frac{4}{1.5} = 2.666$$

Here, $F \neq 0$, although checking the table at $df = 1,4$ we see that $F < F_{\text{critical}}$. To confirm the relationship between $F = 2.666$ and $r^2 = .40$, we go to the conversion formula.

$$F = \frac{r^2(n-2)}{1-r^2} = \frac{.40(6-2)}{1-.40} = \frac{.40(4)}{.60} = \frac{.1.60}{.60} = 2.666$$

Before leaving this topic, we may now shed greater light on the definition of r^2 given in Chapter 13: r^2 is the proportion of variation in the dependent variable that can be explained by the independent variable. If we replace the word *variation* with the more specific SS_{Total} and replace "the independent variable" by "the regression line," then

> r^2 is the proportion of SS_{Total} that can be explained by the regression line.

Or, simply:

$$r^2 = \frac{SS_{\text{Explained}}}{SS_{\text{Total}}} = \frac{\sum (y' - \bar{y})^2}{\sum (y - \bar{y})^2}$$

SS_{Total} is the sum of the squared deviations in the *y*-direction (vertical) of the values of *y* of the points on the scattergram from the mean value of *y*. $SS_{\text{Explained}}$ is that proportion of SS_{Total} accounted for by the regression line: the sum of the squared deviations in the *y*-direction between the predicted (y') values and the mean values of *y*.

SIGNIFICANCE OF *r*

If we know Pearson's *r* and wish to test its significance directly, we can convert the *r* to an *F* using the previously developed formula $F = [r^2(n-2)]/(1-r^2)$ and then compare the *F* to F_{critical} at $df = 1$ and $n - 2$. However, it is easier to use Table 14.1. The logic of this table makes use of an algebraic transformation of the above formula to

$$r^2 = \frac{F}{n-2+F}$$

Table 14.1 Critical Values of the Correlation Coefficient

	Level of significance for one-tailed test			
	.05	*.025*	*.01*	*.005*
	Level of significance for two-tailed test			
df	*.10*	*.05*	*.02*	*.01*
1	.988	.997	.9995	.9999
2	.900	.950	.980	.990
3	.805	.878	.934	.959
4	.729	.811	.882	.917
5	.669	.754	.833	.874
6	.622	.707	.789	.834
7	.582	.666	.750	.798
8	.549	.632	.716	.765
9	.521	.602	.685	.735
10	.497	.576	.658	.708
11	.476	.553	.634	.684
12	.458	.532	.612	.661
13	.441	.514	.592	.641
14	.426	.497	.574	.623
15	.412	.482	.558	.606
16	.400	.468	.542	.590
17	.389	.456	.528	.575
18	.378	.444	.516	.561
19	.369	.433	.503	.549
20	.360	.423	.492	.537
21	.352	.413	.482	.526
22	.344	.404	.472	.515
23	.337	.396	.462	.505
24	.330	.388	.453	.496
25	.323	.381	.445	.487
26	.317	.374	.437	.479
27	.311	.367	.430	.471
28	.306	.361	.423	.463
29	.301	.355	.416	.456
30	.296	.349	.409	.449
35	.275	.325	.381	.418
40	.257	.304	.358	.393
45	.243	.288	.338	.372
50	.231	.273	.322	.354
60	.211	.250	.295	.325
70	.195	.232	.274	.303
80	.183	.217	.256	.283
90	.173	.205	.242	.267
100	.164	.195	.230	.254

SOURCE: Table 14.1 is abridged from Table VII, of R. A. Fisher and F. Yates, *Statistical Tables for Biological, Agricultural and Medical Research* (6th ed.), published by Longman Group UK Ltd., 1974, by permission of the authors and the publisher.

If we plug in for F the value of $F_{critical}$ at the .05 level, solve for r^2, and take the square root of r^2, we will find the *lowest* value of r that would be significant at the .05 level at the designated degrees of freedom. Thus

$$r_{critical} = \pm\sqrt{\frac{F_{critical}}{n - 2 + F_{critical}}}$$

Table 14.1 gives critical values for either one-tailed or two tailed tests. (Since we always only use one tail of the sampling distribution of F, the terminology really should be directional or nondirectional H_1.) In both cases, H_0: $r_{population} = 0$. In the nondirectional ("two-tailed") instance, H_1: $r_{population} \neq 0$. If we could make a directionality assumption, we would have as H_1 either $r_{population} > 0$ or $r_{population} < 0$, whichever is most appropriate.

Let us assume a nondirectional alternative hypothesis for now. If $n = 15$, then $df = 1{,}13$, and at the .05 level, $F_{critical} = 4.67$.

$$r_{critical} = \sqrt{\frac{4.67}{15 - 2 + 4.67}} = \sqrt{\frac{4.67}{17.67}} = \sqrt{.246} = .514$$

At 13 df ($n = 15$), the lowest value of r that would be significant at the .05 level would be .514. Any sample r below that should not lead us to conclude that r in the population differs from zero. By using the formula $df = n - 2$, we get this information from Table 14.1 without needing to calculate F. It shows that at $df = 13$, at the .05 level, $r_{critical} = .514$. We also see that at $df = 13$, the $r_{criticals}$ are .592 and .641 at the .02 and .01 levels, respectively.

If we had been able to make a directionality assumption in advance, we could have used the one-tailed critical values of r. At 13 df we would now only need an r of .441 to reject H_0, as opposed to the .514 needed if no directionality could be assumed.

Before we leave this topic, it is important to consider again the question of when a correlation coefficient is large enough to have practical research significance. In the previous chapter this decision was based on the coefficient of determination and the proportion of variation in the dependent variable explained by the independent variable. From a purely statistical point of view, using the coefficient of determination is an appropriate guide to establishing relevance. In most research we encounter, however, the statistical significance of r is really what is used to "establish" research significance, despite the fact that research significance and statistical significance are quite different concepts.

The reason for this is that most of the correlation coefficients th we actually generate in the social and behavioral sciences are relativel low. We are much more likely to see coefficients in the range of .20 t .40 than in the range of .70 to .90. Social and behavioral phenomena ar complex, so we only account for proportions of variation in snippets a time. Remember, though, that statistical significance is based in pa on sample size, so the larger the *n*, the more statistically significar correlations are likely to be generated. Bear in mind that some of thes coefficients will not have as much research significance as the re searchers preparing such studies perhaps might wish.

PARTIAL CORRELATIONS AND CAUSAL MODELS

Up to this point in our discussion of correlation-regression analysis, w have been dealing with relationships between two variables at a tim We now turn our attention to correlation techniques that involve mor than two variables. Specifically, we are examining the effects of othe variables on the relationship between two given variables. In doing so we revisit the three-variable situations initially discussed in Chapter 6 Consider two variables, designated x_1 and x_2. (In this section we wi refer to each variable as *x* plus a numbered subscript.) Suppose that th correlation between these two variables, r_{12}, is .60. Now suppose tha there is also a third variable, x_3. When we ask about the impact of x_3 o the relationship between x_1 and x_2 we are asking to what extent th correlation r_{12} is dependent on the presence of variable x_3. At on extreme, the presence of x_3 would have no impact on r_{12}. At the othe extreme, the entire correlation between x_1 and x_2 would be due to th presence of x_3; if x_3 were not present, there would be no correlation *Recall that when a correlation between two variables is due to th presence of one or more additional variables, we say that the relationship i* **indirect:** *either* **spurious** *if the control variable is dependent, or providin interpretation if the control variable is an intervening variable.*

For example, suppose we are studying a group of volunteers who ar working for a charitable organization. We survey these volunteers an among other things determine the following for each person.

x_1 = the mean hours per week that the subject is involved in volun teer activities of a charitable nature.

x_2 = the dollars per year that the subject has donated to all charities

x_3 = the mean annual income for that subject's family.

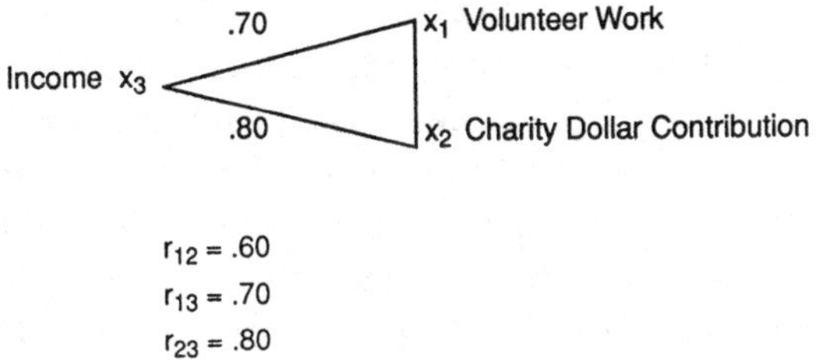

Figure 14.5

We show the three variables and the Pearson's correlations between them in schematic form in Figure 14.5. By convention, the variable presumed to be the independent variable is placed on the left-hand side and the dependent variable(s) on the right.

In a three-variable case, a quick way to determine if a correlation is indirect is to see if it equals the product of the remaining two correlations. If r_{12} is indirect, then $r_{12} = r_{13} \cdot r_{23}$. Here, $r_{13} \cdot r_{23} = (.70)(.80) = .56$, very close to the actual r_{12} of .60. Since r_{12} is probably indirect, we indicate it with a dotted line on the schematic shown in Figure 14.6.

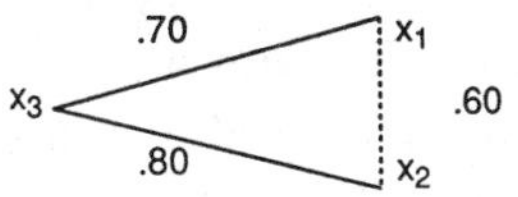

Figure 14.6

Note also that $r_{12} \cdot r_{13} = .42 \neq .80$ and $r_{12} \cdot r_{23} = .48 \neq .70$, so only r_{12} is indirect. By saying r_{12} is *spurious we imply that the correlation is due to the presence of variable* x_3. The relationship between time spent as a charity volunteer and money spent as a charity contributor is due to the existence of the respondent's income. Note that if the original hypothesis had been that higher incomes lead to greater involvement of all kinds in charities, then the finding of r_{12} to be indirect would confirm the hypothesis. If r_{12} had not been indirect, we would have to investigate the possibility that volunteer work motivates one to contribute funds, or vice versa. Or perhaps both activities motivate people to the other activity. Remember also that we are only looking at three variables and many other variables could be able to influence charitable activities.

In all probability, however, income is an independent variable independently influencing the two dependent variables volunteer work time and charity donations. Since in this case all correlations are positive, we can say that the greater one's income, the greater will be one's time and money spent in both activities. Since r_{12} is indirect, volunteer work time is not directly affecting charitable donations. Nor is it going the other way. We say that the independent variable is causing the changes in the two dependent variables. In Figure 14.7, we add arrowheads to the lines in our diagram to indicate the probable direction of causation.

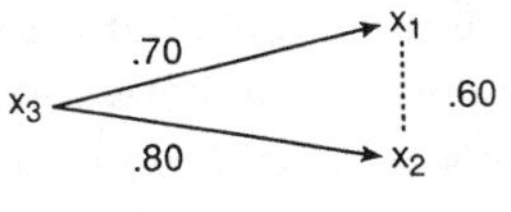

Figure 14.7

Now imagine a situation where the values of the correlations are the same, but x_2 is now the respondent's years of formal schooling. Assume also that in time order, schooling came first, then the acquisition of

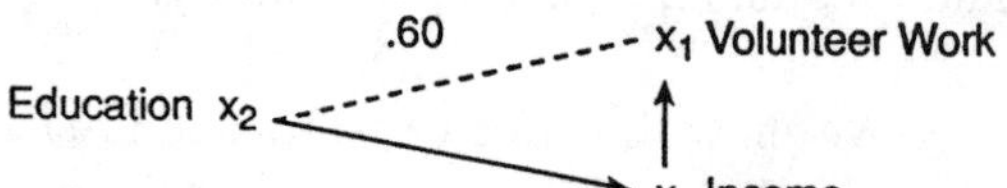

Figure 14.8

income, and then the time devoted to volunteer activity. The arrow now goes from x_2 to x_3, as seen in Figure 14.8.

Here, education is the independent variable, income is the **intervening variable,** and volunteer work time is the dependent variable. Education "causes" volunteer work time, but indirectly, through the intervening variable income. In effect, more education leads to higher income and higher income frees up more time for volunteer activities. Income provides interpretation of the relationship between education and volunteer work.

When we eliminate indirect correlations and at the same time make assumptions about causation, as we have just done, *we are undertaking what is known* as **causal modeling.** The direction of causation (the arrows) are based on logical assumptions, primarily time ordering. Despite the fact that reciprocal causation might also be logical, we usually ignore that possibility and place only one arrow on a given line. (That is, we ignore the prospect that while more education leads to more income, more income may also enable a person to acquire more education.) After setting up the model, we eliminate "paths" by determining which correlations may be indirect.

THE ROLE OF THE PARTIAL CORRELATION COEFFICIENT

In the previous examples, we determined that r_{12} was indirect by comparing it to the products of the other two correlations. Another approach involves the calculation of partial correlations. In simple terms, the **partial correlation** *between two variables is the amount of relationship not attributable to other variables in the system.* In our original problem, the partial correlation between x_1 and x_2 is the amount of correlation *not* due to x_3. It is the amount of direct correlation shared solely by x_1 and x_2. When we find the partial correlation between x_1 and x_2, we say that we are *controlling for the effects of* x_3 *on that relationship and we call* x_3 *the* **control variable.** We indicate the partial correlation as $r_{12.3}$, with the number of the variable to the right of the period being the control variable. Similarly, the partial correlation between x_1 and x_3 controlling for x_2 would be indicated $r_{13.2}$. Thus we have new notation and terminology.

r_{12} The *simple* Pearson's r between x_1 and x_2. *Because we are not controlling for any other variables, this is also called a* **zero-order correlation.**

$r_{12.3}$ The partial correlation between x_1 and x_2 controlling for the effects of x_3. *Since we are controlling for only one variable, we refer to this as a* **first-order partial correlation.**

$r_{12.34}$ The partial correlation between x_1 and x_2 controlling for the combined effects of two other variables x_3 and x_4 (a new variable in our system). *Since we control for two variables here, we call it a* **second-order partial correlation.**

We can have as many control variables as we want, the order being the number of control variables being used.

Calculating partial correlations can be complex. Third-order partials are calculated from second-order partials, second-order partials are calculated from first-order partials, and first-order partials are calculated from the zero-order correlations. Generally we use computer programs for this technique.

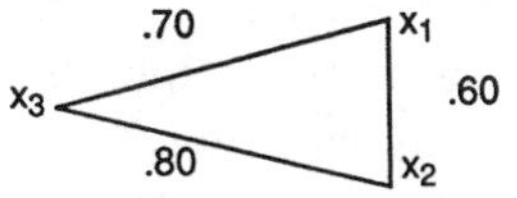

Figure 14.9
Zero-Order (Simple) Correlation

However, in the case of our three-variable problem, the calculations are not hard. We need to find first-order partials from the original zero-order correlations shown in Figure 14.9. The formula for $r_{12.3}$ is

$$r_{12.3}=\frac{r_{12}-r_{13}r_{23}}{\sqrt{1-r_{13}^2}\sqrt{1-r_{23}^2}}=\frac{.60-(.70)(.80)}{\sqrt{1-.70^2}\sqrt{1-.80^2}}=\frac{.60-.56}{\sqrt{1-.49}\sqrt{1-.64}}$$

$$=\frac{.04}{\sqrt{.51}\sqrt{.36}}=\frac{.04}{(.7141)(.6)}=\frac{.04}{.42846}=.09335$$

Rounding, we get $r_{12.3}$ = .09.

Whenever a partial correlation is close to zero, we say that the original relationship was indirect. We went from .60 to .09. (A zero in the tenths place + anything in the hundredths place can be thought of as nearly 0.) Thus, .09 being close to 0, suggests a spurious correlation or other indirect relationship.

We use similar formulas to get the other two partials.

$$r_{13.2}=\frac{r_{13}-r_{12}r_{23}}{\sqrt{1-r_{12}^2}\sqrt{1-r_{23}^2}}=\frac{.70-(.60)(.80)}{\sqrt{1-.60^2}\sqrt{1-.80^2}}=.46$$

$$r_{23.1}=\frac{r_{23}-r_{12}r_{13}}{\sqrt{1-r_{12}^2}\sqrt{1-r_{13}^2}}=\frac{.80-(.60)(.70)}{\sqrt{1-.60^2}\sqrt{1-.70^2}}=.67$$

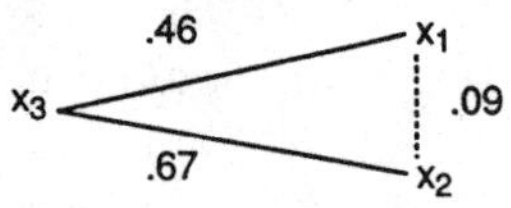

Figure 14.10
First-Order Partials

Summarizing, the first-order partials are presented in Figure 14.10 Only $r_{12.3}$ approaches 0, so only one indirect relationship is discovere here. If $r_{12.3}$ had not been low, we would have concluded that there wer *no* indirect relationships. If that had been the case, we would keep solid line between x_1 and x_2 and put an arrowhead on that line. In th example where x_2 was charity dollar contributions this decision woul be a difficult one: Do contributions lead to volunteer work, or vice versa? (See Figure 14.11.)

Figure 14.11

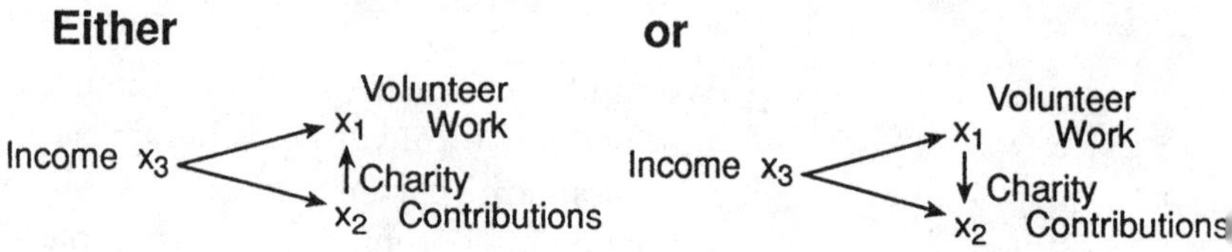

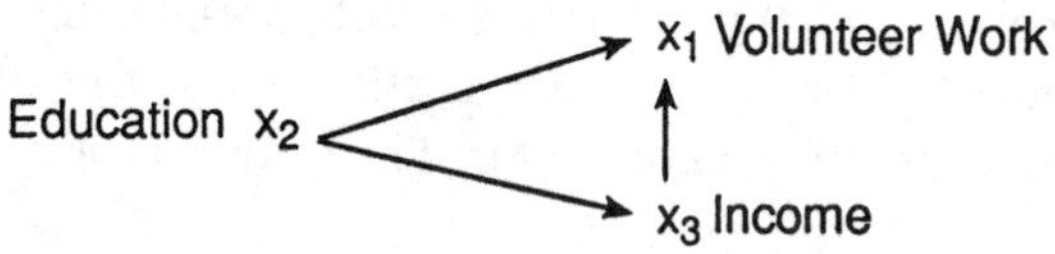

Figure 14.12

In the case where x_2 was education, th decision would have been easier, assumin education occurs first in time. Refer to Figur 14.12. Here, education has a direct impact o volunteer work as well as an indirect impac via income, the intervening variable.

Sometimes when we generate partial cor relations we encounter a situation where th zero-order coefficient is very low, but th partial correlation is not. Here is an example We assign attitude scores to survey respon dents as follows:

x_1: *Pro-choice.* The more pro-choice the respondent, the higher the score.

x_2: *Pro-death penalty.* The higher the score, the greater the respon dent's support for the death penalty.

x_3: *Religiosity.* The higher the score, the more religious the respondent.

In Figure 14.13 we summarize our results.

Although initially it appeared that there was no correlation between pro-choice and pro-death penalty attitudes, when we control for religi osity, a large positive correlation emerges. Such situations could result

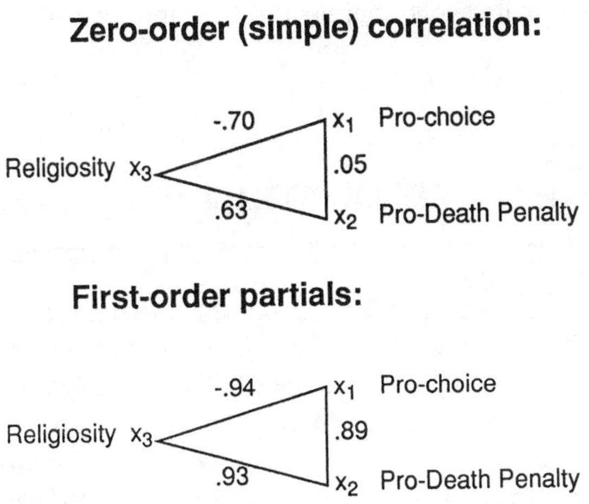

Figure 14.13

from offsetting consequences of religiosity. If, for instance, among the very religious the relationship between x_1 and x_2 was the opposite of the relationship for the less religious, the two effects would offset one another. Suppose for the very religious, respondents are either pro-choice and opposed to the death penalty or pro-life and support the death penalty. For the nonreligious, respondents are either pro-choice and support the death penalty or pro-life and oppose the death penalty. Without knowing religiosity, this strange relationship would not be apparent. It would be suppressed.

We can illustrate this using tables. Suppose 50 people have high religiosity and 50 have low religiosity.

For the Religious

Death Penalty	*Pro-Choice*	*Pro-Life*		
Support	0	25	25	
Oppose	25	0	25	$Q = -1.00$
	25	25	50	

For the Nonreligious

Death Penalty	*Pro-Choice*	*Pro-Life*		
Support	25	0	25	
Oppose	0	25	25	$Q = +1.00$
	25	25	50	

But when we combine both groups and do not differentiate by religiosity, the following table results.

Death Penalty	*Pro-Choice*	*Pro-Life*		
Supports	25	25	50	$Q = 0$
Opposes	25	25	50	
	50	50	100	

We can also use causal modeling with some of the other coefficients in this chapter, such as partial regression slopes and standardized partial regression slopes (known also as path coefficients). These techniques were more commonly applied in the 1960s, seemed to decline in popularity until a few years ago, but appear to be staging a comeback in recent literature.

MULTIPLE CORRELATION AND THE COEFFICIENT OF MULTIPLE DETERMINATION

The concept of the multiple correlation, R, and the coefficient of multiple determination, R^2, is an extension of Pearson's r and r^2 to more than two variables. In fact, some systems of notation and many computer programs use R and R^2 even for the simple bivariate r and r^2. The **multiple correlation** *coefficient measures the correlation between a dependent variable and the combined effect of other designated variables in the system.* The **coefficient of multiple determination** *measures the proportion of variation in the dependent variable accounted for by those other variables.* In the three-variable example where x_1 is the dependent variable, $R_{1.23}$ tells us that x_1 is dependent and that there are two independent variables, x_2 and x_3. If you see $R_{1.234}$, you would know that x_1 is dependent and that there are three independent variables, x_2, x_3, and x_4. If you see $R_{2.13}$, you know that x_2 is dependent and x_1 and x_3 are the two independent variables. Similarly, $R^2_{2.13}$ is the proportion of variation in x_2 accounted for by x_1 and x_3 acting together.

Actually, the multiple correlation is simply a mathematical extension of Pearson's r. In the two-variable case, there were two perpendicular axes and the data points were based on coordinates referenced to the two axes. The regression was the best-fitting line running through the data that could be used to predict the dependent variable. When a third variable is added, a third perpendicular axis for the new variable extends out from the origin. Two and three dimensional coordinate systems are illustrated in Figure 14.14. (The negative-valued axes are deleted for visual clarity in this figure.) As in a hologram or a 3D film, the points that were once on the flat two-dimensional $x_1 + x_2$ coordinate system are now extended out toward the viewer (or back behind the "screen" where x_3 coordinates have negative values). Each point in space is referenced to the three axes by an **ordered triplet:** (x_1, x_2, x_3).

Although this three-dimensional scatter of points in space, like its two-dimensional cousin, could have no pattern (a "blob" of points) or could be curvilinear (note the design on the cover of this book), we are here still looking for linearity. In three dimensions, a linear relationship becomes a plane. Think of the plane as a flat piece of window glass with

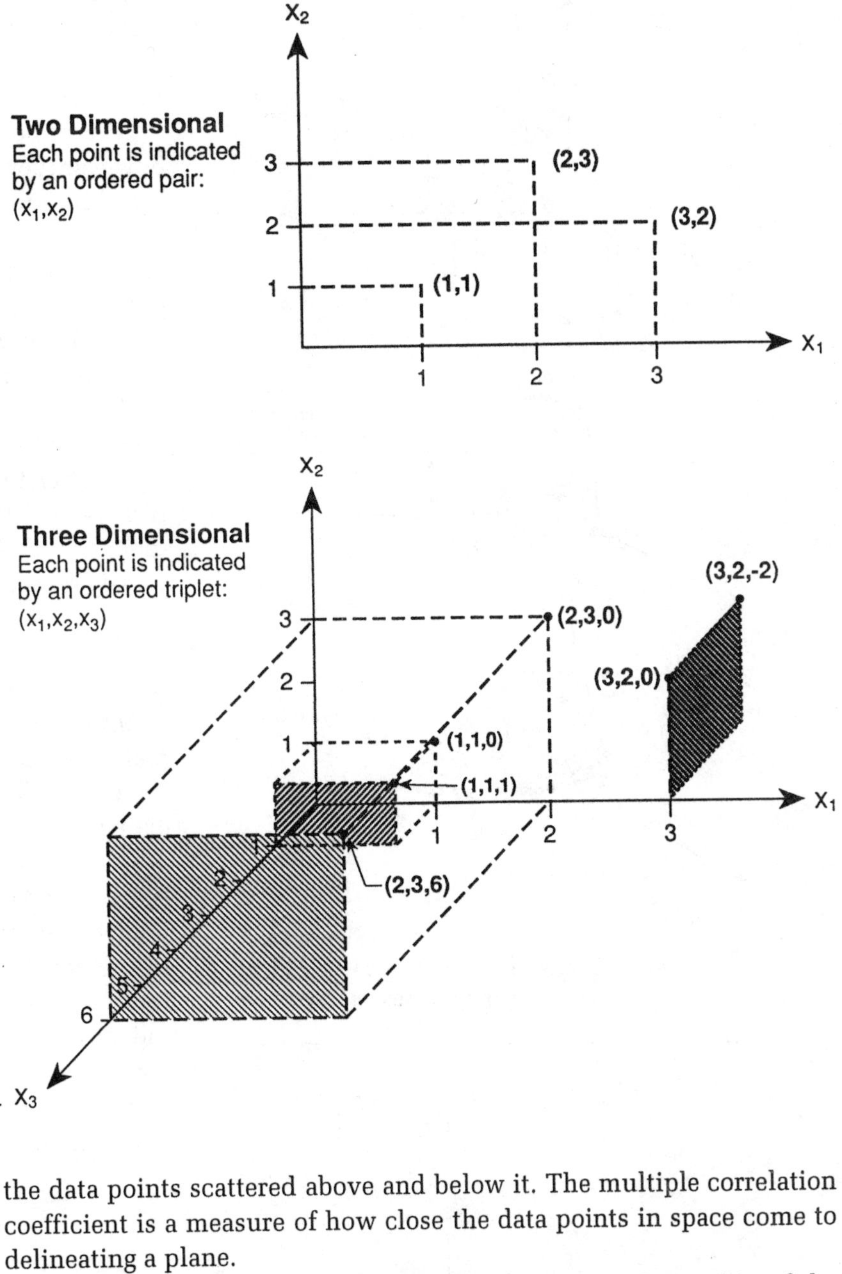

Figure 14.14

the data points scattered above and below it. The multiple correlation coefficient is a measure of how close the data points in space come to delineating a plane.

Figure 14.15 shows, in a simplified fashion, the relationship of the points to the plane. When points are scattered far and unevenly around the plane, R is low. As the scatter becomes closer and more even about the plane, R grows larger. When the points all fall exactly on the plane,

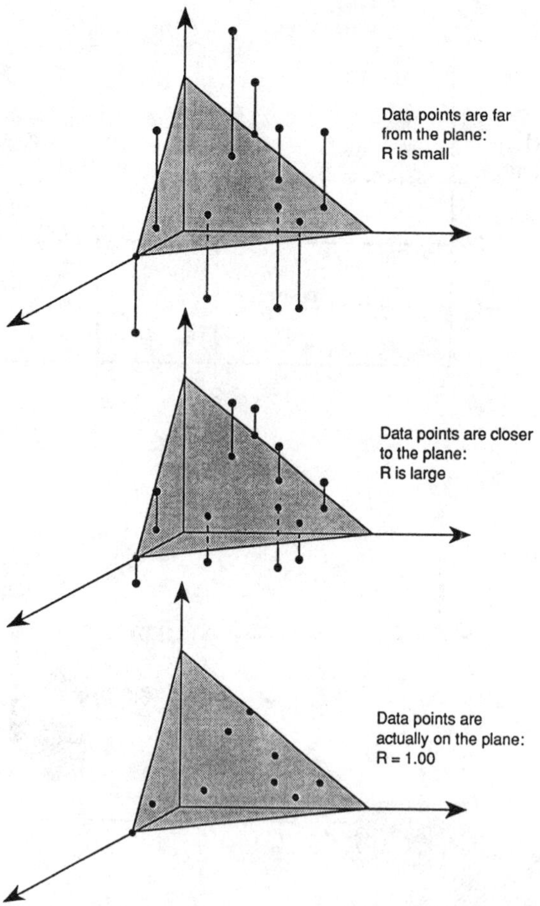
Data points are far
from the plane:
R is small
Data points are closer
to the plane:
R is large
Data points are
actually on the plane:
R = 1.00

Figure 14.15
Relationship of the Points to the Plane as R Increases in Magnitude

R equals 1. Note how this parallels what takes place in two variable-linear regression where as the points become more linear, simple r increases, approaching ±1.0 as a limit.

We are not limited here to three variables; however, beyond three dimensions we lose the ability to graph the structure. *Nevertheless, we can still describe the structure mathematically and call it an* **n-dimensional hyperplane.** No matter how many dimensions, we can still find R and R^2. So we have the freedom to add as many new variables as we wish.

Unlike the bivariate Pearson's r, the multiple correlation is always a positive number. It does not attempt to indicate whether the independent variables are positively or inversely related to the dependent variable, simply because when there is more than one independent variable, some may be positively related, whereas at the same time others may be inversely related. Since R has no sign, its real value comes once it is squared and is indicating the proportion of variation being explained.

As with partial correlations, the calculation of R and R^2 can become complex as more independent variables are added to the soup. In the case of a three-variable problem, the formula with x_1 dependent and x_2 and x_3 independent could be either

$$R^2_{1.23} = r^2_{12} + r^2_{13.2}(1 - r^2_{12})$$

or

$$R^2_{1.32} = r^2_{13} + r^2_{12.3}(1 - r^2_{13})$$

Verbally, using proportion of variation terminology, we take the proportion of variation explained by one independent variable and add to it

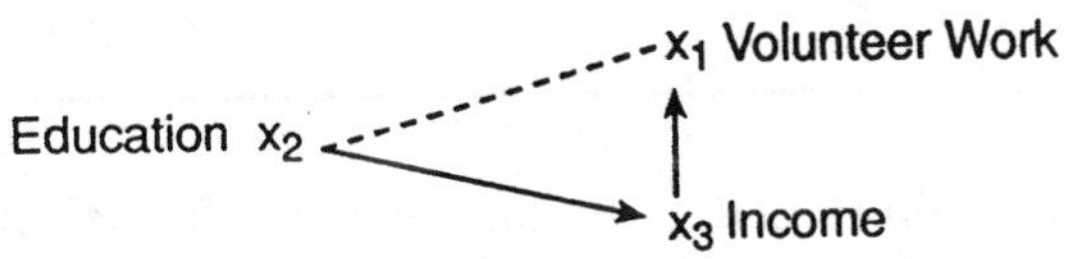

Figure 14.16

the product of the additional proportion of variation explained by the second independent variable times the proportion of variation left unexplained by the first independent variable. (This is not an easy definition to grasp at first presentation.)

In the case of the earlier example where volunteer work time is being explained by education and income together (Figure 14.16):

We were given	*We found*
$r_{12} = .60$	$r_{12.3} = .09$
$r_{13} = .70$	$r_{13.2} = .46$
$r_{23} = .80$	$r_{23.1} = .67$

Using the first of the two multiple R^2 formulas we get

$$\begin{aligned} R^2_{1.23} &= r^2_{12} + r^2_{13.2}(1 - r^2_{12}) \\ &= (.60)^2 + (.46)^2\,[1 - .60)^2] \\ &= .36 + .2116(1 - .36) \\ &= .36 + .1354 \\ &= .4954 \end{aligned}$$

About 49 1/2% of the variation in volunteer work time can be explained by the combined effects of education and income. (The other $R^2_{1.23}$ formula yields .4944. The difference is the result of rounding back the partial correlations to two decimal places.)

Suppose you wanted a variable other than x_1 to be dependent. You can make use of the *same* formulas by temporarily designating the dependent variable as x_1 and the other two variables, in no particular order, as x_2 and x_3.

Before we conclude this topic let us address an often-asked question. Why not add up the simple r^2's to get the multiple R^2? In other words, could we not say:

$$R^2_{1.23} = r^2_{12} + r^2_{23}$$

It turns out that this formula is correct only if x_2 and x_3 are uncorrelated; it only works if $r_{23} = 0$. Since this is generally not the case, $R_{1.23}$ is almost always less than $r_{12}^2 + r_{13}^2$.

MULTIPLE REGRESSION

Multiple regression, a shortened term for its more formal title of **multiple linear regression,** *is the technique of developing predictive equations when there is more than one independent variable present.* Just as we can find R and R^2 where we add variables, we also can develop a predictive equation to predict a score on the dependent variable from the combined effect of the independent variables. Recall that the equation for a straight line followed the format $y = a + bx$, where a is the y-intercept and b is the slope. If we indicate all of our variables as x + a subscript as we have been doing in this chapter, then for x_1 dependent,

$$x_1 = a_1 + b_2x_2$$

The new subscripts for a and b indicate that a_1 is the value of x_1 where the regression line crosses the x_1-axis and b_2 is the slope going with variable x_2.

When we add another independent variable, x_3, our equation becomes the equation for a three-dimensional plane:

$$x_1 = a'_1 + b'_2x_2 + b_3x_3$$

The little prime symbols now above a_1 and b_2 are to remind us that these numbers are *not* necessarily the same as the ones in the first equation. They change each time a variable is added. If we bring in another variable, x_4, the equation for the four-dimensional hyperplane would have the form

$$x_1 = a''_1 + b''_2x_2 + b'_3x_3 + b_4x_4$$

As before, the a's and b's change from what they were prior to the addition of x_4.

Once we are at three dimensions and are dealing with a plane, the a value becomes the intercept of the dependent variable, that is, the value of the dependent variable when the two independent variables' scores are each zero, which is the point where the plane crosses the dependent variable's axis. This is illustrated in Figure 14.17. If x_1 is dependent, we move along the x_1 axis to where the plane crosses it. Then a'_1 is the distance of that point from the origin. (If x_2 were dependent, we would move up the x_2 axis and intercept the plane at a'_2.)

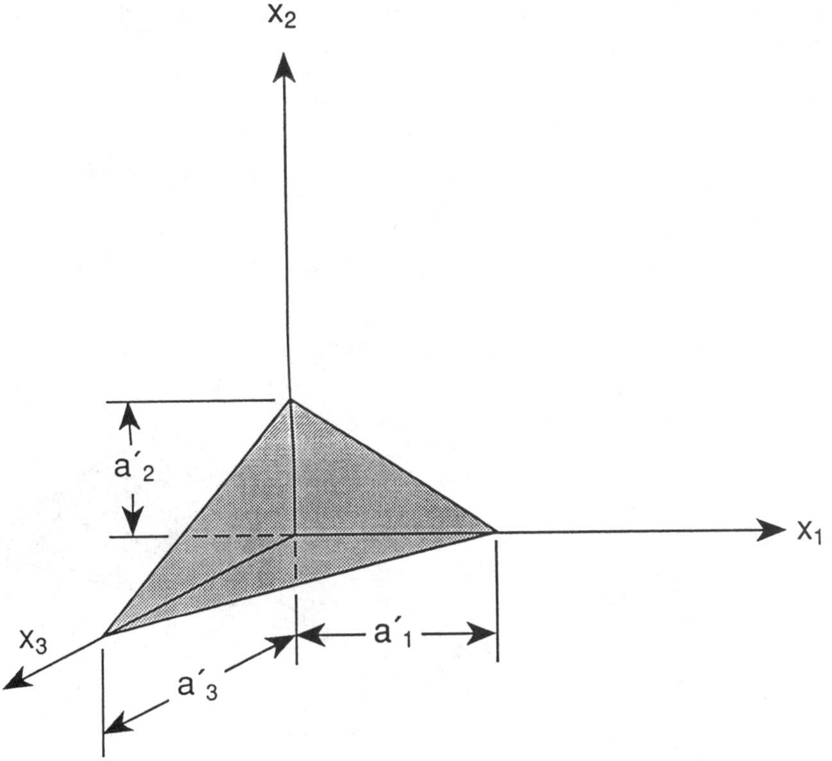

Figure 14.17
The Intercepts for x_1, x_2, and x_3

The slopes in the multiple regression, the *b*'s, are referred to as **partial regression slopes.** We use the same terminology as we did with partial correlations. *The simple two-variable regression slope is a* **zero-order regression slope.** *When we add a third variable the slopes become* **first-order partial regression slopes.** When we add a fourth variable, the slopes become *second-order* partial regression slopes, and so on.

The order's number tells us the number of other variables whose influence is being held constant. To understand what we mean by "holding constant," examine Figure 14.18. If we wanted a simple regression slope, as before with linear regression, we would take the original data points (not shown here) and use them all (ignoring x_3) to find the slope of the line minimizing the squares of x_1 distances. With the partial regression slopes, we are not using all the data points at the same time. Instead, what we do has the effect of taking some specific value of x_3 (it does not matter which value) and using only the data points with that particular value. Notice in Figure 14.18 that if we take *just* the points on

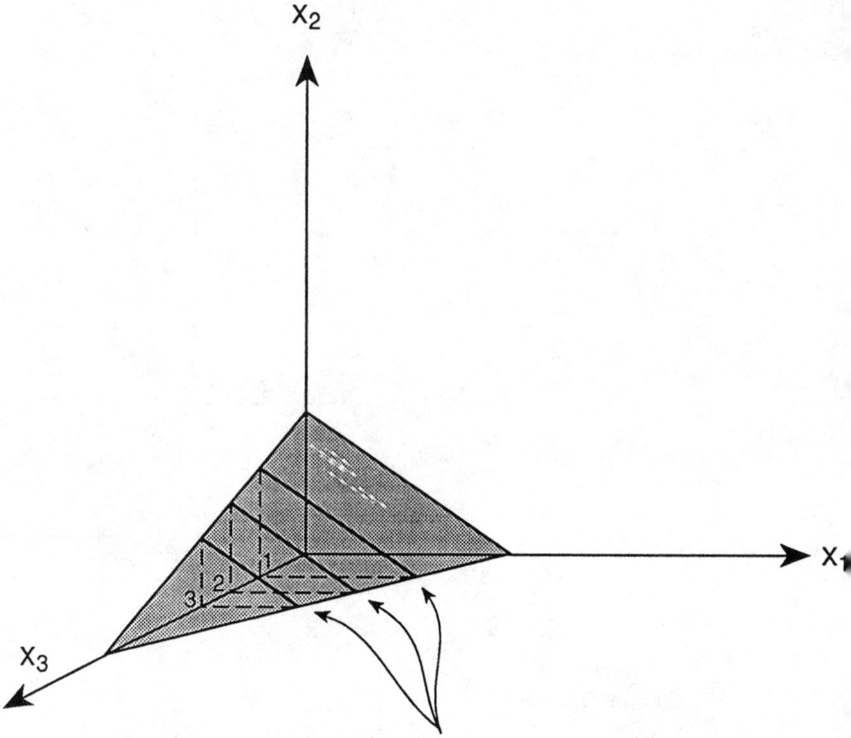

Each one of these cuts in the plane, where x_3=1, x_3=2, and x_3=3, respectively, is parallel to the others, thus having the **same** slope. This is the partial regression slope of x_1 on x_2, controlling for (holding constant) the effects of x_3. It differs from the simple (zero order) b_{12}, because **that** value is based on **all** points on the plane, disregarding their x_3 values.

Figure 14.18
A Partial Regression Slope

the plane where $x_3 = 1$ and reflect them back on the surface where the x_1- and x_2-axes cross, those points will compose a line. If we use only the points where $x_3 = 2$ on the plane and reflect those points back, we will get a second line. Doing the same where $x_3 = 3$, we get a third line (see Figure 14.19). All three lines when reflected back on the surface where the x_1- and x_2-axes cross will be parallel, thus sharing a common slope, which is the partial regression slope. But that slope will not necessarily be the same as the slope established when all the original data points were used to find the simple b_{12}. (Another explanation of b follows later in this chapter.)

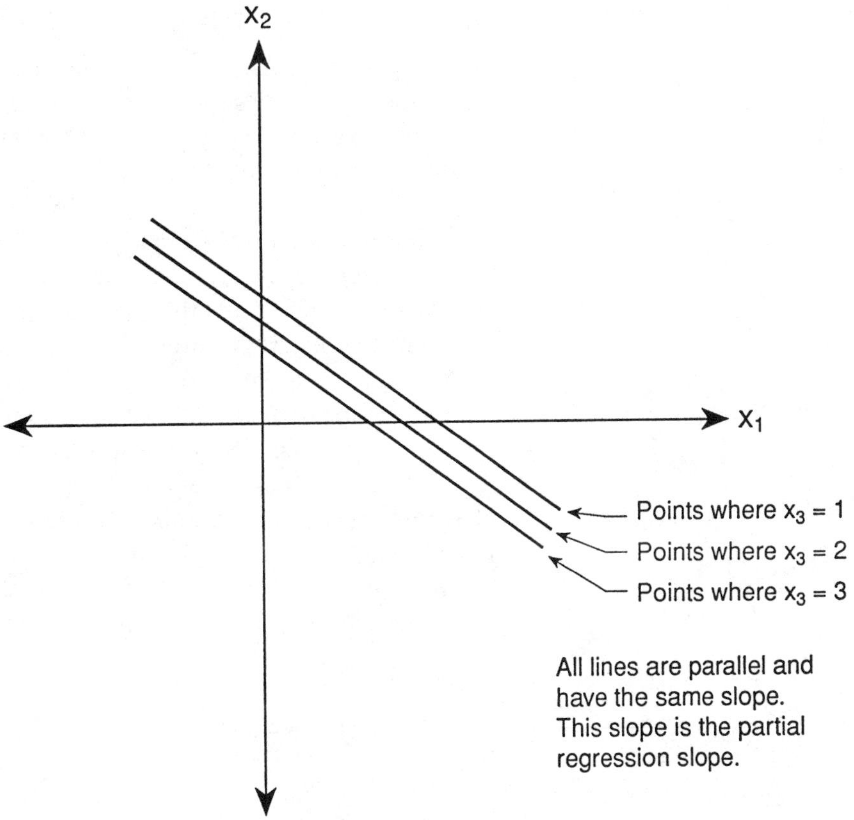

Figure 14.19
Reflections of the Cuts in the Plane (Figure 14.18) on the Surface Where the x_1- and x_2-Axes Cross

All of this notation and graphical representation takes some getting used to. However, even if we have difficulty visualizing all of these *a*'s and *b*'s and what they do, we can still apply multiple regression techniques. We can trust that the equation we obtain will be the best equation, using the least squares criterion, for predicting the value of a dependent variable from the combined effect of several other variables. What makes our job easy is the computer. Because the calculations for multiple regression intercepts and slopes are tedious and complicated, they rarely will be done even with a calculator. Let us, therefore, assume that we will have at our disposal a software package that will calculate all items necessary to build the regression equation. Using a computer also enables us to disregard a lot of the subscripts and other notation that have been used up to this point.

AN EXAMPLE FROM JUDICIAL BEHAVIOR

Suppose some organization wishes to influence the process whereby U.S. Supreme Court justices are selected. The organization has rated each judge currently serving on the Federal Appeals bench or on a state supreme court and has assigned each judge a score designed to measure support for individual civil liberties. The scores range from 5 to 30. The higher the score, the greater the judge's judicial support for the "liberal" decisions of the Warren Court. The lower the score, the more that judge took a "strict constructionist" stance in opposition to the decisions of the Warren Court. Thus the more "conservative" the judge, the lower would be his or her overall score.

You are a scholar in the field of judicial behavior and have been studying many of these same judges. From previous court decisions, you have been able to code their positions along several areas of recent legal contention. For each issue area, you have scored each judge on a scale ranging from 1 (most strict constructionist) to 5 (most civil libertarian/liberal). You have coded the overall score and each issue area score for each judge and entered the data as computer input. The variables are indicated with short names listed below.

Variable	*Short Name*
Dependent:	
Judicial Liberalism	JLIB
Independent:	
Abortion/Pro-Choice	ABOR
Capital Punishment/Opposition	CAPP
Censorship/Freedom of Expression	CENS
Consumer Protection	CONS
"Right to Die" Legislation	RDIE

In most software packages, the *short names* (**mnemonics**) replace algebraic letters such as x_1 or x_2 in identifying the variables.

As the researcher collecting these data, you would not have been operating in a vacuum. Your expertise in judicial behavior would have led you to select independent variables that you would expect to explain overall judicial liberalism. You would generally not code those issue areas that in your experience do not correlate with overall liberalism. If you had not previously noted a liberal/conservative pattern to issues such as import-export regulation, zoning authority, or income tax law, you would exclude them from your study. On the other hand, if there were no experiential or theoretical basis for excluding the latter issue areas, you might include them as well. If they provide little predictive capability, the regression and its accompanying statistics will demonstrate that fact. Finally, if there is little theory but many independent variables, you might make use of a procedure known as stepwise regression, which will be presented shortly.

From the regression program, you determine what later will be shown to be the "best" formula for predicting JLIB. It predicts JLIB from the *first three* independent variables on the list—ABOR, CAPP, and CENS:

$$\text{JLIB} = -.82 + 2.46\ \text{ABOR} + 1.81\ \text{CAPP} + 1.59\ \text{CENS}$$

The intercept (sometimes referred to as *the constant*) is –.82. The partial regression slopes are 2.46 for ABOR, 1.81 for CAPP, and 1.59 for CENS. To predict a judge's overall score from the issue area scores, take the latter scores and substitute the scores for the short names in the equation, just as we did with the *x* values in linear regression. (Note that these coefficients are not based on an actual study, but are from simulated data.)

Let us begin by examining four hypothetical cases. A judge who scored most conservative on each of the issue areas (a score of 1) would have the following predicted overall score.

$$\begin{aligned}\text{JLIB} &= -.82 + 2.46\ \text{ABOR} + 1.81\ \text{CAPP} + 1.59\ \text{CENS}\\ &= -.82 + 2.46(1) + 1.81(1) + 1.59(1)\\ &= -.82 + 2.46 + 1.81 + 1.59\\ &= -.82 + 5.86\\ &= 5.04\end{aligned}$$

A judge who scored most liberal (5) on each issue area would have this predicted overall score.

$$\begin{aligned}\text{JLIB} &= -.82 + 2.46(5) + 1.81(5) + 1.59(5)\\ &= -.82 + 12.3 + 9.05 + 7.95\\ &= 28.48\end{aligned}$$

(Recall that the original scale ranges from a low of 5 to a high of 30.) The next case is a relatively liberal judge. His scores are 3 on ABOR, 4 on CAPP, and 5 on CENS; his overall predicted score follows.

$$\begin{aligned}\text{JLIB} &= -.82 + 2.46(3) + 1.81(4) + 1.59(5)\\ &= -.82 + 7.36 + 7.24 + 7.95\\ &= 21.75\end{aligned}$$

Another more conservative judge whose scores are ABOR = 2, CAPP = 1, and CENS = 3, would have this overall predicted score.

$$\begin{aligned} \text{JLIB} &= -.82 + 2.46(2) + 1.81(1) + 1.59(3) \\ &= -.82 + 4.92 + 1.81 + 4.77 \\ &= 10.68 \end{aligned}$$

Remember that these are the scores *predicted* from the regression formula. The actual overall scores assigned to the latter two judges have been coded into the system. We expect our predictions to come close to the actual scores, though they rarely will be exactly identical. R^2 will tell us approximately how close overall the predicted values come to the values actually observed.

Let us return to the subject of the partial regression slopes in the formula. Earlier we discussed the meaning of partial slopes in terms of a visual example. We can now explain such slopes in terms of their impact on prediction. We start with a two-variable example. Suppose we only wanted to predict JLIB from one variable, ABOR. In that case, the regression formula would be:

$$\text{JLIB} = 10.63 + 2.39 \text{ ABOR}$$

If ABOR = 1,

$$\begin{aligned} \text{JLIB} &= 10.63 + 2.39(1) \\ &= 10.63 + 2.39 \\ &= 13.02 \end{aligned}$$

If we increased ABOR by one unit, so that ABOR = 2, then

$$\begin{aligned} \text{JLIB} &= 10.63 + 2.39(2) \\ &= 10.63 + 4.78 \\ &= 15.41 \end{aligned}$$

Note that the difference between the two predictions is 15.41 – 13.02 = 2.39, the slope of ABOR. Thus b tells us the amount of change in the dependent variable, JLIB, generated by a one-unit increase in the independent variable, ABOR. Note that had the sign of ABOR's slope been negative, so that JLIB = 10.63 – 2.39 ABOR, then a one-unit increase in ABOR would have caused JLIB to decrease by 2.39 units, because the variables would be inversely related.

Suppose we wanted to predict from ABOR plus another independent variable, CAPP. Our formula would be

$$\text{JLIB} = 2.28 + 2.55\ \text{ABOR} + 2.48\ \text{CAPP}$$

The slope on ABOR, 2.55, is now a first-order partial regression slope, telling as that if ABOR increased by one unit while CAPP remained unchanged, JLIB would increase by 2.55 units. (Saying that CAPP remained unchanged is another way of saying that we are "controlling for" that variable.) Also, CAPP's *b* of 2.48 tells us that if ABOR remained unchanged and CAPP increased by one unit, JLIB would increase by 2.48 units.

Returning to our three-independent-variable formula,

$$\text{JLIB} = -.82 + 2.46\ \text{ABOR} + 1.81\ \text{CAPP} + 1.59\ \text{CENS}$$

the slope of ABOR, 2.46, is a second-order partial regression slope that shows the change in JLIB generated by a one-unit increase in ABOR when both CAPP and CENS remain unchanged. The slope is a second-order partial slope because we are controlling for two other variables.

Since the size of the slope measures the impact of the change on the dependent variable, coming from a one-unit increase of the slope's independent variable, it would appear that the larger the slope the more important is that variable in explaining the dependent variable. Thus ABOR is more important than CAPP and CAPP is a bit more important than CENS. Indeed, in this problem that turns out to be the case. Generally, though, we *cannot* reach this conclusion by examining the regression slopes because the independent variables are usually measured along scales of differing ranges and dispersions. Recall that in this problem, each independent variable is measured on a scale ranging from 1 to 5. Suppose instead that ABOR was scored from a low of 10 to a high of 50. Then its slope would have been .246 instead of 2.46, and it would falsely appear to be less important than CAPP or CENS. The easiest way to avoid this pitfall is to use a scale of identical range and standard deviation for each of the independent variables. If this is not possible, we use the alternative approach presented next.

THE STANDARDIZED PARTIAL REGRESSION SLOPE

To compensate for differences in measurement, when we want to determine the relative importance of our independent variables we calculate a new measure known as a **beta coefficient** or a **beta weight** Beta (β) is a **standardized partial regression slope.** By *standardized* we mean converted into standard deviation units of the dependent variable much in the same vein as changing values of x into standard z scores. Beta tells us by how many of its standard deviation units the dependent variable will change when the variable associated with that beta weight increases by one of its standard deviation units and the other independent variables remain unchanged.

To get each beta, we multiply its b by its standard deviation and divide by the standard deviation of the dependent variable.

$$\text{for ABOR,} \qquad \beta_{\text{ABOR}} = b_{\text{ABOR}} \times \frac{s_{\text{ABOR}}}{s_{\text{JLIB}}}$$

$$\text{for CAPP,} \qquad \beta_{\text{CAPP}} = b_{\text{CAPP}} \times \frac{s_{\text{CAPP}}}{s_{\text{JLIB}}}$$

and so on.

In this case, for predicting JLIB,

$$\beta_{\text{ABOR}} = .77$$

$$\beta_{\text{CAPP}} = .33$$

$$\beta_{\text{CENS}} = .26$$

Thus if ABOR increases by one of its standard deviation units and both CAPP and CENS remain unchanged, JLIB will increase by .77 of one of its standard deviation units. This is much larger a change than would result from a standard deviation increase in either of the other two variables.

As a further illustration of the value of betas, note the following data set

$x_1 =$	$x_2 =$	$x_3 =$
7.4	6.7	.75
6.8	6.2	.69
8.2	7.9	.81
7.8	6.9	.80
6.6	6.2	.65

When a regression is run on the above to predict x_1 from x_2 and x_3, the b for x_2 is 0.35 and the b for x_3 is 6.40. However, if we change x_3 from a scale running from 0 to 1.00 to a scale running from 0 to 100 (75, 69, 81, and so on), the b for x_2 is unchanged at 0.35 but the b for x_3 changes from 6.40 to 0.064. In the first example, $b_2 < b_3$ and in the second $b_2 > b_3$. However, the beta coefficients will be the same for both examples, 0.37 for x_2 and 0.66 for x_3. Regardless of the scale used for x_3, it has twice the impact on the dependent variable as x_2. Thus, while a comparison of the unstandardized regression slopes varies with the scales of measurement of the independent variables, this is not true with the standardized partial regression slopes. The betas are unchanged.

Beta will always have the same sign as its b, so if b is negative, beta will tell us by how many standard deviations the dependent variable will *decrease* when that independent variable increases by a standard deviation. Also, we normally never use the betas in a prediction equation. Accordingly, we use the partial b's in the formula so that we are predicting the independent variable value in its original scale. We would rarely want to make predictions in standard deviation units, which use of the betas would do. The betas are used to compare the relative impact of each independent variable on the dependent variable. The b's are used for prediction. Consequently, each coefficient has its own role to play.

USING A REGRESSION PRINTOUT

In Table 14.2 you will see a sample multiple regression printout using SAS's PROC REG procedure. In this case, all five of the original independent variables in our problem were retained. The printout begins with an ANOVA for the overall statistical significance of the regression equation. The F of 72.606 has a probability of .0001—very significant. Below the ANOVA is additional information. Note in the center that an R-SQUARE value of 0.8561 is given. Thus 85.61% of the variation in JLIB can be accounted for by this regression with five independent variables. Below the R-SQUARE, note the ADJ R-SQ which stands for **adjusted R-SQUARE.** The adjustment does for R what $n - 1$ in the denominator does for the standard deviation. Just as with S, an R from sample data tends to overstate the R in the population. With the adjustment, we may estimate that in the population as a whole (all similar judges), 84.43% of the variation in JLIB can be accounted for by this model.

In the next lower part of the printout, you will find the information needed to construct the regression formula. (Why doesn't this procedure

Table 14.2 The SAS REG Procedure

SAS

DEP VARIABLE: JLIB
ANALYSIS OF VARIANCE

SOURCE	DF	SUM OF SQUARES	MEAN SQUARE	F VALUE	PROB > F
MODEL	5	3010.44928	602.08986	72.606	0.0001
ERROR	61	505.84923	8.29261031		
C TOTAL	66	3516.29851			

ROOT MSE	2.879689	R-SQUARE	0.8561
DEP MEAN	19.41791	ADJ R-SQ	0.8443
C.V.	14.83007		

PARAMETER ESTIMATES

VARIABLE	DF	PARAMETER ESTIMATE	STANDARD ERROR	T FOR H0; PARAMETER = 0	PROB > |T|
INTERCEP	1	−0.96098181	1.37008962	−0.701	0.4857
CENS	1	1.40981955	0.34490030	4.088	0.0001
CAPP	1	1.45120569	0.32446883	4.473	0.0001
CONS	1	0.97003442	0.41702892	2.326	0.0234
RDIE	1	−0.02914917	0.01595688	−1.827	0.0726
ABOR	1	2.40255343	0.15804215	15.202	0.0001

VARIABLE	DF	STANDARDIZED ESTIMATE
INTERCEP	1	0
CENS	1	0.23138398
CAPP	1	0.27146741
CONS	1	0.13229489
RDIE	1	−0.09153555
ABOR	1	0.75883123

just generate the formula?) The first column on the left is entitled VARIABLE. Below the title is INTERCEP, meaning, of course, the intercept (a) for the variable JLIB. Below INTERCEP are the names of the variables in the order in which the program was requested to enter them.

To the right of the VARIABLE column is a degrees of freedom column, DF, and to its right, PARAMETER ESTIMATE. In this latter column is the information we need for our equation. The first number is the intercept, −0.96098181, and the number under it is the partial regression slope for the first variable entered, CENS, which is 1.40981955. (From now on, for simplicity's sake, we'll use only the first two decimal places, and in order to ease your reading of the printout, we will *not* round the entries.) Going down this column we can build our regression formula:

$$\text{JLIB} = -0.96 + 1.40\ \text{CENS} + 1.45\ \text{CAPP} + 0.97\ \text{CONS}$$
$$-0.02\ \text{RDIE} + 2.40\ \text{ABOR}$$

After the intercept of −0.96, we could have entered the variables in any order that we preferred: JLIB = −0.96 + 2.40 ABOR + 1.45 CAPP . . . and so on. Note that below these numbers on the printout is a column entitled STANDARDIZED ESTIMATE. These are the beta coefficients. (The intercept, of course, has no beta.) Note that both the *b*'s and betas differ from those used earlier in this chapter because now two *new* variables have been added.

Moving back up to the PARAMETER ESTIMATE column and moving two columns to the right, you'll see T FOR H0: PARAMETER = 0 and to its right a PROB > |T| column. In addition to the overall *F* already done, SAS tests each component of the equation for significance. The first *t* value, −0.701, tests the null hypothesis $a_{\text{population}} = 0$. Since $p = 0.4857$, we cannot reject H_0. This is OK because the printed intercept is small to begin with, −0.96. For the population's regression formula we could substitute 0 for −0.96. (Make no substitution, of course, if $p < .05$.) The rest of the *t* tests are for the null hypothesis: $b_{\text{population}} = 0$. With the exception of RDIE, all *t*'s are significant and their respective variables retained in the equation.

Since the *t* for RDIE is not significant, we must assume $b_{\text{population}} = 0$. If that is the case, remember that $r_{\text{population}} = 0$. It would be preferable to rerun this procedure without using RDIE at all, as shown in Table 14.3. In addition, the independent variables were entered in *decreasing* order of the size of their slopes. Note that just as when we add a variable the coefficients change, the same thing happens where we delete one. Our new equation becomes

$$\text{JLIB} = -1.57 + 2.44\ \text{ABOR} + 1.55\ \text{CAPP} + 1.44\ \text{CENS}$$
$$+ 0.91\ \text{CONS}$$

Since the intercept is not significantly different from zero, we could delete it as well. (However, many programs do not test the intercept for significance, so it is rarely deleted in practice.)

$$\text{JLIB} = 2.44\ \text{ABOR} + 1.55\ \text{CAPP} + 1.44\ \text{CENS} + 0.91\ \text{CONS}$$

Table 14.3 Regression Without RDIE and Independent Variables Reordered

SAS

DEP VARIABLE: JLIB
ANALYSIS OF VARIANCE

SOURCE	DF	SUM OF SQUARES	MEAN SQUARE	F VALUE	PROB > F
MODEL	4	2982.77683	745.69421	86.656	0.0001
ERROR	62	533.52167	8.60518829		
C TOTAL	66	3516.29851			

ROOT MSE	2.93346	R-SQUARE	0.8483
DEP MEAN	19.41791	ADJ R-SQ	0.8385
C.V.	15.10698		

PARAMETER ESTIMATES

VARIABLE	DF	PARAMETER ESTIMATE	STANDARD ERROR	T FOR HO: PARAMETER = 0	PROB > |T|
INTERCEP	1	−1.57963264	1.35236320	−1.168	0.2473
ABOR	1	2.44618610	0.15914391	15.371	0.0001
CAPP	1	1.55217744	0.32569620	4.766	0.0001
CENS	1	1.44292853	0.35085499	4.113	0.0001
CONS	1	0.91434287	0.42367918	2.158	0.0348

VARIABLE	DF	STANDARDIZED ESTIMATE
INTERCEP	1	0
ABOR	1	0.77261233
CAPP	1	0.29035552
CENS	1	0.23681793
CONS	1	0.12469958

STEPWISE MULTIPLE REGRESSION

In the above applications, it was up to the researcher to specify th independent variables for the regression model. *There are a number c techniques* subsumed under the title of **stepwise multiple regressio** *that allow the computer to determine which independent variables t enter and when to enter them.* In theory we are telling the computer t first find the "best" independent variable for predicting the dependen variable. The computer, in step 1 below, finds that "best" variable amon a menu of variables and gives us a simple regression for predicting the dependent variable from the selected independent variable.

Then the computer finds the "second best" independent variable, and in step 2, adds that variable to the model. Now we have two independent variables in a regression. Then the computer finds the "third best" independent variable and adds it to the equation. It continues doing this until it runs out of independent variables, or out of time, or other preselected criteria are met for shutting down the process.

There are a number of techniques for finding the "best," "second best," and so on, independent variables. One common way is as follows:

Step 1. Do a matrix of simple correlations and enter the variable most highly correlated with the dependent variable.

Step 2. Do a matrix of first-order partial correlations (controlling for the variable selected in step 1). Find in this matrix the variable most highly correlated with the dependent variable and use this new variable along with the one selected in step 1 to create an equation for predicting the dependent variable.

Step 3. Find the variable whose second-order partial correlation (controlling for the two previously selected variables) correlates the highest and enter it in step 3 in a new regression equation.

At each step, the program generates an R or an R^2. The first one (hopefully) will be large. The one in the second step will be even larger than the first, since two variables should account for more variation than just one. However, with each subsequent step, the increment in the size of the coefficient will get smaller, as each latter variable only tends to bring about a small increase in explained variation. Watching to see where R^2 "tops off," together with an analysis of the significance of the regression slopes, enables the researcher to decide which model to select.

In Table 14.4, we see the results of the SAS PROC STEPWISE procedure as applied to our judicial scaling problem. Each step resembles, in general, the earlier format in PROC REG.

At the beginning of each step, the printout shows the step number, the independent variable entered, and R SQUARE. (STEPWISE does not find the adjusted R^2 and does not calculate beta coefficients. However, once you have selected the variables for your final model, you could also run PROC REG, as we did previously, to get this information.)

Below this information an F is generated for that model and its probability given. Below the analysis of variance in the column entitled

Table 14.4 SAS Stepwise Regression

```
SAS

STEPWISE REGRESSION PROCEDURE FOR DEPENDENT VARIABLE JLIB

NOTE: SLENTRY AND SLSTAY HAVE BEEN SET TO
      .15 FOR THE STEPWISE TECHNIQUE.

STEP 1       VARIABLE ABOR ENTERED                      R SQUARE = 0.57096438
                                                          C(P) = 118.92309202

                    DF  SUM OF SQUARES   MEAN SQUARE             F   PROB > F

REGRESSION           1   2007.68119830  2007.6811983         86.50     0.0001
ERROR               65   1508.61730917    23.2094971
TOTAL               66   3516.29850746

               B VALUE       STD ERROR    TYPE II SS             F   PROB > F

INTERCEPT  10.63390350
ABOR        2.39239214      0.25722777  2007.6811983         86.50     0.0001

BOUNDS ON CONDITION NUMBER:             1,          1
-----------------------------------------------------------------------------
STEP 2       VARIABLE CAPP ENTERED                      R SQUARE = 0.78386266
                                                          C(P) = 30.64827058

                    DF  SUM OF SQUARES   MEAN SQUARE             F   PROB > F

REGRESSION           2   2756.29511365  1378.1475568        116.05     0.0001
ERROR               64    760.00339381    11.8750530
TOTAL               66   3516.29850746

               B VALUE       STD ERROR    TYPE II SS             F   PROB > F

INTERCEPT   2.28313643
CAPP        2.48235217      0.31264553   748.6139154         63.04     0.0001
ABOR        2.55779778      0.18516922  2265.8458299        190.81     0.0001

BOUNDS ON CONDITION NUMBER:             1.01282,
                                        4.051278
-----------------------------------------------------------------------------
STEP 3       VARIABLE CENS ENTERED                      R SQUARE = 0.83687406
                                                          C(P) = 10.16995879

                    DF  SUM OF SQUARES   MEAN SQUARE             F   PROB > F

REGRESSION           3   2942.69899378  980.89966459        107.73     0.0001
ERROR               63    573.59951368    9.10475419
TOTAL               66   3516.29850746
```

continued

Table 14.4 Continued

	B VALUE	STD ERROR	TYPE II SS	F	PROB > F
INTERCEPT	−0.82773845				
CENS	1.59823760	0.35322219	186.4038801	20.47	0.0001
CAPP	1.81212925	0.31126325	308.5964051	33.89	0.0001
ABOR	2.46070774	0.16355183	2060.9976536	226.36	0.0001

BOUNDS ON CONDITION NUMBER: 1.309335, 10.91348

STEP 4 VARIABLE CONS ENTERED R SQUARE = 0.84827179
C(P) = 7.33700054

	DF	SUM OF SQUARES	MEAN SQUARE	F	PROB > F
REGRESSION	4	2982.77683321	745.69420830	86.66	0.0001
ERROR	62	533.52167426	8.60518829		
TOTAL	66	3516.29850746			

	B VALUE	STD ERROR	TYPE II SS	F	PROB > F
INTERCEPT	−1.57963264				
CENS	1.44292853	0.35085499	145.5441367	16.91	0.0001
CAPP	1.55217744	0.32569620	195.4419285	22.71	0.0001
CONS	0.91434287	0.42367918	40.0778394	4.66	0.0348
ABOR	2.44618610	0.15914391	2033.1026734	236.26	0.0001

BOUNDS ON CONDITION NUMBER: 1.516799, 21.0738

STEP 5 VARIABLE RDIE ENTERED R SQUARE = 0.85614156
C(P) = 6.00000000

	DF	SUM OF SQUARES	MEAN SQUARE	F	PROB > F
REGRESSION	5	3010.44927833	602.08985567	72.61	0.0001
ERROR	61	505.84922913	8.29261031		
TOTAL	66	3516.29850746			

	B VALUE	STD ERROR	TYPE II SS	F	PROB > F
INTERCEPT	−0.96098181				
CENS	1.40981955	0.34490030	138.5578569	16.71	0.0001
CAPP	1.45120569	0.32446883	165.8835156	20.00	0.0001
CONS	0.97003442	0.41702892	44.8676379	5.41	0.0234
RDIE	−0.02914917	0.01595688	27.6724451	3.34	0.0726
ABOR	2.40255343	0.15804215	1916.4236580	231.10	0.0001

BOUNDS ON CONDITION NUMBER: 1.562133, 32.06837

NO OTHER VARIABLES MET THE 0.1500 SIGNIFICANCE LEVEL FOR ENTRY

B VALUE you will find the same information as was in the column PARAMETER ESTIMATES in the REG procedure done earlier.

In step 1, therefore, we see that R SQUARE = 0.57 (we again report only to two decimal places). This 0.57 in the first step is the simple r between JLIB and ABOR. The intercept is 10.63 and the slope of ABOR is 2.39. Also note that in each step each slope (but not the intercept) is tested for significance. STEPWISE does this with F rather than t, but the probabilities are identical to those in REG. We construct from the data the first equation.

$$\text{JLIB} = 10.63 + 2.39\ \text{ABOR}$$

For step 2, CAPP is entered and R^2 increases from .57 in step 1 to .78. The new intercept is 2.28, the slope for CAPP is 2.48, and the new slope for ABOR is 2.55. (In later steps you will see that in the list of independent variables, the variables are not listed in the same order as they were brought into the model; rather they are listed in the order in which they were specified when the run was set up.)

Following is a summary of the regression equations for each step and with the independent variables listed in the order in which they were brought in.

Step	R^2	*Equation*
1	.57	JLIB = 10.63 + 2.39 ABOR
2	.78	JLIB = 2.28 + 2.55 ABOR + 2.48 CAPP
3	.83	JLIB = −0.82 + 2.46 ABOR + 1.81 CAPP + 1.59 CENS
4	.84	JLIB = −1.57 + 2.44 ABOR + 1.55 CAPP + 1.44 CENS + 0.91 CONS
5	.85	JLIB = −0.96 + 2.40 ABOR + 1.45 CAPP + 1.40 CENS + 0.97 CONS − 0.02 RDIE

As before, RDIE in step 5 is not statistically significant, $p = .0726$. Additionally, CONS, while significant in both step 4 ($p = .0348$) and step 5 ($p = .0234$), has a much higher probability of error than the other three independent variables. Also, R^2 reaches .83 in step 3 and grows very little in the subsequent steps. For these reasons, we select the model in step 3 as our "working model" for further analysis. This is why this model was given to you earlier in the chapter as the "best" model for our example.

A word of caution here: Because the stepwise procedure is based on statistical considerations alone and requires no theoretical justification for the independent variables it includes or excludes from entry, we know what was and was not included, but not why. Perhaps consumer protection and right to die issues have not been under judicial consideration long enough for legal scholars to see them in a liberal/conservative continuum. Perhaps these issues are judged on very situation-specific considerations. Whatever the reasons, this procedure may raise more questions than it answers, a weakness of many techniques where many variables are manipulated simultaneously.

At the same time, the questions raised may in turn give rise to theory. How many cases in a given issue area must be considered before the area itself becomes subject to judicial ideology? What other cross-cutting values may be present in the judges' minds? The stepwise procedure is like a fishing expedition. We cast our nets into waters that are murky, because we have little theory to guide us. We collect whatever our nets entangle not knowing until later if the fish we caught are even edible. Still, it is fun to go fishing and sometimes we make a good catch. Many scholars, however, are mistrustful of "fishing expeditions." They would rather the researcher enter in the variables, basing the selection of the variables to be entered on a theoretical foundation rather than the "whims" of the computer.

CONCLUSION

Multiple regression is the first of several techniques for studying large numbers of variables. For that reason it is often excluded from introductory statistics and methodology texts. Yet its usage in political, social, and economic analysis is so widespread that it is to your advantage to know about this procedure.

This means that you are at the stage where you must depend on a computer to do the number crunching for you. Although it is a pleasure to let the calculations go, that pleasure is soon replaced by the pain (or at least the challenge) of figuring out what the data mean. Why, for instance, in our simulated data, was there no correlation to speak of between a judge's position on the right to die issue and the judge's overall judicial ideology? Also, why did the consumer protection factor appear not to relate to JLIB? Sometimes it's easier to grind out answers and have your instructor tell you whether or not they are right than it is to look at the answers on a printout and figure out their meaning!

Beyond multiple regression are a variety of other multivariate techniques with such strange names as cluster analysis, factor analysis,

PROBIT, and LISREL. And there are others on the way. Each new technique opens up more vistas for data analysis but also poses new problems. Some techniques are tried and discarded. Sometimes they are resurrected later and sometimes not. Some, like multiple regression, have staying power. You are likely to encounter these multivariate techniques again should you pursue additional coursework.

If this will be your last encounter with a statistics course, I hope you have gained an appreciation of the fact that much of the content you have covered is a venture into applied logic. The numbers are only the symbols used, the language in which the logic is applied and communicated. Perhaps you will be aware that in many endeavors you are applying the same logical process used here, whether you are buying a car, selecting a sofa, or listening to some expert tell you that some scientific finding is statistically significant. You can now remind that expert that there exists at least some probability that the finding could be wrong.

EXERCISES

Exercise 14.1

Using Table 14.1, test the following correlations for statistical significance. Assume nondirectional alternative hypotheses, unless told otherwise.

1. $r = -.22$, $n = 10$
2. $r = .49$, $n = 25$
3. $r = .83$, $n = 17$
4. $r = -.10$, $n = 80$
5. $r = -.36$, $n = 100$
6. $r = .32$, $n = 45$
7. $r = .32$, $n = 45$, one-tailed H_1
8. $r = -.70$, $n = 6$, directional H_1
9. $r = -.19$, $n = 102$
10. $r = .50$, $n = 12$, directional H_1

Exercise 14.2

Calculate all partial correlations and derive the most logical causal model for each of the following.

1. x_1 = Age, x_2 = Social Conservatism, x_3 = Religiosity. $r_{12} = .93$, $r_{13} = .64$, and $r_{23} = .59$.
2. x_1 = Age, x_2 = Pro-Choice Attitude Scale, x_3 = Religiosity. $r_{12} = -.16$, $r_{13} = .90$, and $r_{23} = -.18$.
3. x_1 = Conservatism, x_2 = Pro-Life Attitude Scale, x_3 = Pacifism (Antiwar Attitude Scale). $r_{12} = .58$, $r_{13} = -.25$, and $r_{23} = .02$.
4. x_1 = Socioeconomic Status, x_2 = Support for Gun Control Legislation, x_3 = Support for Capital Punishment. $r_{12} = .58$, $r_{13} = -.25$, and $r_{23} = -.43$.

Exercise 14.3

Calculate and interpret the coefficients of multiple determination for predicting x_1 from x_2 and x_3 (that is, $R^2_{1.23}$) for each of the four data sets presented in Exercise 14.2.

Exercise 14.4

A researcher is studying the factors that make younger voters support a given candidate in an election. The higher one's affinity score for a candidate, the more likely one will be to vote for the candidate. Following are the variables studied.

x_1 = Candidate Affinity (0, Low to 20, High).

x_2 = Partisanship (0, the voter totally opposes the candidate's party to 20, the voter totally supports the candidate's party).

x_3 = The voter's years of formal schooling.

x_4 = The voter's age.

Here are summary statistics:

$\bar{x}_1 = 10.0$	$s_1 = 2.510$
$\bar{x}_2 = 9.9$	$s_2 = 3.131$
$\bar{x}_3 = 12.5$	$s_3 = 5.252$
$\bar{x}_4 = 20.5$	$s_4 = 17.194$

The multiple regression formula for predicting x_1 from the other variables was obtained by computer.

$$x_1 = 2.756 + .780x_2 - .027x_3 - .007x_4$$

EXERCISES

1. Use the formula to predict x_1 when x_2, x_3, and x_4 are at their respective mean values. How close does it come to $\bar{x}_1$?
2. Predict x_1 when x_2, x_3, and x_4 are all 0.
3. Generate the betas for x_2, x_3, and x_4 and interpret their meanings.
4. Interpret the meaning of $b_2 = .780$ and compare it to your interpretation of the corresponding beta.
5. What is the most important independent variable? What is the least important?
6. What other variables would you personally like to add to the model for possible explanation of candidate affinity? (Remember that these must be subject to interval level measurement to work in a regression.)

Exercise 14.5

An economist is trying to predict annual increases in the CPI, the Consumer Price Index, which indicates the rate of inflatation. The independent variables used are MONEYMKT, the average annualized yields from money market funds; HOME, the change in the average home price from year to year; and WHEAT, the year to year change in wheat futures (an indicator of anticipated changes in the cost of that commodity). A computer generates the table on the following page.

In effect, the model shows the "contribution" to the CPI of three different types of commodities: the "cost" of money (interest), the cost of housing, and the cost of food.

1. Using only the first two decimal places, write out the regression formula for predicting CPI.
2. What will be the predicted change in the CPI in a year when money market yields, home prices, and wheat future prices do not change?
3. Predict CPI for a year when money market yields are up 7% (plug in 7, not .07), housing prices increase by 4%, and wheat futures decline 3%.
4. Predict CPI for a year when money market yields decline 3%, housing prices increase 2%, and wheat futures rise 10%.
5. Examine the betas (look at the column titled STANDARDIZED ESTIMATE) and interpret them. What has the greatest impact on CPI? The second greatest? The least?

Table E14.5 The SAS System

MODEL: MODEL 1
DEPENDENT VARIABLE: CPI

ANALYSIS OF VARIANCE

SOURCE	DF	SUM OF SQUARES	MEAN SQUARE	F VALUE	PROB > F
MODEL	3	139.80704	46.60235	103.832	0.0001
ERROR	6	2.69296	0.44883		
C TOTAL	9	142.50000			

ROOT MSE	0.66994	R-SQUARE	0.9811
DEP MEAN	6.50000	ADJ R-SQ	0.9717
C.V.	10.30684		

PARAMETER ESTIMATES

VARIABLE	DP	PARAMETER ESTIMATE	STANDARD ERROR	T FOR H0: PARAMETER = 0	PROB > |T|	STANDARDIZED ESTIMATE
INTERCEP	1	-3.839252	0.74232459	-5.172	0.0021	0.00000000
MONEYMKT	1	0.687741	0.07243392	9.495	0.0001	0.60121698
HOME	1	0.459485	0.07263228	6.326	0.0007	0.52155505
WHEAT	1	0.088931	0.01873714	4.746	0.0032	0.40916409

Appendix 1 Continued—Proportions of Area Under Standard Normal Curve

A	B	C	A	B	C	A	B	C
z			z			z		
0.00	.0000	.5000	0.42	.1628	.3372	0.84	.2995	.2005
0.01	.0040	.4960	0.43	.1664	.3336	0.85	.3023	.1977
0.02	.0080	.4920	0.44	.1700	.3300	0.86	.3051	.1949
0.03	.0120	.4880	0.45	.1736	.3264	0.87	.3078	.1922
0.04	.0160	.4840	0.46	.1772	.3228	0.88	.3106	.1894
0.05	.0199	.4801	0.47	.1808	.3192	0.89	.3133	.1867
0.06	.0239	.4761	0.48	.1844	.3156	0.90	.3159	.1841
0.07	.0279	.4721	0.49	.1879	.3121	0.91	.3186	.1814
0.08	.0319	.4681	0.50	.1915	.3085	0.92	.3212	.1788
0.09	.0359	.4641	0.51	.1950	.3050	0.93	.3238	.1762
0.10	.0398	.4602	0.52	.1985	.3015	0.94	.3264	.1736
0.11	.0438	.4562	0.53	.2019	.2981	0.95	.3289	.1711
0.12	.0478	.4522	0.54	.2054	.2946	0.96	.3315	.1685
0.13	.0517	.4483	0.55	.2088	.2912	0.97	.3340	.1660
0.14	.0557	.4443	0.56	.2123	.2877	0.98	.3365	.1635
0.15	.0596	.4404	0.57	.2157	.2843	0.99	.3389	.1611
0.16	.0636	.4364	0.58	.2190	.2810	1.00	.3413	.1587
0.17	.0675	.4325	0.59	.2224	.2776	1.01	.3438	.1562
0.18	.0714	.4286	0.60	.2257	.2743	1.02	.3461	.1539
0.19	.0753	.4247	0.61	.2291	.2709	1.03	.3485	.1515
0.20	.0793	.4207	0.62	.2324	.2676	1.04	.3508	.1492
0.21	.0832	.4168	0.63	.2357	.2643	1.05	.3531	.1469
0.22	.0871	.4129	0.64	.2389	.2611	1.06	.3554	.1446
0.23	.0910	.4090	0.65	.2422	.2578	1.07	.3577	.1423
0.24	.0948	.4052	0.66	.2454	.2546	1.08	.3599	.1401
0.25	.0987	.4013	0.67	.2486	.2514	1.09	.3621	.1379
0.26	.1026	.3974	0.68	.2517	.2483	1.10	.3643	.1357
0.27	.1064	.3936	0.69	.2549	.2451	1.11	.3665	.1335
0.28	.1103	.3897	0.70	.2580	.2420	1.12	.3686	.1314
0.29	.1141	.3859	0.71	.2611	.2389	1.13	.3708	.1292
0.30	.1179	.3821	0.72	.2642	.2358	1.14	.3729	.1271
0.31	.1217	.3783	0.73	.2673	.2327	1.15	.3749	.1251
0.32	.1255	.3745	0.74	.2704	.2296	1.16	.3770	.1230
0.33	.1293	.3707	0.75	.2734	.2266	1.17	.3790	.1210
0.34	.1331	.3669	0.76	.2764	.2236	1.18	.3810	.1190
0.35	.1368	.3632	0.77	.2794	.2206	1.19	.3830	.1170
0.36	.1406	.3594	0.78	.2823	.2177	1.20	.3849	.1151
0.37	.1443	.3557	0.79	.2852	.2148	1.21	.3869	.1131
0.38	.1480	.3520	0.80	.2881	.2119	1.22	.3888	.1112
0.39	.1517	.3483	0.81	.2910	.2090	1.23	.3907	.1093
0.40	.1554	.3446	0.82	.2939	.2061	1.24	.3925	.1075
0.41	.1591	.3409	0.83	.2967	.2033	1.25	.3944	.1056
A	B	C	A	B	C	A	B	C
−z			−z			−z		

continued

Appendix 1 Continued—Proportions of Area Under Standard Normal Curve

A z	B	C	A z	B	C	A z	B	C
1.26	.3962	.1038	1.68	.4535	.0465	2.10	.4821	.0179
1.27	.3980	.1020	1.69	.4545	.0455	2.11	.4826	.0174
1.28	.3997	.1003	1.70	.4554	.0446	2.12	.4830	.0170
1.29	.4015	.0985	1.71	.4564	.0436	2.13	.4834	.0166
1.30	.4032	.0968	1.72	.4573	.0427	2.14	.4838	.0162
1.31	.4049	.0951	1.73	.4582	.0418	2.15	.4842	.0158
1.32	.4066	.0934	1.74	.4591	.0409	2.16	.4846	.0154
1.33	.4082	.0918	1.75	.4599	.0401	2.17	.4850	.0150
1.34	.4099	.0901	1.76	.4608	.0392	2.18	.4854	.0146
1.35	.4115	.0885	1.77	.4616	.0384	2.19	.4857	.0143
1.36	.4131	.0869	1.78	.4625	.0375	2.20	.4861	.0139
1.37	.4147	.0853	1.79	.4633	.0367	2.21	.4864	.0136
1.38	.4162	.0838	1.80	.4641	.0359	2.22	.4868	.0132
1.39	.4177	.0823	1.81	.4649	.0351	2.23	.4871	.0129
1.40	.4192	.0808	1.82	.4656	.0344	2.24	.4875	.0125
1.41	.4207	.0793	1.83	.4664	.0336	2.25	.4878	.0122
1.42	.4222	.0778	1.84	.4671	.0329	2.26	.4881	.0119
1.43	.4236	.0764	1.85	.4678	.0322	2.27	.4884	.0116
1.44	.4251	.0749	1.86	.4686	.0314	2.28	.4887	.0113
1.45	.4265	.0735	1.87	.4693	.0307	2.29	.4890	.0110
1.46	.4279	.0721	1.88	.4699	.0301	2.30	.4893	.0107
1.47	.4292	.0708	1.89	.4706	.0294	2.31	.4896	.0104
1.48	.4306	.0694	1.90	.4713	.0287	2.32	.4898	.0102
1.49	.4319	.0681	1.91	.4719	.0281	2.33	.4901	.0099
1.50	.4332	.0668	1.92	.4726	.0274	2.34	.4904	.0096
1.51	.4345	.0655	1.93	.4732	.0268	2.35	.4906	.0094
1.52	.4357	.0643	1.94	.4738	.0262	2.36	.4909	.0091
1.53	.4370	.0630	1.95	.4744	.0256	2.37	.4911	.0089
1.54	.4382	.0618	1.96	.4750	.0250	2.38	.4913	.0087
1.55	.4394	.0606	1.97	.4756	.0244	2.39	.4916	.0084
1.56	.4406	.0594	1.98	.4761	.0239	2.40	.4918	.0082
1.57	.4418	.0582	1.99	.4767	.0233	2.41	.4920	.0080
1.58	.4429	.0571	2.00	.4772	.0228	2.42	.4922	.0078
1.59	.4441	.0559	2.01	.4778	.0222	2.43	.4925	.0075
1.60	.4452	.0548	2.02	.4783	.0217	2.44	.4927	.0073
1.61	.4463	.0537	2.03	.4788	.0212	2.45	.4929	.0071
1.62	.4474	.0526	2.04	.4793	.0207	2.46	.4931	.0069
1.63	.4484	.0516	2.05	.4798	.0202	2.47	.4932	.0068
1.64	.4495	.0505	2.06	.4803	.0197	2.48	.4934	.0066
1.65	.4505	.0495	2.07	.4808	.0192	2.49	.4936	.0064
1.66	.4515	.0485	2.08	.4812	.0188	2.50	.4938	.0062
1.67	.4525	.0475	2.09	.4817	.0183	2.51	.4940	.0060
A −z	B	C	A −z	B	C	A −z	B	C

continued

Appendix1 Continued—Proportions of Area Under Standard Normal Curve

A	B	C	A	B	C	A	B	C
z			z			z		
2.52	.4941	.0059	2.80	.4974	.0026	3.08	.4990	.0010
2.53	.4943	.0057	2.81	.4975	.0025	3.09	.4990	.0010
2.54	.4945	.0055	2.82	.4976	.0024	3.10	.4990	.0010
2.55	.4946	.0054	2.83	.4977	.0023	3.11	.4991	.0009
2.56	.4948	.0052	2.84	.4977	.0023	3.12	.4991	.0009
2.57	.4949	.0051	2.85	.4978	.0022	3.13	.4991	.0009
2.58	.4951	.0049	2.86	.4979	.0021	3.14	.4992	.0008
2.59	.4952	.0048	2.87	.4979	.0021	3.15	.4992	.0008
2.60	.4953	.0047	2.88	.4980	.0020	3.16	.4992	.0008
2.61	.4955	.0045	2.89	.4981	.0019	3.17	.4992	.0008
2.62	.4956	.0044	2.90	.4981	.0019	3.18	.4993	.0007
2.63	.4957	.0043	2.91	.4982	.0018	3.19	.4993	.0007
2.64	.4959	.0041	2.92	.4982	.0018	3.20	.4993	.0007
2.65	.4960	.0040	2.93	.4983	.0017	3.21	.4993	.0007
2.66	.4961	.0039	2.94	.4984	.0016	3.22	.4994	.0006
2.67	.4962	.0038	2.95	.4984	.0016	3.23	.4994	.0006
2.68	.4963	.0037	2.96	.4985	.0015	3.24	.4994	.0006
2.69	.4964	.0036	2.97	.4985	.0015	3.25	.4994	.0006
2.70	.4965	.0035	2.98	.4986	.0014	3.30	.4995	.0005
2.71	.4966	.0034	2.99	.4986	.0014	3.35	.4996	.0004
2.72	.4967	.0033	3.00	.4987	.0013	3.40	.4997	.0003
2.73	.4968	.0032	3.01	.4987	.0013	3.45	.4997	.0003
2.74	.4969	.0031	3.02	.4987	.0013	3.50	.4998	.0002
2.75	.4970	.0030	3.03	.4988	.0012	3.60	.4998	.0002
2.76	.4971	.0029	3.04	.4988	.0012	3.70	.4999	.0001
2.77	.4972	.0028	3.05	.4989	.0011	3.80	.4999	.0001
2.78	.4973	.0027	3.06	.4989	.0011	3.90	.49995	.00005
2.79	.4974	.0026	3.07	.4989	.0011	4.00	.49997	.00003
A	B	C	A	B	C	A	B	C
−z			−z			−z		

SOURCE: Appendix 1 is abridged from Table III of R. A. Fisher and F. Yates, *Statistical Tables for Biological, Agriculture and Medical Research* (6th ed.), published by Longman Group UK Ltd., 1974, by permission of the authors and the publisher.

Appendix 2 Distribution of t

	Level of significance for one-tailed test					
	.10	*.05*	*.025*	*.01*	*.005*	*.0005*
	Level of significance for two-tailed test					
df	*.20*	*.10*	*.05*	*.02*	*.01*	*.001*
1	3.078	6.314	12.706	31.821	63.657	636.619
2	1.886	2.920	4.303	6.965	9.925	31.598
3	1.638	2.353	3.182	4.541	5.841	12.941
4	1.533	2.132	2.776	3.747	4.604	8.610
5	1.476	2.015	2.571	3.365	4.032	6.859
6	1.440	1.943	2.447	3.143	3.707	5.959
7	1.415	1.895	2.365	2.998	3.499	5.405
8	1.397	1.860	2.306	2.896	3.355	5.041
9	1.383	1.833	2.262	2.821	3.250	4.781
10	1.372	1.812	2.228	2.764	3.169	4.587
11	1.363	1.796	2.201	2.718	3.106	4.437
12	1.356	1.782	2.179	2.681	3.055	4.318
13	1.350	1.771	2.160	2.650	3.012	4.221
14	1.345	1.761	2.145	2.624	2.977	4.140
15	1.341	1.753	2.131	2.602	2.947	4.073
16	1.337	1.746	2.120	2.583	2.921	4.015
17	1.333	1.740	2.110	2.567	2.898	3.965
18	1.330	1.734	2.101	2.552	2.878	3.922
19	1.328	1.729	2.093	2.539	2.861	3.883
20	1.325	1.725	2.086	2.528	2.845	3.850
21	1.323	1.721	2.080	2.518	2.831	3.819
22	1.321	1.717	2.074	2.508	2.819	3.792
23	1.319	1.714	2.069	2.500	2.807	3.767
24	1.318	1.711	2.064	2.492	2.797	3.745
25	1.316	1.708	2.060	2.485	2.787	3.725
26	1.315	1.706	2.056	2.479	2.779	3.707
27	1.314	1.703	2.052	2.473	2.771	3.690
28	1.313	1.701	2.048	2.467	2.763	3.674
29	1.311	1.699	2.045	2.462	2.756	3.659
30	1.310	1.697	2.042	2.457	2.750	3.646
40	1.303	1.684	2.021	2.423	2.704	3.551
60	1.296	1.671	2.000	2.390	2.660	3.460
120	1.289	1.658	1.980	2.358	2.617	3.373
∞	1.282	1.645	1.960	2.326	2.576	3.291

SOURCE: Appendix 2 is abridged from Table III of R. A. Fisher and F. Yates, *Statistical Tables for Biological, Agriculture and Medical Research* (6th ed.), published by Longman Group UK Ltd., 1974, by permission of the authors and the publisher.

Appendix 3 Critical Values of F for $p = .05$

$n_2 \backslash n_1$	1	2	3	4	5	6	8	12	24	∞
1	161.40	199.50	215.70	224.60	230.20	234.00	238.90	243.90	249.00	254.30
2	18.51	19.00	19.16	19.25	19.30	19.33	19.37	19.41	19.45	19.50
3	10.13	9.55	9.28	9.12	9.01	8.94	8.84	8.74	8.64	8.53
4	7.71	6.94	6.59	6.39	6.26	6.16	6.04	5.91	5.77	5.63
5	6.61	5.79	5.41	5.19	5.05	4.95	4.82	4.68	4.53	4.36
6	5.99	5.14	4.76	4.53	4.39	4.28	4.15	4.00	3.84	3.67
7	5.59	4.74	4.35	4.12	3.97	3.87	3.73	3.57	3.41	3.23
8	5.32	4.46	4.07	3.84	3.69	3.58	3.44	3.28	3.12	2.93
9	5.12	4.26	3.86	3.63	3.48	3.37	3.23	3.07	2.90	2.71
10	4.96	4.10	3.71	3.48	3.33	3.22	3.07	2.91	2.74	2.54
11	4.84	3.98	3.59	3.36	3.20	3.09	2.95	2.79	2.61	2.40
12	4.75	3.88	3.49	3.26	3.11	3.00	2.85	2.69	2.50	2.30
13	4.67	3.80	3.41	3.18	3.02	2.92	2.77	2.60	2.42	2.21
14	4.60	3.74	3.34	3.11	2.96	2.85	2.70	2.53	2.35	2.13
15	4.54	3.68	3.29	3.06	2.90	2.79	2.64	2.48	2.29	2.07
16	4.49	3.63	3.24	3.01	2.85	2.74	2.59	2.42	2.24	2.01
17	4.45	3.59	3.20	2.96	2.81	2.70	2.55	2.38	2.19	1.96
18	4.41	3.55	3.16	2.93	2.77	2.66	2.51	2.34	2.15	1.92
19	4.38	3.52	3.13	2.90	2.74	2.63	2.48	2.31	2.11	1.88
20	4.35	3.49	3.10	2.87	2.71	2.60	2.45	2.28	2.08	1.84
21	4.32	3.47	3.07	2.84	2.68	2.57	2.42	2.25	2.05	1.81
22	4.30	3.44	3.05	2.82	2.66	2.55	2.40	2.23	2.03	1.78
23	4.28	3.42	3.03	2.80	2.64	2.53	2.38	2.20	2.00	1.76
24	4.26	3.40	3.01	2.78	2.62	2.51	2.36	2.18	1.98	1.73
25	4.24	3.38	2.99	2.76	2.60	2.49	2.34	2.16	1.96	1.71
26	4.22	3.37	2.98	2.74	2.59	2.47	2.32	2.15	1.95	1.69
27	4.21	3.35	2.96	2.73	2.57	2.46	2.30	2.13	1.93	1.67
28	4.20	3.34	2.95	2.71	2.56	2.44	2.29	2.12	1.91	1.65
29	4.18	3.33	2.93	2.70	2.54	2.43	2.28	2.10	1.90	1.64
30	4.17	3.32	2.92	2.69	2.53	2.42	2.27	2.09	1.89	1.62
40	4.08	3.23	2.84	2.61	2.45	2.34	2.18	2.00	1.79	1.51
60	4.00	3.15	2.76	2.52	2.37	2.25	2.10	1.92	1.70	1.39
120	3.92	3.07	2.68	2.45	2.29	2.17	2.02	1.83	1.61	1.25
∞	3.84	2.99	2.60	2.37	2.21	2.09	1.94	1.75	1.52	1.00

SOURCE: Appendix 3 is abridged from Table V of R. A. Fisher and F. Yates, *Statistical Tables for Biological, Agricultural and Medical Research* (6th ed.), published by Longman Group UK, Ltd., 1974, by permission of the authors and the publisher.
NOTE: Values of n_1 and n_2 represent the degrees of freedom associated with the larger and smaller estimates of variance, respectively.

Appendix 3 Continued—Critical Values of *F* for $p = .01$

$n_2 \backslash n_1$	1	2	3	4	5	6	8	12	24	∞
1	4052	4999	5403	5625	5764	5859	5981	6106	6234	6366
2	98.49	99.01	99.17	99.25	99.30	99.33	99.36	99.42	99.46	99.50
3	34.12	30.81	29.46	28.71	28.24	27.91	27.49	27.05	26.60	26.12
4	21.20	18.00	16.69	15.98	15.52	15.21	14.80	14.37	13.93	13.46
5	16.26	13.27	12.06	11.39	10.97	10.67	10.27	9.89	9.47	9.02
6	13.74	10.92	9.78	9.15	8.75	8.47	8.10	7.72	7.31	6.88
7	12.25	9.55	8.45	7.85	7.46	7.19	6.84	6.47	6.07	5.65
8	11.26	8.65	7.59	7.01	6.63	6.37	6.03	5.67	5.28	4.86
9	10.56	8.02	6.99	6.42	6.06	5.80	5.47	5.11	4.73	4.31
10	10.04	7.56	6.55	5.99	5.64	5.39	5.06	4.71	4.33	3.91
11	9.65	7.20	6.22	5.67	5.32	5.07	4.74	4.40	4.02	3.60
12	9.33	6.93	5.95	5.41	5.06	4.82	4.50	4.16	3.78	3.36
13	9.07	6.70	5.74	5.20	4.86	4.62	4.30	3.96	3.59	3.16
14	8.86	6.51	5.56	5.03	4.69	4.46	4.14	3.80	3.43	3.00
15	8.68	6.36	5.42	4.89	4.56	4.32	4.00	3.67	3.29	2.87
16	8.53	6.23	5.29	4.77	4.44	4.20	3.89	3.55	3.18	2.75
17	8.40	6.11	5.18	4.67	4.34	4.10	3.79	3.45	3.08	2.65
18	8.28	6.01	5.09	4.58	4.25	4.01	3.71	3.37	3.00	2.57
19	8.18	5.93	5.01	4.50	4.17	3.94	3.63	3.30	2.92	2.49
20	8.10	5.85	4.94	4.43	4.10	3.87	3.56	3.23	2.86	2.42
21	8.02	5.78	4.87	4.37	4.04	3.81	3.51	3.17	2.80	2.36
22	7.94	5.72	4.82	4.31	3.99	3.76	3.45	3.12	2.75	2.31
23	7.88	5.66	4.76	4.26	3.94	3.71	3.41	3.07	2.70	2.26
24	7.82	5.61	4.72	4.22	3.90	3.67	3.36	3.03	2.66	2.21
25	7.77	5.57	4.68	4.18	3.86	3.63	3.32	2.99	2.62	2.17
26	7.72	5.53	4.64	4.14	3.82	3.59	3.29	2.96	2.58	2.13
27	7.68	5.49	4.60	4.11	3.78	3.56	3.26	2.93	2.55	2.10
28	7.64	5.45	4.57	4.07	3.75	3.53	3.23	2.90	2.52	2.06
29	7.60	5.42	4.54	4.04	3.73	3.50	3.20	2.87	2.49	2.03
30	7.56	5.39	4.51	4.02	3.70	3.47	3.17	2.84	2.47	2.01
40	7.31	5.18	4.31	3.83	3.51	3.29	2.99	2.66	2.29	1.80
60	7.08	4.98	4.13	3.65	3.34	3.12	2.82	2.50	2.12	1.60
120	6.85	4.79	3.95	3.48	3.17	2.96	2.66	2.34	1.95	1.38
∞	6.64	4.60	3.78	3.32	3.02	2.80	2.51	2.18	1.79	1.00

SOURCE: Appendix 3 is abridged from Table V of R. A. Fisher and F. Yates, *Statistical Tables for Biological, Agricultural and Medical Research* (6th ed.), published by Longman Group UK, Ltd., 1974, by permission of the authors and the publisher.
NOTE: Values of n_1 and n_2 represent the degrees of freedom associated with the larger and smaller estimates of variance, respectively.

Appendix 3 Continued—Critical Values of F for $p = .001$

$n_2 \backslash n_1$	1	2	3	4	5	6	8	12	24	∞
1	405284	500000	540379	562500	576405	585937	598144	610667	623497	636619
2	998.5	999.0	999.2	999.2	999.3	999.3	999.4	999.4	999.5	999.5
3	167.5	148.5	141.1	137.1	134.6	132.8	130.6	128.3	125.9	123.5
4	74.14	61.25	56.18	53.44	51.71	50.53	49.00	47.41	45.77	44.05
5	47.04	36.61	33.20	31.09	29.75	28.84	27.64	26.42	25.14	23.78
6	35.51	27.00	23.70	21.90	20.81	20.03	19.03	17.99	16.89	15.75
7	29.22	21.69	18.77	17.19	16.21	15.52	14.63	13.71	12.73	11.69
8	25.42	18.49	15.83	14.39	13.49	12.86	12.04	11.19	10.30	9.34
9	22.86	16.39	13.90	12.56	11.71	11.13	10.37	9.57	8.72	7.81
10	21.04	14.91	12.55	11.28	10.48	9.92	9.20	8.45	7.64	6.76
11	19.69	13.81	11.56	10.35	9.58	9.05	8.35	7.63	6.85	6.00
12	18.64	12.97	10.80	9.63	8.89	8.38	7.71	7.00	6.25	5.42
13	17.81	12.31	10.21	9.07	8.35	7.86	7.21	6.52	5.78	4.97
14	17.14	11.78	9.73	8.62	7.92	7.43	6.80	6.13	5.41	4.60
15	16.59	11.34	9.34	8.25	7.57	7.09	6.47	5.81	5.10	4.31
16	16.12	10.97	9.00	7.94	7.27	6.81	6.19	5.55	4.85	4.06
17	15.72	10.66	8.73	7.68	7.02	6.56	5.96	5.32	4.63	3.85
18	15.38	10.39	8.49	7.46	6.81	6.35	5.76	5.13	4.45	3.67
19	15.08	10.16	8.28	7.26	6.61	6.18	5.59	4.97	4.29	3.52
20	14.82	9.95	8.10	7.10	6.46	6.02	5.44	4.82	4.15	3.38
21	14.59	9.77	7.94	6.95	6.32	5.88	5.31	4.70	4.03	3.26
22	14.38	9.61	7.80	6.81	6.19	5.76	5.19	4.58	3.92	3.15
23	14.19	9.47	7.67	6.69	6.08	5.65	5.09	4.48	3.82	3.05
24	14.03	9.34	7.55	6.59	5.98	5.55	4.99	4.39	3.74	2.97
25	13.88	9.22	7.45	6.49	5.88	5.46	4.91	4.31	3.66	2.89
26	13.74	9.12	7.36	6.41	5.80	5.38	4.83	4.24	3.59	2.82
27	13.61	9.02	7.27	6.33	5.73	5.31	4.76	4.17	3.52	2.75
28	13.50	8.93	7.19	6.25	5.66	5.24	4.69	4.11	3.46	2.70
29	13.39	8.85	7.12	6.19	5.59	5.18	4.64	4.05	3.41	2.64
30	13.29	8.77	7.05	6.12	5.53	5.12	4.58	4.00	3.36	2.59
40	12.61	8.25	6.60	5.70	5.13	4.73	4.21	3.64	3.01	2.23
60	11.97	7.76	6.17	5.31	4.76	4.37	3.87	3.31	2.69	1.90
120	11.38	7.31	5.79	4.95	4.42	4.04	3.55	3.02	2.40	1.56
∞	10.83	6.91	5.42	4.62	4.10	3.74	3.27	2.74	2.13	1.00

SOURCE: Appendix 3 is abridged from Table V of R. A. Fisher and F. Yates, *Statistical Tables for Biological, Agricultural and Medical Research* (6th ed.), published by Longman Group UK, Ltd., 1974, by permission of the authors and the publisher.
NOTE: Values of n_1 and n_2 represent the degrees of freedom associated with the larger and smaller estimates of variance, respectively.

Appendix 4 Critical Values of Chi-Square

df	*.10*	*.05*	*.01*	*.001*
1	2.71	3.84	6.64	10.83
2	4.60	5.99	9.21	13.82
3	6.25	7.81	11.34	16.27
4	7.78	9.49	13.28	18.47
5	9.24	11.07	15.09	20.52
6	10.64	12.59	16.81	22.46
7	12.02	14.07	18.48	24.32
8	13.36	15.51	20.09	26.12
9	14.68	16.92	21.67	27.88
10	15.99	18.31	23.21	29.59
11	17.28	19.68	24.72	31.26
12	18.55	21.03	26.22	32.91
13	19.81	22.36	27.69	34.53
14	21.06	23.68	29.14	36.12
15	22.31	25.00	30.58	37.70
16	23.54	26.30	32.00	39.25
17	24.77	27.59	33.41	40.79
18	25.99	28.87	34.80	42.31
19	27.20	30.14	36.19	43.82
20	28.41	31.41	37.57	45.32
21	29.62	32.67	38.93	46.80
22	30.81	33.92	40.29	48.27
23	32.01	35.17	41.64	49.73
24	33.20	36.42	42.98	51.18
25	34.38	37.65	44.31	52.62
26	35.56	38.88	45.64	54.05
27	36.74	40.11	46.96	55.48
28	37.92	41.34	48.28	56.89
29	39.09	42.56	49.59	58.30
30	40.26	43.77	50.89	59.70
40	51.80	55.76	63.69	73.40
50	63.17	67.50	76.15	86.66
60	74.40	79.08	88.38	99.61
70	85.53	90.53	100.42	112.32

SOURCE: Appendix 4 is abridged from Table V of R. A. Fisher and F. Yates, *Statistical Tables for Biological, Agricultural and Medical Research* (6th ed.), published by Longman Group UK Ltd., 1974, by permission of the authors and the publisher.

Appendix 5 Critical Values of the Correlation Coefficient

	Level of significance for one-tailed test			
	.05	*.025*	*.01*	*.001*
	Level of significance for two-tailed test			
df	*.10*	*.05*	*.02*	*.01*
1	.988	.997	.9995	.9999
2	.900	.950	.980	.990
3	.805	.878	.934	.959
4	.729	.811	.882	.917
5	.669	.754	.833	.874
6	.622	.707	.789	.834
7	.582	.666	.750	.798
8	.549	.632	.716	.765
9	.521	.602	.685	.735
10	.497	.576	.658	.708
11	.476	.553	.634	.684
12	.458	.532	.612	.661
13	.441	.514	.592	.641
14	.426	.497	.574	.623
15	.412	.482	.558	.606
16	.400	.468	.542	.590
17	.389	.456	.528	.575
18	.378	.444	.516	.561
19	.369	.433	.503	.549
20	.360	.423	.492	.537
21	.352	.413	.482	.526
22	.344	.404	.472	.515
23	.337	.396	.462	.505
24	.330	.388	.453	.496
25	.323	.381	.445	.487
26	.317	.374	.437	.479
27	.311	.367	.430	.471
28	.306	.361	.423	.463
29	.301	.355	.416	.456
30	.296	.349	.409	.449
35	.275	.325	.381	.418
40	.257	.304	.358	.393
45	.243	.288	.338	.372
50	.231	.273	.322	.354
60	.211	.250	.295	.325
70	.195	.232	.274	.303
80	.183	.217	.256	.283
90	.173	.205	.242	.267
100	.164	.195	.230	.254

SOURCE: Appendix 5 is abridged from Table VII, p. 63 of R. A. Fisher and F. Yates, *Statistical Tables for Biological, Agricultural and Medical Research* (6th ed.), published by Longman Group UK Ltd., 1974, by permission of the authors and the publisher.

Answers to Selected Exercises

CHAPTER 1

Exercise 1.3 (There could be many other hypotheses here.)

1. There is a relationship between gender and salary, such that—in like occupations—men earn more than women. Dependent variable: salary (in appropriate national currency). Independent variable: gender (male, female). Unit of analysis: individual. I might seek data from a source with well-defined job categories (civil service jobs, for instance, or academic rankings in similar universities). Be careful to consider the length of time each person has held his or her job.
2. There is a relationship between race and access to housing, such that people of African origin have less access than others. Dependent variable: access to housing (high, medium, low). Independent variable: race (African origin, other). Unit of analysis: individual. Look at statistics on integration levels in residential areas.
3. There is a relationship between the language studied by students and the existence of minority linguistic groups in the student's country, such that a greater proportion of students will study the minority language than will study other languages. Dependent variable: language

selected. Independent variable: existence of minority linguistic groups. Units of analysis: individual. Compare the number of U.S. students studying Spanish to the number studying other languages.

4. Interest groups and PACs contribute a lower percentage of campaign contributions where such contributions are restricted by law than they do elsewhere. Independent: campaign contribution regulations (yes, no). Dependent: contributions by interest groups (PACs as a proportion of total campaign contributions). Unit: probably country (assuming those data are available).
5. There is a positive relationship between the size of a country's armed forces and the number of chemical and biological weapons it possesses. Independent: size of armed forces. Dependent: number of chemical and biological weapons. Unit: country. Look at defense statistics collected by NATO or by private organizations.

Exercise 1.4

1. Women is not a variable; it is a category of the variable gender. Also, there is no comparison with men being made. There is a relationship between gender and math anxiety, such that women have higher levels of math anxiety than do men.
2. Don't state the hypothesis as a question. Age and need for social services are positively related.
3. Babies have the lower birth weights, not their smoking mothers. There is a relationship between birth weight of babies and their mother's smoking habit, such that babies of mothers who smoke have lower birth weights than babies of mothers who do not smoke.
4. There is *no* relationship posited. All this says is that *all* British parties support national health insurance. There is a relationship between political party and support for national health insurance, such that the Labour Party supports a wider range of such benefits than do the other parties.

Exercise 1.5

1. Countries.
2. Geographic area.
3. British counties.
4. The kindergarten students in Mrs. Smith's class at Apple Valley Elementary School registered last February 7.
5. Employees.

CHAPTER 2

Exercise 2.1

1. Ordinal, class intervals are of differing sizes with the upper and lower ones also being open-ended.
2. Ratio. All class intervals are closed-ended and all are equal in size.
3. Nominal. This would have been a Likert-type ordinal scale, except that the sequence was broken by listing "strong disagree" above "agree." To make this ordinal, reorder the categories correctly as follows:

Response	$f =$
Strong Agree	25
Agree	20
Unsure	5
Disagree	15
Strongly Disagree	5
Total	70

4. Ordinal. Categories range from most to least ideal. Note that we do not know what they meant by idealism and how it was measured. (More on that in the next chapter.)
5. Ratio. All categories are equal-sized closed-ended class intervals.
6. Ordinal. Upper class interval is open-ended and the other class intervals are of unequal size.
7. Ordinal. The categories are intended to range from most to least censorship. Note that the unit of analysis here is probably country or nation-state.
8. Nominal. Types of races are listed with no ordering.

Exercise 2.2

1. Ratio level with individual people as the units of analysis.
2. Ordinal. The units of analysis are individuals, probably high school students.
3. Probably nominal level, because not all of Canada's parties fit neatly into an easily defined ideological spectrum. (There is room for disagreement here.) Individual Canadians are the unit of analysis.
4. Ordinal. The numbers are population rankings. Country is the unit of analysis.

CHAPTER 3

There are many possible ways to operationalize these variables. Th following are thoughts or suggestions.

Exercise 3.1

1. Age. You could merely request the respondent to fill in his or he age. But if you fear incorrect information or suspect that olde subjects may lie about their ages, consider asking them to chec an appropriate class interval, such as:

   Age (Please indicate)
   ____ Above 70
   ____ 60 – 69
   ____ 50 – 59
   ____ 40 – 49
   etc.

2. Religion. Make sure you include all relevant faiths appropriate t your audience, as well as categories for "other" and "none." Als consider whether you really want to know the subject's religion c birth or current religion, if they are not the same.
3. One could check an appropriate category, such as:

   ____ Single
   ____ Married
   ____ Divorced
   ____ Widowed

Among your subjects, there may be people cohabiting but not formall married. How will you count them? Are you interested in marriag as a legal status or as a sociological status? What about "marriage between gay men or lesbian women?

4. One could check the appropriate party or (in the United States) a part plus intensity scale. Make sure there are categories for independent and (if needed) supporters of smaller parties, such as the Greens.
5. Pick environmental problems that you deem important: acid rair ozone depletion, nuclear waste, other waste, tropical rain forests, an so on. Also, what do you think "Attitude on environmental problems means? Attitudes about what exists? What should exist? What shoul be done about what exists? Try to differentiate the energy issue fron the pollution issue. Where are they the same? Where do they differ

Exercise 3.2

1. Look for arrests (or lack thereof) for speeches or articles expressing politically unpopular topics. Look for societal intolerance, such as inability to publish articles with unpopular views in the private media. How many political parties and interest groups function in the polity? To what extent are they governmentally controlled? Is there blacklisting or other evidence of persecution?
2. Take several similar crimes and compare the legal penalties from one country to another. See if you can find out for each country the average prison term actually applied in sentencing or the actual years of incarceration by type of crime. To what crimes is the death penalty applied, if any? Is there evidence of torture (legal or not)? Other cruel and unusual penalties?

CHAPTER 4

Exercise 4.1

For the oil exporters, $\bar{x} = 410/7 = 58.57$ and for the nonexporters, $\bar{x} = 266/5 = 53.20$. Men born in the oil-exporting countries have longer life expectancies.

Exercise 4.2

Reordering the scores from high to low:

Oil Exporters		*Nonexporters*	
Qatar	72	Syria	65
Bahrain	68	Lebanon	63
Kuwait	68	Jordan	60
Iraq	56	S. Yemen	41
S. Arabia	54	N. Yemen	37
Oman	51		
U.A.E.	41		

For the oil exporters, *Md. Pos.* $= (n + 1)/2 = (7 + 1)/2 = 8/2 = 4$. The fourth country in the array is Iraq with a male life expectancy of 56. Therefore, *Md.* $= 56$.

For the nonexporters, *Md. Pos.* $= (5 + 1)/2 = 6/2 = 3$. The third country in the array is Jordan with a figure of 60. Thus, *Md.* $= 60$.

ANSWERS

Since 60 is greater than 56, it is the *nonexporting* countries which hav the greater life expectancies, in complete contradiction to th results in Exercise 4.1! (Frustrating, isn't it!)

Exercise 4.3

The figures for the nonexporters are unaffected. For the exporters, th mean now becomes 369/6 or 61.5 years. The median position nov becomes (6 + 1)/2 = 7/2 = 3.5. This position is shared by the thir and fourth countries, Kuwait and Iraq, whose male life expectancie are 68 and 56, respectively. Thus, *Md.* = (68 + 56)/2 = 124/2 = 62 Summarizing:

	Exporters	*Nonexporters*
Mean	61.5	53.2
Median	62.0	60.0

Consistent results are obtained regardless of the measure used. Mal life expectancies in the oil-exporting countries are greater than i the nonexporting countries.

Exercise 4.4

Exporters

$x =$	$f =$	$fx =$	$cf =$
147	1	147	5
127	1	127	4
116	2	232	3
96	1	96	1
	$n = \sum f = 5$	$\sum fx = 602$	

$$\bar{x} = \frac{\sum fx}{n} = \frac{\sum fx}{\sum f} = \frac{602}{5} = 120.40$$

$$Md.\ Pos. = \frac{n+1}{2} = \frac{\sum f + 1}{2} = \frac{5+1}{2} = \frac{6}{2} = 3$$

$$Md. = 116.00$$

Nonexporters

$x =$	$f =$	$fx =$	$cf =$
120	1	120	8
116	1	116	7
115	1	115	6
102	1	102	5
99	2	198	4
86	1	86	2
76	8	76	1
	$\sum f = 8$	$\sum fx = 813$	

$$\bar{x} = \frac{813}{8} = 101.63$$

$$Md.\,Pos. = \frac{8+1}{2} = \frac{9}{2} = 4.5$$

$$Md. = \frac{99+102}{2} = \frac{201}{2} = 100.50$$

Summarizing:

	Exporters	*NonExporters*
Mean	120.40	101.63
Median	116.00	100.50

On the average, citizens of Arab oil-exporting countries have a greater caloric intake (that is, they eat better) than their nonexporting cousins.

Exercise 4.5

1. III
2. VI; this would be a bimodal, symmetric distribution.
3. I
4. V
5. IV
6. II

Exercise 4.6

a. Median
b. Mean

c. Median
d. Mean
e. Mean
f. Median
g. Mean
h. Median

Exercise 4.13

Extroversion
50 - 59 | 2 7 9
40 - 49 | 1 3 5 7 8
30 - 39 | 0 0 3 4 6 7 8 9
20 - 29 | 0 1 2 4 8 8 9
10 - 19 | 0 2 4 7
0 - 9 | 7 8 9

Exercise 4.14

Md. Pos. = 15.5

For 1st Quartile, *Md. Pos.* = 8
For 3rd Quartile, *Md. Pos.* = 8

Md. = 30

1st Quartile = 20
3rd Quartile = 41

CHAPTER 5

Exercise 5.1

1. $\bar{x} = 18.3$
2. *Md.* = 5
3. *M.D.* = 19.63
4. and 5. $s^2 = 5300.01/9 = 588.89$
6. $s = 24.27$

Exercise 5.2

1. $\overline{x} = 79.375$
2. $Md. = 90$
3. $s^2 = 5343.756/16 = 333.985$
4. $s^2 = 333.984$ (Due to rounding in 3 above, this is slightly different.)
5. $s = 18.275$
6. $\overline{x}_{pr} < \overline{x}_{pu}$ Public employees rank it higher.
   $s_{pr} > s_{pu}$ Private employees have greater variability and public employees less variability.

Exercise 5.3

1. $\overline{x} = 61.86$
2. $Md. = 68$
3. $M.D. = 29.02$
4. and 5. $s^2 = 7274.86/7 = 1039.27$
6. $s = 32.24$

CHAPTER 6

Exercise 6.3

Use the middle percentage in each cell, adding to 100% for each *row.* An inverse relationship.

Exercise 6.4

A positive relationship.

Exercise 6.5

GNP is independent. It follows "by" in the title and is the column variable. Phones is dependent. A positive relationship.

Exercise 6.6

Inconsistent. Excluding the VL Deaths row, a positive relationship. But VL Deaths are not in O PDEMS but in M. Later on, we will discuss relationships such as this, which are called curvilinear relationships.

CHAPTER 7

Exercise 7.1

1. H_0: $\mu_{\text{men}} = \mu_{\text{women}}$ and H_1: $\mu_{\text{men}} \neq \mu_{\text{women}}$
3. H_0: $\mu_{\text{France}} = \mu_{\text{USA}}$ and H_1: $\mu_{\text{France}} \neq \mu_{\text{USA}}$

Exercise 7.2

1. Do a one-sample *z* test.
3. Two-sample *t* test.
6. This is a table. Do chi-square.

Exercise 7.3

1. You have population data for both groups. No sampling, so no test.
3. H_1 says all political science majors, but should say all political science majors in the honors program.

Exercise 7.4

1. $z = -2.000$
3. $z = 1.500$
5. $z = -2.286$

Exercise 7.5

1. $p < .001$
3. $p < .01$
5. $p < .01$ Here, actually, $p = .01$.
7. $p < .01$

Exercise 7.6

2. $p < .05$
4. $p < .05$
6. $p < .05$
8. Not significant

Exercise 7.9

1. $z = 12.35$ $p < .001$

CHAPTER 8

Exercise 8.1

1. $z = 1.65$ Area = .0495
3. Area = .5000 + .4505 = .9505
5. Area = .0517
7. Area = .0495 + .4483 = .4978

Exercise 8.2

1. Sigma known. OK to relax normality assumption. Directional H_1.
   $z = 1.655$ $p < .05$
3. Normality assumption may be relaxed, but at some risk.
   $z = 2.48$ $p < .01$
5. Normality assumption may be relaxed with only slight risk. Sigma unknown. Do *t* test. $t = 2.000$, $df = 80$ (use $df = 60$ on the critical values table), $p = .05$. If you had the critical values at 80 degrees of freedom you could say $p < .05$.

Exercise 8.3

1. For VIO, $z = -9.16$. $p < .001$

Exercise 8.4

1. For REM, $t = 1.440$. $df = 49$. Use 40 *df* on the table. Not significant.

Exercise 8.5

1. $z = 2.97$, $p < .01$
2. $z = 4.67$, $p < .001$
3. $z = -1.56$, not significant.

Exercise 8.6

1. The interval ranges from 45.3 to 54.7.
3. The interval ranges from 45.5 to 54.5.

Exercise 8.7

1. The interval ranges from .49 to .61.

Exercise 8.8

1. .0475
2. .1587
3. .2062
4. .1293

Exercise 8.9

2. 336 and 56

CHAPTER 9

Exercise 9.1

$F = 1.11$ cannot reject H_0. Assume equal population variances.

$t = -2.058$. $df = 18$. Not significant.

Exercise 9.3

$F = 1.74$ cannot reject H_0. Assume equal population variances.

$t = -2.429$. $df = 18$. Reject H_0. $p < .05$.

Exercise 9.5

$F = 15.53$. Reject H_0. Do not assume equal population variances.
$t = -3.315$. df estimated at 10. Reject H_0. $p < .01$.

Exercise 9.7

In each case the t values for equal and unequal variances are the same. But df differs and thus the conclusions about p differ.

Exercise 9.9

$t = 1.561$. $df = 9$. Cannot reject H_0.

CHAPTER 10

Exercise 10.1

Source	*SS*	*df*	*MS*	*F*	*p*
Between	11.28	1	11.28	3.96	> .05
Within	25.63	9	2.85		

Not Significant. (The t test in Chapter 9 had a directional H_1.)

Exercise 10.3

Source	*SS*	*df*	*MS*	*F*	*p*
Between	27.20	2	13.60	11.06	< .01
Within	18.41	15	1.23		

H_0	$\lvert\bar{x}_i - \bar{x}_j\rvert =$		*Scheffé's Critical Value*	*Conclusion*
$\mu_1 - \mu_2$	0.93	<	1.67	Cannot reject H_0.
$\mu_1 - \mu_3$	3.03	>	1.76	Reject H_0.
$\mu_2 - \mu_3$	2.10	>	1.82	Reject H_0.

Exercise 10.5

Source	*MS*	*F*	*p*
Between	21,462.25	46.38	< .001
Within	462.77		

Exercise 10.7

Source	*SS*	*df*	*MS*	*p*
Total	6235.36	55		
Between			4093.87	< .001
Within		23	93.11	

Exercise 10.9

Having only the printout, all we can say is that we may reject H_0. $p = .0032$ (between .01 and .001).

CHAPTER 11

Exercise 11.1

$Q = -.680$, $\phi = -.382$.

Exercise 11.2

$Q = .818$, $\phi = .459$

Exercise 11.3

$P_s = 6$, $P_d = 102$, $\gamma = -.889$
Agric. Dep. $E_1 = 14$, $E_2 = 10$, $\lambda = .286$
GNP Dep. $E_1 = 13$, $E_2 = 7$, $\lambda = .462$
$\lambda_{symmetric} = .374$

Exercise 11.4

$P_s = 121$, $P_d = 3$, $\gamma = .952$
λ Either Way, $E_1 = 12$, $E_2 = 5$
$\lambda = .583$, $\lambda_{symmetric} = .583$
$\gamma_{\text{Most Appropriate}}$: ordinal × ordinal

Exercise 11.5

$P_s = 106$, $P_d = 2$, $\gamma = .963$

Exercise 11.6

$P_s = 61$, $P_d = 42$, $\gamma = .184$ low
$E_1 = 13$, $E_2 = 10$, $\lambda = .231$, fairly large for a lambda. Hence, a curvilinear relationship.

CHAPTER 12

Exercise 12.1

$\chi^2 = 25.86$, Reject H_0, $p < .001$
$\phi = 1.14$, $C = .75$, $V = .80$

Exercise 12.3

$\chi^2 = 20.00$, Reject H_0, $p < .001$
$\phi = +1.00$, $C = +.71$, $V = +1.00$

Exercise 12.5

For Agriculture, combine High and Medium into one category and Low and Very Low into another. $\chi^2 = 20.95$.

Exercise 12.7

For the four-category variable, make it two categories by combining High with Medium and combining Low with Very Low. For the three-category variable, either combine High with Medium or Medium with Low. Either way, two f_e's will be below 5. Chi-square is invalid. (We would have to do Fisher's Exact Test in this situation.)

CHAPTER 13

Exercise 13.1

1. $r = -.962$

Exercise 13.2

1. $b = -.730$, $a = 97.66$, CHEM = 97.66 − .73 DRAMA
3. Emile: 87 predicted, 93 actual
   Dimitri: 92 predicted, 93 actual
   Deborah: 31 predicted, 29 actual

Exercise 13.4

1. ALGEBRA = 30, LITER = 71.94
   ALGEBRA = 80, LITER = 26.94
2. ALGEBRA = 81, LITER = 26.04
   CHEM = 75, LITER = 21.97
3. GEOM = 0, LITER = 87.34
   DRAMA = 100, LITER = 85.89

Exercise 13.6

3. $r^2 = .9025$
4. For Jack, predict LITER = 44
   For Jill, predict LITER = 87

Exercise 13.7

For exercise 10.3, $r_i = .627$ and $E^2 = .596$.

CHAPTER 14

Exercise 14.1

1. Not significant.
3. Reject H_0. $p < .01$
5. Reject H_0. $p < .01$
7. Reject H_0. $p < .025$
9. Not significant.

Exercise 14.2

1. r_{23} is spurious.
3. No correlations are spurious.

Exercise 14.3

1. $R^2 = .877$
3. $R^2 = .405$

Exercise 14.4

1. $x_1 = 9.996$
3. $\beta_2 = .973$, $\beta_3 = -.057$

Exercise 14.5

1. CPI = − 3.83 + .68 MONEYMKT + .45 HOME + .08 WHEAT
3. Up 2.49
5. MONEYMKT: .60 the most.
   WHEAT: .40 the least.

Index

About the Author

Mark Sirkin, Ph.D. is Assistant Professor of Political Science at Wright State University in Dayton, Ohio. He earned his M.A. and Ph.D. in Political Science at Pennsylvania State University. In the past, in addition to teaching, he served as Assistant Dean in the College of Liberal Arts at Wright State University. Among other responsibilities, he undertook studies and provided data analysis on such subjects as course enrollment patterns, grade distributions, and other topics pertaining to the college's students and faculty.

After that, he was Assistant Dean and later Associate Dean of Wright State's School of Graduate Studies. As Associate Dean, he was also Director of the Office of University Research Services, which administered sponsored programs—grants and contracts—as well as supervising the use of human subjects in research. During that period, he prepared an accreditation self-study for the University and was involved in the analysis of data on graduate students and faculty research throughout the University, including its schools of Medicine and Professional Psychology. He also administered the Institutional Review Board for human subjects in research. These positions gave him a great deal of experience in academic research, not only in the social and behavioral sciences, but also in the biological, physical, and medical sciences.

Currently, he is teaching full time. His courses include quantitative methods, empirical analysis, and statistical applications in the social sciences. He also teaches courses on the Middle East.